The Pineal Gland and Its Hormones

Fundamentals and Clinical Perspectives

NATO ASI Series

Advanced Science Institutes Series

A series presenting the results of activities sponsored by the NATO Science Committee, which aims at the dissemination of advanced scientific and technological knowledge, with a view to strengthening links between scientific communities.

The series is published by an international board of publishers in conjunction with the NATO Scientific Affairs Division

A	Life Sciences	Plenum Publishing Corporation
B	Physics	New York and London
C	Mathematical and Physical Sciences	Kluwer Academic Publishers
		Dordrecht, Boston, and London
D	Behavioral and Social Sciences	
E	Applied Sciences	
F	Computer and Systems Sciences	Springer-Verlag
G	Ecological Sciences	Berlin, Heidelberg, New York, London,
H	Cell Biology	Paris, Tokyo, Hong Kong, and Barcelona
I	Global Environmental Change	

PARTNERSHIP SUB-SERIES

1. Disarmament Technologies	Kluwer Academic Publishers
2. Environment	Springer-Verlag
3. High Technology	Kluwer Academic Publishers
4. Science and Technology Policy	Kluwer Academic Publishers
5. Computer Networking	Kluwer Academic Publishers

The Partnership Sub-Series incorporates activities undertaken in collaboration with NATO's Cooperation Partners, the countries of the CIS and Central and Eastern Europe, in Priority Areas of concern to those countries.

Recent Volumes in this Series:

Volume 275 — Neural Development and Schizophrenia: Theory and Research
edited by Sarnoff A. Mednick and J. Meggin Hollister

Volume 276 — Thrips Biology and Management
edited by Bruce L. Parker, Margaret Skinner, and Trevor Lewis

Volume 277 — The Pineal Gland and Its Hormones: Fundamentals and
Clinical Perspectives
edited by Franco Fraschini, Russel J. Reiter, and Bojidar Stankov

Volume 278 — Obesity Treatment: Establishing Goals, Improving Outcomes, and
Reviewing the Research Agenda
edited by David B. Allison and F. Xavier Pi-Sunyer

Series A: Life Sciences

The Pineal Gland and Its Hormones

Fundamentals and Clinical Perspectives

Edited by

Franco Fraschini

University of Milan
Milan, Italy

Russel J. Reiter

University of Texas
San Antonio, Texas

and

Bojidar Stankov

University of Milan
Milan, Italy

Springer Science+Business Media, LLC

Proceedings of a NATO Advanced Study Institute on
The Pineal Gland and Its Hormones: Fundamentals and Clinical Perspectives,
held June 7–13, 1994,
in Erice, Italy

NATO-PCO-DATA BASE

The electronic index to the NATO ASI Series provides full bibliographical references (with keywords and/or abstracts) to about 50,000 contributions from international scientists published in all sections of the NATO ASI Series. Access to the NATO-PCO-DATA BASE is possible in two ways:

—via online FILE 128 (NATO-PCO-DATA BASE) hosted by ESRIN, Via Galileo Galilei, I-00044 Frascati, Italy

—via CD-ROM "NATO Science and Technology Disk" with user-friendly retrieval software in English, French, and German (©WTV GmbH and DATAWARE Technologies, Inc. 1989). The CD-ROM also contains the AGARD Aerospace Database.

The CD-ROM can be ordered through any member of the Board of Publishers or through NATO-PCO, Overijse, Belgium.

Library of Congress Cataloging-in-Publication Data

On file

ISBN 978-1-4613-5781-0   ISBN 978-1-4615-1911-9 (eBook)
DOI 10.1007/978-1-4615-1911-9

© 1995 Springer Science+Business Media New York
Originally published by Plenum Press New York in 1995
Softcover reprint of the hardcover 1st edition 1995

10 9 8 7 6 5 4 3 2 1

PREFACE

This volume contains the written contributions to the proceedings of a workshop related to the pineal gland and its hormones, which was held in Erice, Italy, on June 7 - June 13, 1994. This series of workshops, which began in 1982 and which have been held at four-year intervals since that time, has provided important continuity for advancing the state of knowledge relating to this very important investigative area. The enthusiasm for these conferences has increased steadily, as reflected in the number of individuals applying to attend and in the input of individuals who participate in the meeting.

The 1994 meeting was important because of its timeliness. In the two years preceding the meeting a number of revolutionary discoveries were made relative to the actions of the pineal hormone melatonin. The *Xenopus* melatonin receptor was cloned, melatonin was demonstrated to be a potent antioxidant, the significance of melatonin receptors at the level of pars tuberalis in the regulation of the hypothalamo-pituitary-gonadal axis was questioned, a number of melatonin receptor analogues were discovered and successfully utilized, the mechanisms by which melatonin retards initiation and promotion of cancer was further elucidated, the clinical aspects of the pineal gland was re-scrutinized. Reviews relating to each of these subjects, as well as many others, are contained in this proceedings book. This volume represents an up-to-date repository for the most recent information related to this rapidly advancing field. The book contains contributions that will appeal to both basic scientists and clinicians in virtually any discipline, but certainly in the fields of physiology, endocrinology, free radical biology, oncology, pharmacology circadian biology, immunology, and cellular and molecular biology.

The organizers are highly grateful for the overwhelming support they received in the organization of this noteworthy workshop. The invited speakers enthusiastically accepted the invitations to present their newest data and the attendees participated actively. The enthiusiam was obvious during the meeting and hopefully will linger in the pages of this proceedings volume. We thank all those who contributed to or were associated with this important event.

Franco Fraschini

Milano, Italy

Russel J. Reiter

San Antonio, Texas

Bojidar Stankov

Milano, Italy

CONTENTS

PINEAL GLAND REGULATION

MELATONIN RECEPTORS

MELATONIN ANALOGS

MELATONIN IN HUMANS

EXPRESSION OF NEUROTRANSMITTER RECEPTOR SUBTYPES AND SUBUNITS IN THE MAMMALIAN PINEAL GLAND

M. Møller[1], P. Phansuwan-Pujito[2], and G. Mick[3]

[1]Institute of Medical Anatomy, University of Copenhagen, Denmark
[2]Dept. of Anatomy, Faculty of Medicine, Srinakarinwirot, University of
Prasarnmit, Bangkok, Thailand
[3]G. Mick, Unité 94, Institut national de la Santé et de la Recherche
Médicale, Bron, France

INTRODUCTION

The mammalian pineal gland receives multiple pinealopetal nervous projections (for surveys see Korf and Møller, 1984; Korf and Møller, 1985; Møller et al., 1991). In all investigated species, the gland is innervated by sympathetic nerve fibres originating from perikarya located in the superior cervical ganglia (Kappers, 1960). Further, parasympathetic nerve fibres have been found to innervate the gland in the monkey (Kenny, 1965; Nielsen and Møller, 1975), rabbit (Romijn, 1975), and gerbil (Shiotani et al., 1986). Also nuclei located in the forebrain project to the pineal gland, via the pineal stalk, connecting the gland directly with the optic system and the hypothalamus (central innervation) (Møller and Korf, 1983a,b; Møller and Korf, 1987; Mikkelsen and Møller, 1990; Fink-Jensen and Møller, 1991; Mikkelsen et al., 1991; Larsen et al., 1991). In the rodents, these central pinealopetal fibres mostly terminate in the rostral part of the pineal complex, e.g. the deep pineal gland and the pineal stalk (Møller, 1992)

The pinealopetal nerve fibres contain different neurotransmitters/modulators (see Møller and Mikkelsen, 1991). Noradrenaline is present in the sympathetic fibres (Møller and van Veen, 1981) colocalized in most of the fibres with neuropeptide Y (Schröder, 1986; Zhang et al., 1991; Cozzi et al., 1992). However, several studies have shown NPY to be present also in fibres of the central innervation (Zhang et al., 1991; Cozzi et al., 1992). Vasoactive intestinal peptide (VIP) and peptide histidine isoleucine (PHI), two peptides encoded from the same gene, are probably present in the parasympathetic pinealopetal nerve fibres the perikarya of which are located in the pterygopalatine ganglion (Cozzi et al., 1994). Recently, in the rat, it has been shown that histaminergic nerve fibres project from the magnocellular mammillary nuclei to the pineal (Mikkelsen et al., 1992). In the bovine pineal, cholinergic nerve fibres (Phansuwan-Pujito et al., 1991b) as well as fibres immunoreactive to substance P (Møller et al., 1993) have been demonstrated.

Biochemical binding studies have verified the presence of multiple binding sites for neurotransmitters on the pinealocyte cell membrane in several species (Ebadi et al., 1989). Thus, studies have shown binding sites for VIP in the rat pineal gland (Yuwiler, 1987; Møller

The Pineal Gland and Its Hormones
Edited by F. Fraschini *et al.*, Plenum Press, New York, 1995

1

et al., 1985) and NPY (subtype Y_1) (Olcese, 1991, Simonneaux et al., 1994) in the pineal of the cow. In the same species, binding sites for substance P (Ebadi et al., 1989) and acetylcholine (Phansuwan-Pujito et al, 1991b) have been pharmacologically demonstrated.

The excitatory amino acid glutamate has recently been suggested to play a role in pineal function. Glutamate is present at high concentration in rat pinealocytes, mainly at the level of synaptic ribbons, and to a lesser degree in glial cells, neural elements and endothelial cells (McNulty et al., 1990). In a perifusion system, this neuroactive substance reduces the norepinephrine-induced melatonin synthesis in the rat (Govitrapong and Ebadi, 1988; Kus et al., 1994). However, nerve fibres containing glutamate have never been observed in the mammalian pineal gland.

Cloning of many receptors has made it possible to visualize mRNA encoding the receptor proteins by use of **in situ** hybridization (ISH). We have by use of this powerful new techniques demonstrated the presence of mRNA's for several receptors in the pineal gland. We describe in this paper experiments demonstrating the presence of mRNA encoding β_1-receptor and the muscarinic receptor of subtype m_1.

Recent **in situ** hybridization studies have been reporting an expression of glutamate receptor subunits in the rat pineal gland (Sato et al., 1993; Wisden and Seeburg, '93; Tolle et al., 1993). In this paper we therefore also describe our results of our **in situ** experiments investigating glutamate receptors in both rodent and primate pineal glands.

METHODS

Animals and Tissue Preparation

The animals were anaesthetized with tribromethanol (250 mg/kg i.p.) and perfused transcardially with 0.1 M phosphate-buffered saline (pH 7.4) with heparin for 15 min. Then, the brains were rapidly removed and frozen in crushed dry ice. After freezing the tissue blocks were be stored at -70°C until further processing. Cryostat sections, 20 µm in thickness, were cut and thaw-mounted on gelatin-coated slides.

Probes

cDNA antisense oligonucleotide hybridization probes were either obtained commercially or synthesized. Generally, we used a probe mixture of three oligonucleotides (length 40-50 mer) complementary to the base sequences encoding amino acid sequences located on the extacellular N-terminal part of the receptor protein. The oligonucleotides were 3'end-labelled with ^{35}S-dATP (1,211 Ci/mmol, NEN) using terminal deoxynucleotidyl transferase (Boehringer Mannheim) to a specific activity of 1.2×10^{18} dpm/mole.

In situ Hybridization

Prior to hybridization, frozen tissue sections were brought to room temperature, fixed for 5 min in 4% paraformaldehyde in PBS (pH 7.4) and washed twice in PBS, 2 min each. Then, the sections were acetylated in 0.25% acetic anhydride diluted in 0.1 M triethanolamine and 0.9% NaCl for 10 min, dehydrated in a graded series of ethanols, i.e., 70% (1 min), 80% (1 min), 95% (2 min), 100%(1 min), delipidated in 100% chloroform (5 min), rehydrated in 100% and 95% ethanol (1 min each), and air dried.

For hybridization, labelled probes were diluted in hybridization buffer (pH 7.2) containing 25% (V/V) formamide, 4 x SSC (sodium saline citrate), 1 x Denhardt solution, 10% (W/V) dextran sulphate, 10 mM DTT (dithiotreitol), 0.5 mg/ml salmon sperm DNA and 250 µg/ml yeast tRNA, and pipetted onto each section. The sections were covered with parafilm and incubated in humid chamber overnight at 37 °C. The slides were then washed in 1 x SSC for 4 x 15 min at 55 °C, 2 x 30 min at room temperature and dipped twice in distilled water.

After drying of the sections, they were either exposed to an Amersham Hyperfilm[R] or X-ray film for 3 weeks, or dipped in Amersham LM-1[R] nuclear emulsion and exposed for 5 weeks at 4 °C. The autoradiographs were analyzed densitometrically using a Macintosh II computer equipped with an image capture board running the programme "Image 1.42" by Wayne Rasband, NIH, USA. The mean optical density of autoradiographs was measured in dpm/mg tissue by comparison with simultaneously exposed ^{35}S brain paste standards. The localization of grains was studied with in dark- and bright field microscope on emulsion-coated slides.

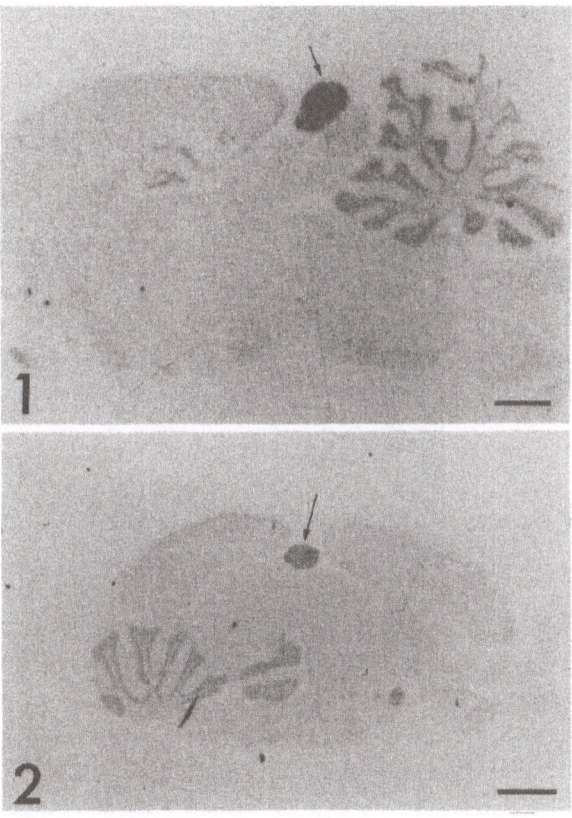

Figs.1, 2.   **In situ** hybridization images of sagittal (fig.1) and frontal (fig.2) rat brain sections showing strong β_1-receptor signals in the pineal gland (arrows). X-ray film. Bar=0.2 mm.

β₁-RECEPTOR

For detection of the β₁-receptor a mixture of cDNA oligonucleotide probes, complementary to the base sequences encoding bases 121-168, 1261-1308 and 1381-1428 of the human β₁-receptor, and sequences encoding bases 7-54, 19-66 of the rat β₁-receptor were used. For detection of β₂-receptor a mixture of three cDNA oligonucleotide probes, complementary to the base sequences encoding bases 4-51, 661-708, and 772-820 were used. Hybridization with the probe recognizing the β₁-mRNA revealed a strong hybridization signal in the pineal gland (Figs. 1, 2, and 3), the hippocampus, dentate gyrus (fig. 1, and 3), and the granular layer of the cerebellum (Fig. 1, and 3). Contrarily, when the probe mixture recognizing the β₂-receptor mRNA was used, no hybridization signal could be detected in the rat pineal. Thus, the present results suggests that mRNA encoding the β₁-adrenoceptor, but not the β₂-adrenoceptor is expressed by pinealocytes of the rat pineal gland.

The presence of mRNA encoding the β₁-receptor is in accord with the biochemical demonstration of β₁-receptors in the rat pineal. Stimulation of this receptor activates a G-protein, which via an elevation of cAMP (Auerbach et al., 1981) stimulates the rate limiting enzyme, serotonin N-acetyl transferase, of the melatonin synthesis (Klein and Weller, 1973). Due to the resolution of the film, the present experiments did not resolve, whether the receptors are located on specific cells in the pineal.

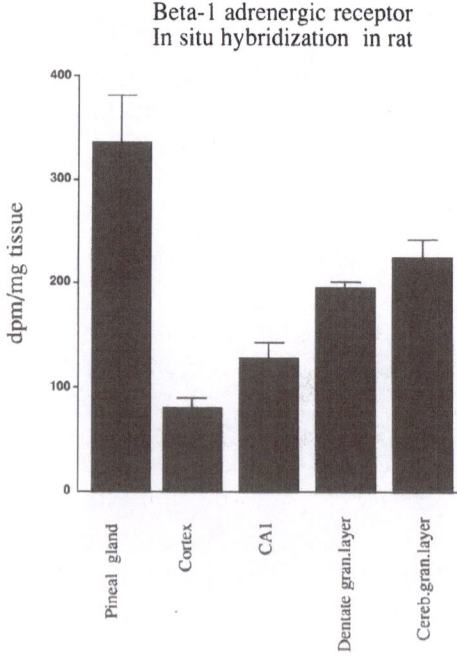

Fig.3   Bar graph of the densitometric measurements of the β₁-receptor signal in the rat brain.

MUSCARINIC RECEPTOR (subtype m₁)

Three oligonucleotide hybridization probes were purchased from Dupont Co., (NEP 561 FP 052 Lot# WF 2289) complementary to the base sequences encoding bases 4-51, 721-768, 811-858 of human muscarinic m_1 receptor cDNA **In situ** hybridization histochemistry revealed a moderate signal in the rat pineal gland (Fig.3). À strong signal was observed in the neocortical layer II and hippocampus with a lower signal in cortical layers III-VI (Fig.4). Within the hippocampus, the signal was strong in the dentate gyrus and the CA3, CA1 and CA2 of Ammon's horn (Fig.4). A weak signal was also present in the superior colliculus (confer Fig.3 in Phansuwan-Pujito et al., 1994).

Muscarinic receptors have been demonstrated by biochemical receptors binding studies in the pineal gland (Taylor et al., 1980; Govitrapong et al., 1989; Finocchiaro et al., 1989). In a study by Laitinen et al. (1989) the muscarinic receptor was demonstrated to stimulate the phosphoinositide system indicating the receptor to be of the M1 subtype.

With regard to function some studies have described a stimulation of the pinealocyte by the muscarinic receptor (Wartman et al., 1969; Finocchiaro et al., 1989; Laitinen et al., 1992). However, in the cow muscarinic agonist inhibits serotonin N-acetyl transferase (Phansuwan-Pujito et al., 1991a). We have previously demonstrated a cholinergic innervation of the cow pineal by use of an antibody against choline acetyltransferase by use of which immunoreactive nerve fibres were shown (Phansuwan-Pujito et al., 1991b). However, in the rat such cholinergic fibres could not be demonstrated.

Fig.4   **In situ** hybridization image of a rat brain sections showing a moderate m_1-receptor signal in the rat pineal gland (arrow). x-ray film. Bar = 0.2 mm.

GLUTAMATE RECEPTORS

The pineal gland of the rat is characterized by the presence of a complex mosaic of glutamate receptor subunits in pinealocytes, glial-like and neuronal-like cells. Typically, for each pharmacological type of ionotropic glutamate receptors (N-methyl-D-aspartate (NMDA), a-amino-3-hydroxy-5-methyl-4-isoxazole propionate (AMPA) and kainate) and for the metabotropic family of glutamate receptors (mGluR), only a single mRNA encoding an ionotropic receptor subunit or a mGluR subtype is predominantly expressed in pinealocytes (table 1). A striking feature is the high expression of NR2C subunit of the NMDA receptor subtype distributed in a lobular pattern. Immunohistochemical staining confirms the presence of the single subunit GluR4 of the AMPA glutamate receptor subtype in pinealocytes in the rat but not in the macaque, and reveals the presence of GluR1 and GluR4 subunits on glial-like subpopulations in both rat and macaque and in neuronal-like cells only in the macaque (table 2). Strikingly in the macaque, AMPA subunits are differentially distributed

TABLE 1		
Subunit		**mRNA expression**
NMDA type	NR1	++
	NR2A	(+)
	NR2B	0
	NR2C	+++
	NR2D	(+)
AMPA type	GluR1	(+)
	GluR2	0
	GluR3	0
	GluR4	+
Kainate type	GluR5	(+)
	GluR6	0
	GluR7	0
	KA1	(+)
	KA2	++
δ subfamily	δ1	(+)
	δ2	0
metabotropic gluR type	mGluR1	+
	mGluR2	0
	mGluR3	0
	mGluR4	0
	mGluR5	0

according to proximal and distal areas of the gland. These findings raise a number of questions regarding the possible mechanisms of action of glutamate in the mammalian pineal gland. First, according to the present knowledge on glutamate receptor structure (Seeburg, 1993), the eventual subunit combinations that may characterize the subtypes of ionotropic glutamate receptors in the pineal gland might be not the same as those commonly found in the nervous and endocrine systems. Particularly, the KA2 subunit of the kainate subtype alone does not constitute a functional receptor. Second, all other receptor types represent channels that are highly permeable to calcium, a second messenger that plays a crucial role in the modulation of melatonin synthesis (Klein, 1993). Third, none of the pharmacological agonist of glutamate receptors has been reported to reproduce glutamate effects in vitro (Kus et al. 1994), except eventually AMPA (Govitrapong et al., 1988). Only a single type of glutamate binding site has been pharmacologically characterized in the rat pineal gland (Kus et al., 1993), for which binding of the ligand is preferentially displaced by AMPA. The bovine pineal gland similarly exhibits a glutamate binding site that is insensitive to NMDA and kainate but displays a high affinity for quisqualate, a strong agonist of the

TABLE 2		
Rat		
Antibody	pinealocytes	glial-like cells
GluR1	0	Superficial and rostral located cells
GluR2/3/4c	0	Few superficial located cells
GluR4	Throughout the gland	Superficial and rostral located cells
Macaque		
GluR1	0	Cells in the distal part
GluR2/3/4c	0	few cells
GluR4	0	cells in the rostral part and processes in the distal part

AMPA-gated glutamate receptors (Govitrapong et al., 1986). Together with our anatomical data, this indicates that glutamate plays a specific role via glutamate receptors of the AMPA subtype in the control of the secretory activity of the mammalian pineal gland, involving both pinealocytes and glial cells. In addition, the differential distribution of AMPA subunits in the rat and in the macaque strongly suggests that sites of glutamate action and hence the role of the transmitter differ with respect to mammalian species. The contribution of the other glutamate receptor subunits and subtypes in pineal function remains totally enigmatic.

ACKNOWLEDGEMENTS

The authors are grateful to J.L. BORACH for excellent technical assistance. This investigation was supported by NOVO's Fond, the Danish Medical Research Foundation (grants no. 12-0236, 12-1642-1, 12-1486-1), and the Danish Biotechnology Programme with a grant to Biotechnology Centre for Signal Peptide Research.

REFERENCES

Auerbach, D.A., Klein, D.C.., Woodward, B, and Auerbach, G.D., 1981, Neonatal rat pinealocytes: Typical and atypical characteristics of ^{125}I-iodohydroxybenzylpindolol binding and adenosine 3',5'-monophosphate accumulation. *Endocrinolgy*, 108:559.

Cozzi, B., Mikkelsen, J.D., Ravault, J.-P., and Møller, M., 1992, Neuropeptide Y (NPY) and C-flanking peptide of NPY in the pineal gland of normal and ganglionectomized sheep. *J.Comp.Neurol.*, 316:238.

Cozzi, B., Mikkelsen, J.D., Ravault, J.-P., Locatelli, A., Fahrenkrug, J., Zhang, E.-T., and Møller, M., 1994, The distribution of peptide histidine-isoleucine (PHI)- and vasoactive instestinal peptide (VIP)-immunoreactive nerve fibres in the sheep pineal gland is not affected by superior cervical ganglionectomy. *J.Comp.Neurol.*, 343:72

Ebadi, M., Hexum, T.D., Pfeiffer, R.F., and Govitrapong, P., 1989, Pineal and retinal peptides and their receptors. *Adv.Pineal Res.* 7:1.

Fink-Jensen, A., and Møller, M., 1991, A direct neuronal projection from the lateral hypothalamic area to the rostral part of the pineal complex in the rat. An anterograde neuron-tracing study using Phaseolus vulgaris leucoagglutinin. *Adv.Pineal Res.* 5:21.

Finocchiaro, L.M.E., Scheucher, A., Finkelman, S., Nahmod, V.E., and Pirola, C.J., 1989, Muscarinic effects on the hydroxy- methoxyindole pathway in the rat pineal gland. *J.Endocrinology*, 123;205

Govitrapong P., and Ebadi M, 1988, The inhibition of pineal arylalkylamine N-acetyltransferase by glutamic acid and its analogues. *Neurochem Int.* 13:223.

Govitrapong P., Ebadi M., and Murrin L.C., 1986, Identification of a Cl-/Ca2+ dependent glutamate (quisqualate) binding site in bovine pineal organ. *J.Pineal Res.* 3, 223.

Govitrapong, P., Phansuwan-Pujito, P., and Ebadi, M., 1989, Studies on the properties of muscarinic cholinergic receptor sites in bovine pineal gland. *Comp.Biochem.Physiol.* 94:159.

Kappers, J.A., 1960, The development, topographical relations, and innervation of the epiphysis cerebri in the albino rat. *Z.Zellforsch.* 52:163.

Kenny, G.C.T., 1965, The innervation of the mammalian pineal body (A comparative study). *Proc.Aust.Assoc.Neurol.* 3:133.

Klein, D.C., 1993, The mammalian melatonin rhythm generating system, *in*: "Light and biological rhythms in man" Pergamon Press, New York.

Klein, D.C., and Weller, J.L., 1973, Adrenergic-adenosine 3',5'-monophosphate regulation of serotonin N-acetyltransferase activity and the temporal relationship of serotonin N-acetyltransferase activity to synthesis of [4]H-melatonin in the cultured rat pineal gland. *J.Pharmacol.Exp.Ther.* 186:516.

Korf, H.-W., and Møller, M., 1984, The innervation of the mammalian pineal gland with special reference to central pinealopetal projections. *Pineal Res.Rev*, 2:41.

Korf, H,-W., and Møller, M., 1985, The central innervation of the mammalian pineal organ. *In*: "The Pineal Gland," B. Mess, Cs.Rúzsas, L. tima, P. Pévet, eds., Akademia Kiado, Budapest.

Kus, L., Handa, R.J., and McNulty, J.A., 1993, Characterization of a [3H]glutamate binding site in rat pineal gland: enhanced affinity following superior cervical ganglionectomy. *J. Pineal Res.* 14:39.

Kus, L., Handa, R.J., and McNulty, J.A., 1994, Glutamate inhibition of the adrenergic-stimulated production of melatonin in rat pineal gland in vitro. *J.Neurochem.* 62:2241.

Laitinen, J.T., Torda, T., and Saavedra, J.M., 1989, Cholinergic stimulation of phosphoinositide hydrolysis in the rat pineal gland. *Eur.J.Pharmacol.*, 161:237.

Laitinen, J.T., Vakkuri, O., and Saavedra, J.M. 1992, Pineal muscarinic phosphoinositide responses: age-associated sensitization, agonist-induced desensitization and increase in melatonin release from cultured pineal gland. *Neuroendocrin.*, 55:492.

Larsen, P.J., Møller, M., and Mikkelsen, J.D., 1991), Efferent projections from the periventricular and medial parvocellular subnuclei of the hypothalamic paraventricular nucleus to circumventricular organs of the rat. A Phaseolus-vulgaris leucoagglutinin (PHA-L) tracing study. *J.Comp.Neurol.*, 306:462.

McNulty, J., MacReynolds, H.D., and Bowman, D.C., 1990, Pineal gland free amino acids and indoles during postnatal development of the rat: correlation in individual glands. *J. Pineal Res.* 9:65.

McNulty, J.A., Kus, L., Ottersen, O.P., 1992, Immunocytochemical and circadian biochemical analysis of neuroactive amino acids in the pineal gland of the rat: effect of superior cervical ganglionectomy. *Cell Tiss.Res.* 269:515.

Mikkelsen, J.D, and Møller, M., 1990, A direct neuronal projection from the intergeniculate leaflet of the lateral geniculate nucleus to the deep pineal gland of the rat, demonstrated with Phaseolus vulgaris leucoagglutinin (PHA-L). *Brain Res.*, 520:342.

Mikkelsen, J.D, Cozzi, B., and Møller, M., 1991, Efferent projections from the lateral geniculate nucleus to the pineal complex of the Mongolian gerbil (Meriones unguiculatus). *Cell Tissue Res.*, 264:95.

Mikkelsen, J.D., Panula, P., and Møller, M., 1992, Histamine-immunoreactive nerve fibres in the rat pineal gland: evidence for a histaminergic central innervation. *Brain Res.*, 597:200.

Møller, M., 1992, The fine structure of the pinealopetal innervation of the mammalian pineal gland. *J.Microsc.Res.Techn.*, 21(3)188.

Møller, M., and van Veen, Th. ,1981, Fluorescence Histochemistry of the Pineal Gland. *In*: The Pineal Gland, vol. 1. Anatomy and Biochemistry, R.J. Reiter, Ed., CRC Press, West Palm Beach.

Møller, M., and Korf, H.-W., 1983a, Central innervation of the pineal organ of the Mongolian gerbil. A histochemical and lesion study. *Cell Tissue Res.*, 230:259.

Møller, M., and Korf, H.-W. ,1983b, The origin of central pinealopetal nerve fibres in the Mongolian gerbil as demonstrated by the retrograde transport of horseradish peroxidase. *Cell Tissue Res.* 230:273.

Møller, M., and Korf, H.-W., 1987, Nervous connections between the brain and the pineal gland of the golden hamster (Mesocricetus auratus). A horseradish peroxidase study. *Cell Tissue Res.*, 247:145.

Møller, M., and Mikkelsen, J.D., 1991, Molecular messengers in brain-pineal interactions. *In*: Recent Advances in Cellular and Molecular Biology. Vol.3.: Neurobiochemical transmitter pathways, adrenoceptors and muscarinic receptors,(R.J.Wegmann and M.A. Wegman, eds., Peeters Press; Leuven.

Møller, M., Mikkelsen, J.D., Fahrenkrug, J., and Korf, H.-W., 1985, The presence of vasoactive intestinal polypeptide (VIP)-like-immunoreactive nerve fibres and VIP-receptors in the pineal gland of the Mongolian gerbil (Meriones unguiculatus). An Immunohistochemical and receptor-autoradiographic study. *Cell Tissue Res.* 241:333.

Møller, M., Ravault, J.-P., Cozzi, B., Zhang, E., Phansuwan-Pujito, Larsen, P.J., and Mikkelsen, J.D., 1991, The multineuronal input to the mammalian pineal gland. *Adv.Pineal Res.*, 6:3.

Møller, Phansuwan-Pujito, P., Govitrapong, P., and Schmidt, P., 1993, Indications for a central innervation of the bovine pineal gland with substance P-immunoreactive nerve fibres. *Brain Res.*, 611:347.

Nielsen, J.T., and Møller, M., 1975, Nervous connections between the brain and the pineal gland in the cat (Felis catus) and the monkey (cercopithecus aethiops). *Cell Tissue Res.* 161:293.

Olcese, J., 1991, Neuropeptide Y: An endogenous inhibitor of norepinephrine stimulated melatonin secretion in the rat pineal gland. *J.Neurochem.* 57:943.

Phansuwan-Pujito, P., Govitrapong, P., and Ebadi, M., 1991a, Cholinergic receptor agonists inhibit the activity of serotonin N-acetyltrasnferase in bovine pineal explants. *Neurochem.Res.*, 16:885

Phansuwan-Pujito, P., Mikkelsen, J.D., Govitrapong, P., and Møller, M., 1991b, Cholinergic innervation of the bovine pineal gland visualised by immunohistochemical detection of choline acetyltransferase (ChAT)-immunoreactive nerve fibres. *Brain Res.*, 545:49.

Phansuwan-Pujito, P., Larsen, P.J., and Møller, M., 1994, Expression of muscarinic receptors of subtype m_1 in the rat pineal gland. *Adv.Pineal Res.*, vol.8. pp.

Romijn, H.J., 1975, Structure and innervation of the pineal gland of the rabbit, Oryctolagus cuniculus (L.). III. An electron microscopic investigation of the innervation. *Cell Tissue Res.* 157:25.

Sato, K., Kiyama, H., Shimada, S., and Tohyama, M., 1993, Gene expression of kainate (KA)-type and NMDA receptors, and of a glycine transporter in the rat pineal gland. *Neuroendocrinololy* 58:77.

Schröder, H-J., 1986, Neuropeptide Y (NPY)-like immunoreactivity in the peripheral and central nerve fibres of the golden hamster (mesocricetus auratus) with special respect to pineal gland innervation. *Histochemistry* 85:321.

Seeburg, P.H., 1993, The molecular biology of mammalian glutamate receptor channels. *Trends in Neurosci.* 16(9):359.

Shiotani, Y., Yamano, M., Shiosaka, S., Emson, P.C., Hillyard, C.J., Girgis, S., and McIntyre, I., 1986, Distribution and origins of substance P (SP)-, calcitonin gene related peptide (CGRP)-, vasoactive intestinal polypeptide (VIP)- and neuropeptide Y (NPY)-containing nerve fibres in the pineal gland of gerbils. *Neurosci. Lett.* 70:187.

Tayler, R.L., Albuquerque, M.L.C., and Burt, D.R., 1980, Muscarinic receptors in pineal. *Life Sci.*, 26:2195.

Tölle, T.R., Berthele, A., Zielgngsberger, W., Seeburg, P.H., and Wisden, W., 1993, The differential expression of 16 NMDA and non-NMDA receptor subunits in the rat spinal cord and in periaqueductal gray. *J. Neurosci.* 13:5009.

Simonneaux, V., Ouichou, A., Craft, C., and Pévet, P., 1994, Presynaptic and postsynaptic effects of neuropeptide Y in the rat pineal gland. *J.Neurochem.* 62:2464.

Wartman, A.S., Branch, B., George, R., and Tayler, A.N., 1969, Evidence for a cholinergic influence in pineal HIOMT activity with changes in environmental lighting. *Life Sci.*, 8:1263.

Wisden, W., Seeburg, P.H., 1993, A complex mosaic of high-affinity kainate receptors in rat brain. *J.Neurosci.* 13:3582.

Yuwiler, A., 1987, Synergistic action of postsynaptic α-adrenergic receptor stimulation on vasoactive intestinal polypeptide-induced increase in pineal N-acetyltransferase activity. *J.Neurochem* 49:806.

Zhang, E., Mikkelsen, J.D.,and Møller, M., 1991, Tyrosine hydroxylase-and neuropeptide Y-immunoreactive nerve fibres in the pineal complex of untreated rats and rats following removal of the superior cervical ganglia. *Cell Tissue Res.*, 265:63.

TRANSCRIPTION FACTOR ICER: REGULATION IN THE RAT

PHOTONEUROENDOCRINE SYSTEM

Jörg H. Stehle[1,2], Nicholas S. Foulkes[3], Marjan Rikkers[2], Paul Pevet[2], and
Paolo Sassone-Corsi[3]

[1]Klinikum der Johann Wolfgang Goethe-Universität, Zentrum der
Morphologie, AG Neurobiologie, Theodor-Stern-Kai 7, 60590 Frankfurt
am Main, Germany
[2]Neurobiologie des Fonctions Rythmiques et Saisonnieres, CNRS URA
1332, Université Louis Pasteur, 67000 Strasbourg, France
[3]Institut de Genetique et de Biologie Moleculaire Cellulaire, CNRS
INSERM, Université Louis Pasteur, 67404 Illkirch, France

Development and survival of living organisms are tightly coupled to their ability to react appropriately to exogenous signals (Darwin, 1880; Aschoff, 1981). Environmental stimuli are perceived by membrane bound receptors, amplified by means of signal transduction cascades and finally translated into a specific gene expression. Receptor-stimulated signal transduction pathways regulate outgrowth, plasticity and survival in neuronal as well as in neuroendocrine tissues (Collins et al., 1992; Karin, 1992). Neuroendocrine cell systems in particular represent excellent model systems to explore the temporal course and tissue-specificity of gene expression during ontogeny or following stimulus perception.

The mammalian **photoneuroendocrine system** is comprised of three basic structures: the **retina** receives cues from ambient lighting conditions (photoperiod) and translates them into a neural signal; these signals adjust the phase of the endogenous oscillator in the hypothalamic **suprachiasmatic nucleus (SCN)**; circadian rhythmicity is relayed from the SCN to the **pineal gland** as a clock target. The pineal transduces the incoming neural information into a hormonal signal, the synthesis of melatonin (Klein 1985). Thus, the photoneuroendocrine system acts as an interface to make the body sense photoperiod through a humoral message (Axelrod, 1974; Reiter, 1991).

The pineal gland of vertebrates as the retina and the SCN is a derivate of the diencephalic anlage. Despite a profound phylogenetic transformation of the directly photosensitive pineal of lower vertebrates into a neuroendocrine gland in mammals (Oksche, 1971; Vollrath, 1981 Korf and Ekström, 1987), secretion of the hormone melatonin is an important pineal effector mechanism (Axelrod, 1974).

The Pineal Gland and Its Hormones
Edited by F. Fraschini *et al.*, Plenum Press, New York, 1995

The nocturnally elevated production of melatonin depends on the integrity of the clock in the SCN (Klein et al., 1991). This clock is active already prenatally (Reppert et al., 1989; Moore et al., 1989), but as shown for the rat, the afferent and efferent neuronal connections of the SCN are established only in the second postnatal week. In two-weeks-old rats the circadian pacemaker can be entrained to the photoperiod and in clock targets there occurs a progressive maturation of robust circadian rhythmicity (Moore et al., 1989). This holds in particular true for the pineal gland, where a circadian rhythm in melatonin synthesis is initially detectable at the beginning of the second postnatal week (Duncan et al., 1986). Hormone production is initiated upon the functional completion of the sympathetic innervation leading to a nocturnally elevated release of norepinephrine (NE) from these nerve endings. NE stimulates adenylyl cyclase via membrane bound β_1-adrenergic receptors and thus causes a rise in intracellular cyclic AMP levels (Klein 1985). This effect on adenylyl cyclase is potentiated via α_1-adrenergic stimulation of the Ca^{2+}-phospholipid-dependent protein kinase C (PKC) mechanism (Sugden et al., 1985). Elevated levels of cAMP activate the rate-limiting enzyme of melatonin synthesis, N-acetyltransferase (NAT) at multiple levels by initiating transcription, stimulating translation and maintaining enzyme activity (Klein 1985). The modulation of NAT activity by changes in intracellular cAMP concentrations is characteristic for a regulated expression of cAMP-responsive genes mediated by a family of specific transcription factors, the so-called cAMP responsive element (CRE) binding proteins (Borrelli et al., 1992). Members of this family are CREB (CRE-binding protein; Montminy et al., 1990) and CREM (CRE modulator; Lalli and Sassone-Corsi, 1994) both of which bind to CRE promoter elements. Amongst the nuclear targets of the cAMP signalling pathway in endocrine/neuroendocrine tissues, members of the CREM gene family appear to occupy a privileged position (Foulkes et al., 1992, 1993; Stehle et al., 1993; Stehle et al., in press).

Particularly, in the rat pineal gland there exists a profound day-night switch in the expression of a novel CREM product, ICER (*I*nducible *c*AMP *E*arly *R*epressor; Stehle et al., 1993). ICER expression peaks in the second half of the night and is induced by the SCN-controlled release of NE from the sympathetic nerve endings terminating in the pineal gland. Furthermore, ICER expression is rapidly and transiently induced via activation of the cAMP pathway. Translation of this surprisingly small novel CREM message generates the powerful transcriptional repressor of cAMP-inducible genes, ICER, the activity of which is solely determined by its intracellular concentrations (Stehle et al., 1993).

The maturation of the cAMP signalling pathway is a prerequisite for the generation and maintenance of a rhythmic melatonin production (Ellison et al., 1972; Klein et al., 1981). Moreover, within the mammalian photoneuroendocrine system, the cAMP signalling pathway is not only of importance in the effector, the pineal gland, but as well in both, the retina as well as in the SCN (Skene, 1992; Klein et al., 1991). The common diencephalic origin of retina, SCN and pineal gland (Korf, 1994), forwarded the idea that the rhythmic ICER expression may not be restricted to the pineal gland. The central role of photoreceptor cells and the endogenous oscillator in the mammalian photoneuroendocrine system suggested that ICER expression may be rhythmic in retina and SCN as well. *In situ* hybridization and RNase protection analysis with an ICER specific riboprobe (P75; Stehle et al., 1993) revealed ICER expression to be low although above background labelling in both, rat retina and SCN. A systematic investigation throughout a day-night cycle failed to detect any significant fluctuations of the CREM signal in both structures, (Figure 3; Stehle et al., in press). Thus, within the photoneuroendocrine system, a rhythmic ICER expression is absent in the retina as the photoneuronal transducer of photoperiodic changes and in the SCN as the site of the endogenous clock. However, as a new result a strong ICER signal was detected by *in situ* hybridization in the lamina intercalaris, the so-called deep pineal, in animals killed during the late dark period (Figure 1). Both, ganglionectomy (SCGX) as well as lesioning the clock (SCNX), reduced ICER expression in the deep pineal to background levels as it does in the superficial part of the pineal gland (Figure 2; Stehle et al., 1993). Thus, ICER expression is in both, the superficial and deep portions of the pineal organ, under clock control via sympathetic pinealopetal nerve fibres (Stehle et al., in press). Taken together, these observations demonstrate that endogenous circadian ICER expression in rat brain is restricted to the two parts of the pineal gland.

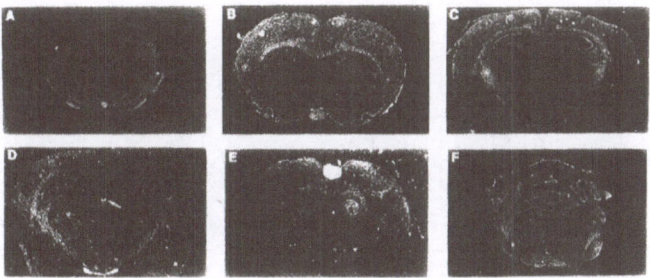

Figure 1. *In situ* hybridization analysis of ICER distribution in the brain of a rat killed at nighttime (02:00h). Coronal sections (A-F) were hybridized with an ^{35}S-labelled antisense P75 probe (Stehle et al., 1993). The sections were exposed to X-ray film for 8 days. As a control, adjacent sections were hybridized with a sense P75 probe. Methods were as described (Stehle et al., 1992). Arrow in D: lamina intercalaris; labelling of base of section D represents residual material from the pituitary gland.

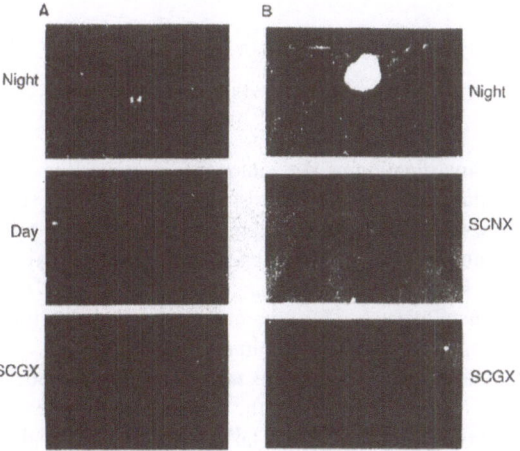

Figure 2. ICER expression in the deep (A) and the superficial (B) part of the pineal gland is under clock-control. Abbreviations: Night: animals were sacrificed at 02:00h, 7h after lights off; Day: animals were killed at 12:00h, 5h after lights on; SCGX: animals had the superior cervical ganglia chronically removed; SCNX: animals received a chronic lesion of the site of the endogenous clock, the suprachiasmatic nucleus (SCN). Operations were performed at least two weeks prior sacrifice of animals. Methods were as described (Stehle et al., 1993; see Figure 1).

Circadian rhythms are already detectable early in development (Reppert et al., 1989). In rat SCN glucose metabolism oscillates from gestational day 19 (G19) on (Reppert and Schwartz, 1984) and a circadian rhythm in electrical activity of SCN cells can be recorded as early as G21 (J.H.S.and S.M. Reppert, unpublished observations). Pineal melatonin synthesis, however, is below detection limit until the end of the first postnatal week (Duncan et al., 1989). Later there is a gradual increase in night-time melatonin values which finally leads to the profound nocturnal melatonin surge as observed in all vertebrates (Vollrath, 1981). This developmental maturation of the pineal

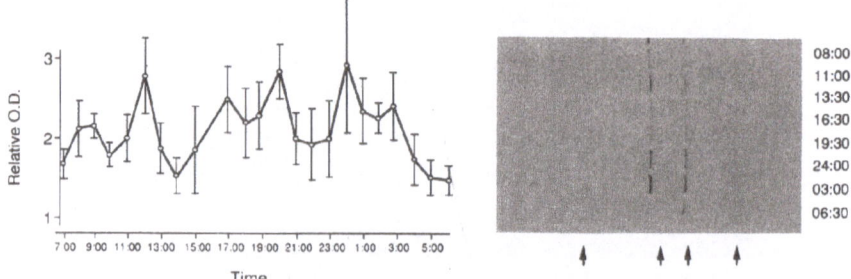

Figure 3. *Left panel:* relative optical density (O.D.) of ICER hybridization signal in the SCN as assessed at indicated timepoints. Error bars indicate SEM. *Right panel:* RNase protection assay of CREM expression using 5µg of RNA extracted from rat retinae. Tissues were taken at indicated timepoints. Arrows indicate specific protected fragments obtained using the mouse p6N/1 probe with rat CREM mRNA (Stehle et al., 1993). Statistical analysis of O.D. values of the specific band showed no difference between samples.

gland forwarded an investigation of the ontogenetic appearance of ICER in rat brain during different gestational and postnatal stages (Stehle et al., in press). By *in situ* hybridization (Figure 4) and RNase protection analysis (not shown) the day/night switch in pineal ICER expression was found to be absent until postnatal day 8 (P8). Subsequently, the night-time expression rises sharply and adult ICER mRNA levels are reached aproximately at postnatal day 15. Thus, the onset in regulated ICER expression is tightly correlated with the onset in pineal melatonin synthesis.

Rat melatonin synthesis is undetectable and also uninducible during the first postnatal week. This is in contrast to the readily inducible NAT activity at that developmental stage (Klein et al., 1981). The absence of a day/night ICER switch during this initial postnatal period gave rise to the question at what developmental stage pineal ICER expression is inducible.

The injection of the ß-adrenergic agonist isoproterenol (5mg/kg b.wt.) in the late morning, i.e. at a time when the CREM gene is inducible (Stehle et al., 1993; N.S.F. and J.H.S. unpublished observations), demonstrated that ICER gains inducibility during postnatal development between day 5 and day 8, concomitantly with the normal onset of the day-night ICER switch (Figure 4, 5). These effects cannot be attributed to a ligand-dependent fluctuation of pineal ß-adrenergic receptors, since the same ontogenetic course of ICER inducibility was observed when dBcAMP was used instead of ISO as a stimulating agent (not shown).

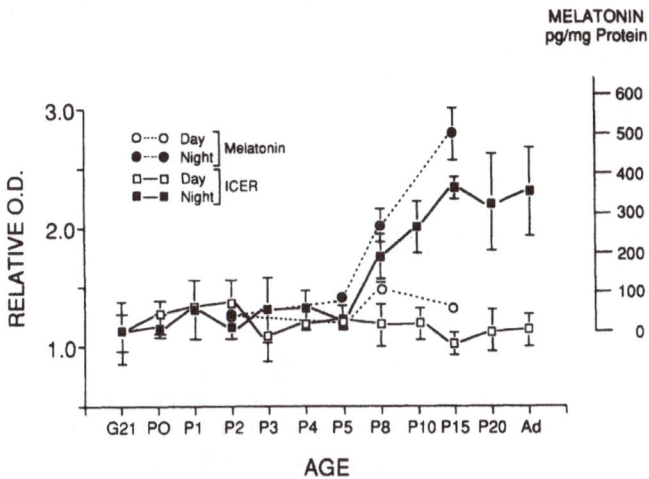

Figure 4. Developmental onset of the day-night switch in ICER expression. Analysis of relative optical density (O.D.) of ICER hybridization signal in rat pineal gland (solid line) obtained in several independent *in situ* hybridization experiments and superimposed pineal melatonin values (dashed line). Significant differences between night- and daytime values of the ICER hybridization signal were first observed in 8 day old animals (P8: $p<0.05$; P15: $p<0.01$). Error bars indicate SEM.

The cAMP signalling pathway is of pivotal importance for mammalian melatonin synthesis (Vanecek et al., 1985). Moreover, cAMP can modulate via CRE binding proteins, such as CREB and CREM (Borrelli et al., 1992) the transcriptional rate of genes expressed in the pineal gland. The CREM gene product ICER is a paradigm for a clock-controlled and transmitter-activated gene of the cAMP-inducible class (Stehle et al., 1993). The crucial role of the pinealopetal sympathetic innervation and its neurotransmitter NE for cAMP-mediated initiation of pineal melatonin synthesis (Klein, 1985) can now be extended to the level of the transcriptional activation of a cAMP-inducible gene, ICER. In addition, the ontogenetic maturation of the cAMP-signalling pathway in the rat pineal gland shapes the appearance of a molecular switch, the onset of a differential day-night expression of ICER. Furthermore, a circadian expression of ICER in rat photoneuroendocrine system is not detected in the retina or the SCN, but it is rather restricted to the superficial and deep portion of the rat pineal gland (Stehle et al., in press).

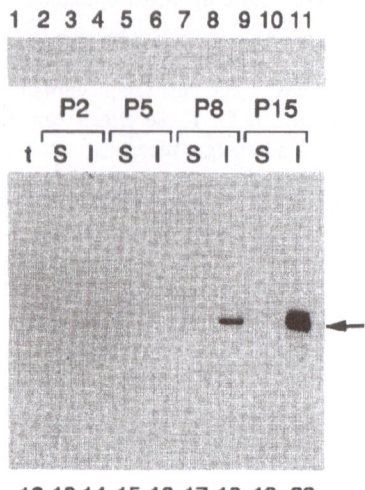

Figure 5. Analysis of ICER expression in rat pineal gland performed in animals sacrificed at postnatal day 2 (P2), P5, P8 and P15, 2 hours following injections of saline (S) or isoproterenol (I). An arrowhead indicates the specific ICER protected fragment. t: tRNA control.

Both, the superficial and the deep part of the rat pineal, are derived from a common diencephalic anlage (Vollrath, 1981). Fluorescence-microscopic investigations have revealed that both parts are innervated by sympathetic postganglionic fibres originating from the SCG (Björklund et al., 1972). The function of this innervation is well characterized for the superficial part of the gland, as the site of melatonin synthesis. The precise function of the deep pineal remains an enigma so far (Korf and Møller, 1985). However, the ICER expression in the deep pineal gland demonstrates the presence of a gene regulated by the clock-driven release of NE from pinealopetal nerve fibres reaching this structure.

CREM is the first gene in the pineal whose differential inducibility has been demonstrated during ontogeny. Development of inducibility coincides temporally with the naturally occurring onset of its stimulation (Stehle et al., in press). Thus, ICER gains inducibility just when the sympathetic innervation of rat pineal gland starts to function during the second week postnatally. The uninducibility of the ICER promoter at early postnatal stages together with the low amount of the ICER message in the pineal gland suggests a transcriptional block, which is exempted upon establishment of a circadian rhythm in NE release from the matured sympathetic innervation. Interestingly, a similar pattern of inducibility and of functional activation has been described for HIOMT (Sugden and Klein, 1983), the final enzyme converting N-acetylserotonin into melatonin. The fact that HIOMT activity can be first demonstrated at the beginning of the second postnatal week provides another example of a temporal gene blockade during a restricted developmental period. For ICER this may very well be explained by the autoregulatory feedback inhibition of the ICER gene (Molina et al., 1993). It is tempting to assume that the HIOMT gene comprises CRE elements as well, which could potentially be targeted by the inhibitory action of ICER. Low amounts of ICER expressed during the early postnatal period may be sufficient to inhibit the transcription of selective cAMP-inducible genes, thus providing a protective mechanism against inappropriate melatonin production. At the molecular level, the complete maturation of cAMP inducibility of gene expression in the pineal gland may thus represents a final step by which pups gain independence from circadian cues initially provided by their mother. Once the neuronal circuits needed for a complete melatonin-rhythm generating system are established, the massive nocturnal release of NE from the sympathetic nerve endings may stimulate transcription of previously blocked cAMP-inducible genes.

REFERENCES

Aschoff, J., 1981, Biological Rhythms, in: "Handbook of Behavioral Neurobiology", Vol. 4, Plenum Press, New York,

Axelrod, J., 1974, The pineal gland: a neurochemical transducer, *Science* 184:1341-1348

Björklund, A., Owman, C.H., and West, K.A., 1972, Peripheral sympathetic innervation and serotonin cells in the habenular region of the rat brain, *Z. Zellforsch.* 127:570-579

Borrelli, E., Montmayeur, J., Foulkes, N., and Sassone-Corsi, P., 1992, Signal transduction and gene control: the cAMP pathway, *CRC Reviews Oncogene* 3:321-338

Collins, S., Caron, M., and Lefkovitz, R., 1992, From ligand binding to gene expression: new insights into the regulation of G-protein-coupled receptors, *TIPS* 17:37-39

Darwin, C., 1880, The power of movement in plants. Murray, London

Duncan, M.J., Banister, M.J., and Reppert S.M., 1986, Developmental appearance of light dark entrainment in the rat, *Brain Research* 369:326-330

Ellison, N., Weller J.L., and Klein D.C., 1972, Development of a circadian rhythm in the activity of pineal serotonin N-acetyltransferase, *J. Neurochem.* 19:1335-1341

Foulkes, N., Mellström, B., Benusiglio, E., and Sassone-Corsi, P., 1992, Developmental switch of CREM function during spermatogenesis: from antagonist to activator, *Nature* 355:80-84

Foulkes, N.S., Schlotter, F., Pévet, P., and Sassone-Corsi, P., 1993, Pituitary hormone FSH directs the CREM functional switch during spermatogensis, *Nature* 362:264-267

Karin, M., 1992, Signal transduction from cell surface to nucleus in development and disease, *FASEB J* 6:2581-2590

Klein, D.C., 1985, Photoneural regulation of the mammalian pineal gland, *in*: "Photoperiodism, Melatonin and the Pineal", Pitman, London

Klein, D.C., Moore R.Y., and Reppert, S.M., 1991, "Suprachiasmatic Nucleus: The Mind's Clock", Oxford University Press, Oxford

Klein, D.C., Auerbach, D., Namboodiri, A., and Wheeler, G., 1981, Indole metabolism in the mammalian pineal gland, *in*: "The Pineal Gland: Anatomy and Biochemistry", CRC Press, Boca Raton

Korf. H.-W., 1994, The pineal organ as a component of the biological clock. Phylogenetic and ontogenetic considerations, *Ann. N. Y. Acad. Sci.* 719:13-42

Korf, H.-W., and Møller, M., 1985, The innervation of the mammalian pineal gland with special reference to central pinealopetal projections, *Pineal Res. Rev.* 2:41-86

Korf, H.-W., and Eckström, P., 1987, Photoreceptor differentiation and neuronal organization of the pineal organ, *in:* "Fundaments and Clinics in Pineal Research", Raven, New York

Lalli, E., and Sassone-Corsi, P., 1994, Signal transduction and gene regulation: the nuclear response to cAMP. *J. Biol. Chem.* 269:17359-17362

Molina, C.A., Foulkes, N.S., Lalli, E., and Sassone-Corsi, P., 1993, Inducibility and negative autoregulation of CREM: an alternative promoter directs the expression of ICER, an early response repressor, *Cell* 75:875-886

Montminy, M., Gonzalez, G., and Yamamoto, K., 1990, Regulation of cAMP-inducible genes by CREB, *TINS* 13:184-188

Moore, R.Y., Shibata, S., and Bernstein, M., 1989, Developmental anatomy of the circadian system, *in*: "Development of Circadian Rhythmicity and Photoperiodism in Mammals", S.M. Reppert, ed., Perinatology Press, New York

Oksche, A., 1971, Sensory and glandular elements of the pineal organ, *in*: "The Pineal Gland", Churchill-Livingstone, Edinburgh

Reiter, R.J., 1991, Pineal gland. Interface between photoperiodic environment and the endocrine system. *Trends Endocr. Met.* 1:13-19

Reppert, S.M., and Schwartz, W.J., 1984, The suprachiasmatic nuclei of the fetal rat: characterization of a functional clock using ^{14}C-labelled deoxyglucose, *J. Neurosci.* 4:1677-1682

Reppert, S.M., Weaver, D.R., and Rivkees, S.A., 1989, Prenatal function and entrainment of circadian clock, in: "Development of Circadian Rhythmicity and Photoperiodism in Mammals", S.M. Reppert, ed., Perinatology Press, New York

Skene, D.J., N-acetyltransferase and melatonin in the retina: regulation, function and mode of action, *Biochem. Soc. Transact.* 20:16-19 (1992)

Stehle, J., Rivkees, S., Lee, J., Weaver, D., Deeds, J., and Reppert, S., 1992, Molecular cloning and expression of a cDNA for a novel A_2-adenosine receptor subtype, *Mol Endocrinol* 6:384-393

Stehle, J., Foulkes, N., Molina, C., Simmoneaux, V., Pévet, P., and Sassone-Corsi, P., 1993, Adrenergic signals direct rhythmic expression of transcriptional repressor CREM in the pineal gland, *Nature* 365:314-320

Stehle, J., Foulkes, N., Pevet, P., and Sassone-Corsi, P., Developmental maturation of pineal gland function: synchronized CREM inducibility and adrenergic stimulation, *Mol Endocrinol,* in press

Sugden, D., and Klein, D.C., 1983, Regulation of rat pineal hydroxyindole-O-methyltransferase in neonatal and adult rats, *J. Neurochem.* 40:1647-1653

Sugden, D., Vanecek, J., Klein D.C., Thomas, T.P., and Anderson, W., 1985, Action of protein kinase C potentiates isoprenaline-induced cyclic AMP accumulation in rat pinealocytes, *Nature* 314:359-361

Vanecek, J., Sugden, D., Weller, J., and Klein, D.C., 1985, Atypical synergistic α_1- and β_1-adrenergic regulation of adenosine 3′, 5′-monophosphate in cultured rat pinealocytes, *Endocrinology* 116:2167-2173

Vollrath, L., The pineal organ, 1981, *in*: "Handbuch der mikroskopischen Anatomie des Menschen", A. Oksche, L. Vollrath, ed., Springer, Berlin

INTRACELLULAR ACTIONS OF MELATONIN WITH A SUMMARY OF ITS INTERACTIONS WITH REACTIVE OXYGEN SPECIES

Russel J. Reiter
Department of Cellular and Structural Biology
University of Texas Health Science Center
7703 Floyd Curl Drive
San Antonio, Texas 78284 USA

INTRODUCTION

Until recently it was assumed that all of melatonin's actions in mammals relied on the binding of melatonin to specific membrane receptors in select areas of the central nervous system as well as to some cells outside of the brain, e.g., the pars tuberalis of the anterior pituitary gland (Stankov et al., 1991; Morgan, 1993). However, recently intracellular binding sites have been identified in the nuclei of hepatic cells (Acuña-Castroviejo et al., 1993, 1994), leaving open the possibly that similar intranuclear binding sites may be found in other cells as well. Additionally, the high lipid solubility of melatonin allows it to enter cells readily where, in the cytosol, it may form a complex with calmodulin (CaM) thereby influencing the actions of this widely acting intracellular molecule (Benitz-King and Anton-Tay, 1993). Finally, melatonin is now known to be a potent oxygen radical scavenger throughout the cell (Tan et al., 1993a). This being the case, melatonin will obviously influence any cellular event which depends on the redox state of the cell.

Since the actions of melatonin via membrane receptors is well accepted and has been reviewed in several recent publications (Dubocovich, 1993; Stankov et al., 1993), the current brief report will summarize non-receptor-mediated actions of melatonin and, furthermore, the findings will be put into clinical perspective.

OXYGEN, REACTIVE OXYGEN SPECIES AND OXIDATIVE STRESS

The best evidence available suggests that life on the plant Earth began about 4 billion years ago, a time at which the atmosphere was devoid of oxygen. Sometime during the next 2 billion years, photosynthetic processes began thereby increasing the oxygen (O_2) content of the atmosphere probably to a level approaching that of the present day (about 21%). While organisms obviously adapted to the increased O_2 levels, nevertheless, its appearance must have been a major insult to organisms that had been totally anaerobic. Obviously, they adapted well since they evolved mechanisms to take advantage of the aerobic environment for the production of energy, a process during which O_2 is reduced to form water.

The fact that the "outer" electrons of the O_2 molecule are in a triplet spin state is an important feature since it prevents the molecule from completely oxidizing all biological material. Molecular oxygen is, in a technical sense, a free radical; thus, the two "outer" electrons are unpaired but they exist in separate orbitals. This is an important property of O_2 because molecules having triple state electrons will not interact with molecules having singlet state electrons. The large majority of biological molecules are in a singlet state and because of this, molecular ground state oxygen (which is in the triplet state) will not interact with and oxidize them. On the other hand, if molecular oxygen is energized to form singlet oxygen by the addition of 23 kcal, the higher energy state allows it to readily oxidize biological molecules or, as usually stated, to induce oxidative stress.

Besides its conversion to singlet oxygen, in order for O_2 to oxidize other molecules it must be converted to reactive oxygen species which have reduction states between O_2 and water. The reactive oxygen species, also referred to a oxygen free radicals, are usually formed at low levels in biological systems and they mediate oxidative damage. A free radical is a molecule or a part of a molecule which has an unpaired valence electron in its outer orbital. In this state, the unparied valence electron of an atomic constitutent does not contribute to the bonding within a molecule and, in this sense, it is "free".

The free radicals of oxygen are the superoxide anion ($O_2 \div$) and the hydroxyl radical ($\cdot OH$) while hydrogen peroxide (H_2O_2) is an intermediate between $O_2 \div$ and $\cdot OH$ (Fig. 1).

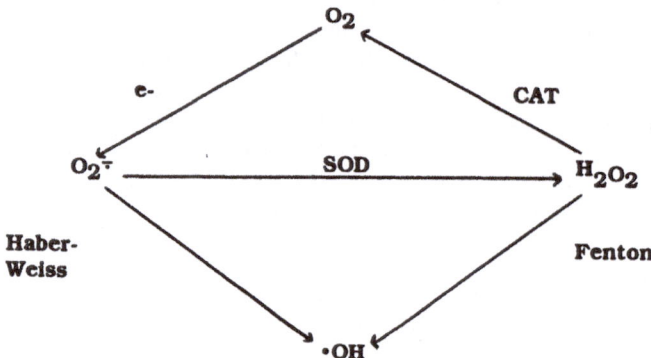

Figure 1. A single electron reduction of molecular oxygen (O_2) results in the generation of the superoxide anion radical ($O_2\div$) which, in a reaction catalyzed by superoxide dismutase (SOD), leads to the synthesis of hydrogen peroxide (H_2O_2). Via the Fenton reaction, which requires a transition metal, H_2O_2 is converted to the hydroxyl radical ($\cdot OH$). The Haber-Weiss reaction involves both $O_2\div$ and H_2O_2 leading to $\cdot OH$ generation. Catalase (CAT) can metabolize H_2O_2 to O_2.

Whereas H_2O_2 is not *per se* a free radical it, like $O_2 \div$ and $\cdot OH$, is a reactive oxygen species and, therefore, it too can induce biological damage. The $O_2 \div$, which is the result of the addition of a single electron to O_2, is negatively charged at physiological pH and it readily passes through membranes on the anion channel (Fridovich, 1978). H_2O_2 is formed in a variety of biological reactions and it easily passes through cellular membranes (Floyd, 1990). $\cdot OH$ once formed react very rapidly (within nanoseconds) very near (several Å) to where they were formed (Girotti, 1990); these interactions are indiscriminate. The circumscribed area in which the •OH radical reacts is sometimes referred to its "reaction cage" (Borg, 1993). •OH react at essentially diffusion mediated rates by adding to or abstracting a hydrogen atom from molecules; in so doing, they form another free radical. Thus, free radical chain reactions are set in motion leading to the formation of progressively more radicals. These radical chain reactions are especially pertinent to lipid peroxidation where polyunsaturated fatty acids of membrane phospholipids are oxidized (Girotti, 1985).

When H_2O_2 reacts with localized Fe on a protein, the resulting •OH will damage specific amino acids in the protein, in particular histidine, lysine, proline and arginine (Stadtman, 1990). This localized damage is referred to as being site specific and it is an important aspect of oxidative damage to DNA, RNA and proteins. Site specific damage is dictated by site specificity of the binding of the transition metal Fe (or Cu) to the macromolecule.

More than 95% of the O_2 that enters mammalian cells is used for the production of energy in the form of adenosine triphosphate (ATP). However, the remaining O_2 is divert-ed to the formation of oxygen free radicals or reactive oxygen species (Fig. 1) as summarized above. The most toxic of these is the •OH. The damage inflicted by oxygen radicals is generally referred to as oxidative stress (Pryor, 1993).

ANTIOXIDATIVE DEFENSE MECHANISMS

Oxidative stress is an important concept since it provides scientists with a means of evaluating oxidative damage. Even though oxygen is unequivocally required for life of aerobics organisms, it presence also ensures that organisms will experience damage by reactive oxygen species. Floyd (1993a) refers to the imposed oxidative stress as oxidative damage potential. Ideally, oxidative damage potential induced by free radicals should be balanced by the antioxidative defense capacity of the organism (Fig. 2). In reality, however, it is usual that the oxidative damage potential exceeds that of the antioxidative defense capacity resulting in cellular damage, disease, aging, and possibly death.

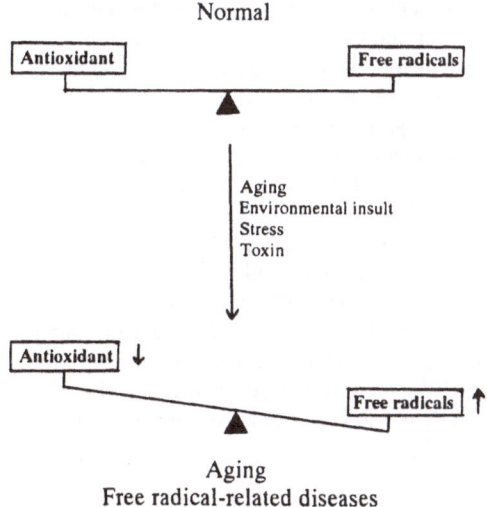

Figure 2. Under ideal conditions the antioxidant capacity of the organism would be equal to or greater than the free radical production within the organism. A number of factors including aging, etc., tilts the balance in favor of free radicals thereby overwhelming the antioxidative defense system. This may lead to aging and the development of age-related diseases.

23

To limit the damage induced by free radicals and reactive oxygen species, organisms have evolved a series of defense mechanisms. Thus, aerobic organisms have develop mechanisms to not only avert damage but to repair it when it occurs. To defend against oxidative damage, organisms rely on both antioxidant enzymes and biological antioxidants, i.e., molecules that neutralize free radicals. Some of the important antioxidative enzymes include superoxide dismutase (SOD, of which there are at least two forms, i.e., copper/zinc and manganese SOD); these enzymes dismutate the $O_2 \div$ to O_2 and H_2O_2. Although generally classified as an antioxidant enzyme, when it is over-expressed such as in Down syndrome (trisomy 21), it results in the eventual increased production of $\bullet OH$ which are highly damaging (Yatham et al, 1988). Other antioxidantive enzymes include catalase which decomposes H_2O_2 to O_2 and water and glutathione peroxidase (GPx) which reduces H_2O_2 to water. GPx also metabolizes hydroperoxides, reducing them to alcohols. Glutathione S-transferase also metabolizes hydoperoxides to alcohols.

There are a large number of biological antioxidants which scavenge oxygen free radicals. Some of these are water soluble and include reduced glutathione (GSH), ascorbic acid (vitamin C) and uric acid. Some of the lipid soluble antioxidants include vitamin E (especially α-tocopherol), carotinoids and ubiquinones. The antioxidant melatonin (Tan et al., 1993a) is in a special category as a scavenger of the $\bullet OH$ because while being highly lipophilic (Reiter, 1991a) is also quite hydrophilic (Shida et al., 1994). Thus the ease with which this small indole passes through cellular membranes and fluids makes it accessible to all parts of the cell.

GSH is one of the most abundant biological reductants and acts as both a thiol reagent and as a substrate by antioxidative enzymes including GPx and glutathione S-transferase (Meister, 1993). Ascorbic acid is likewise abundant in cells and acts to protect against oxidative damage especially in the cytosol. Carotenoids and uric acid probably behave as singlet oxygen quenchers and as free radical scavengers. Vitamin E and ubiquinones are most highly localized in cellular membranes because of their lipid solubility were they function as free radical scavengers (Yu, 1993).

The repair of oxidative damage is also important in cellular homeostasis in organisms. Some of the DNA oxidized by free radicals is repaired by enzymes such as endonuclease and glycosylase (Bernstein and Gensler, 1993). Oxidized proteins are removed by the action of proteases (Stadtman, 1988). Finally, oxidized lipids are directly reduced by glutathione peroxidase or after hydrolysis with phospholipase (Yu, 1993). The role of melatonin, if any, in repairing oxidatively damaged molecules remains univestigated.

It is obvious from many *in vivo* studies that oxidative damage to cellular macromolecules occurs daily and that it goes unrepaired (Floyd, 1993b). Judging from the generation of $O_2 \div$ *in vitro* by mitochondria and microscomes from lung and hepatic tissue, the formation of oxygen free radicals may occur at the rate of 50 nmol/g tissue/min or approximately 10^{11} radicals/cell/day (Fridovich and Freeman, 1986). Furthermore, it has been estimated that there are 8.6×10^4 oxidized DNA residues formed/cell/day and, comparing the quantity of oxidized DNA residues and the number of oxygen radicals produced, it is estimated that one oxidized DNA residue is a consequence of the production of 7.6×10^5 oxygen radicals (Fraga et al., 1990). The role of the antioxidative defense system is to keep oxidative damage to a minimum.

INTERACTION OF MELATONIN WITH REACTIVE OXYGEN SPECIES

The first work to suggest that melatonin may be a free radical scavenger *in vivo* was that of Chen et al (1993). In this study, we found a pineal-mediated rhythm in cardiac Ca^{2+} + stimulated + Mg^{2+}-dependent ATPase (calcium pump) activity in rats. Since the rhythm obviously required an intact pineal gland, we surmised that the hormone responsible was melatonin. Thus, we purified sarcolemmal membranes to which were added melatonin and, in a dose-dependent manner, the activity of the calcium pump increased.

The mechanism by which the addition of melatonin induced a rise in the activity of the calcium pump, however, remained unknown. It had been previously shown that activity of this enzyme was directly related to the redox state of the cell; in particular increased free radicals had been shown to reduce Ca^{2+} + Mg^{2+} ATPase activity (Kaneko et al., 1989). Therefore, Chen and colleagues (1993) theorized that melatonin may in fact be a free radical scavenger. This idea was especially attractive because, if so, it would explain many of the other subtle actions of melatonin in the organism (Reiter, 1991b).

These studies were quickly followed by those of Tan et al (1993a) which directly tested the ability of melatonin to scavenge •OH in an *in vitro* hydroxyl radical generating system. We exposed H_2O_2 to ultraviolet light (at 254nm) in the presence of 5,5-dimethylpyrroline N-oxide (DMPO). DMPO is a spin trapping reagent that forms an adduct with •OH. It was necessary to trap the •OH because, by themselves, they have an extremely short half life and, therefore, they are impossible to directly measure (Finkelstein et al., 1980). The reaction was as follows

$$H_2O_2 + DMPO \xrightarrow[254nm]{UV} DMPO - •OH$$

The adducts (DMPO-•OH) that were formed were quantitified by high performance liquid chromotography with electrochemical detection and verified with electron spin resonance spectroscopy (Floyd et al., 1984). Having developed this system, Tan et al (1993a) then added either melatonin, 5-methoxytryptamine, glutathione or mannitol (the latter two compounds were known free radical scavengers) and compared them in terms of their ability to quench the most toxic oxygen radical, the •OH. Melatonin not only proved to be an efficient scavenger, based on its IC_{50} (the concentration required to quench 50% of the radicals) melatonin was more effective by factors of 5 and 15 compared to glutathione and mannitol, respectively, in neutralizing the •OH. Methoxytryptamine was a rather weak scavenger in this system. These findings were remarkable since, prior to this, glutathione was thought to be the most efficient, endogenously-produced free radical scavenger, and melatonin appeared to the 5 times better.

The next step was to determine if melatonin was an equally efficient antioxidant *in vivo*. To test this, Tan et al (1993b) selected a model system which was well characterized. This entailed the treatment of rats with the chemical carcinogen safrole. Safrole damages cellular macromolecules because it and its metabolites induce the widespread production of toxic oxygen radicals. The endpoint in these studies was damaged DNA which was identified and quantitified using the method of Reddy and Randerath (1986). Twenty-four hours following an injection of safrole (300 mg/kg), massive DNA damage occurred in the livers of safrole only treated rats. When melatonin was given concurrently with the carcinogen, DNA damage was significantly reduced. Thus, when rats were given a dose of melatonin (0.2 mg/kg) that was 1500 times less than that of safrole, the amount of damage to DNA was reduced by about 40% (Fig. 3).

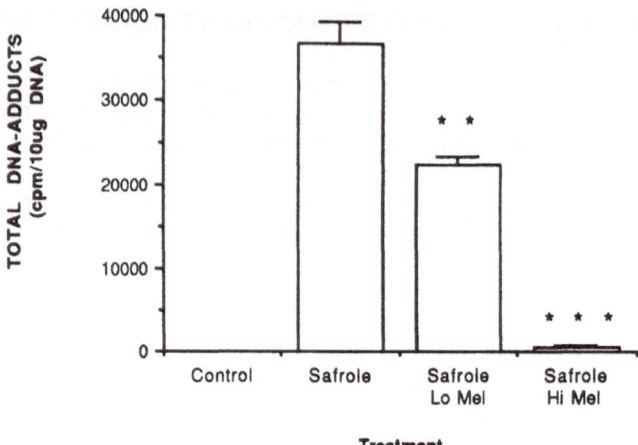

Figure 3. Total DNA-adducts (damaged DNA species) in the livers of untreated control rats and in rats treated with the chemical carcinogen safrole (300 mg/kg) with and without melatonin administration. A low (LoMel, 0.2 mg/kg) and a high (HiMel, 0.4 mg/kg) dose of melatonin were used; both melatonin doses greatly reduced DNA damage. From Tan et al (1993b).

Furthermore, when the dosage of melatonin was doubled to 0.4 mg/kg (but a dose still 750 times less than that of safrole) the number of DNA adducts was reduced by more than 90%. Thus, melatonin, or a metabolite, clearly provided a high degree of protection for DNA against attack by toxic oxygen free radicals. These findings are consistent with melatonin being an important aspect of the antioxidative defense system *in vivo*.

Whereas these findings were significant, the doses of melatonin used certainly caused blood melatonin levels to exceed those that are measured physiologically. Hence, whether physiological concentrations of melatonin could protect against reactive oxygen species remained unknown, but was soon checked. These studies depended on the endogenous rhythm of melatonin (Reiter, 1991a), i.e., low levels during the day and high values at night. Rats were given an injection of safrole (100 mg/kg) either early in the 14h light period or early in the 10h dark period. Livers were collected from the animals 8 hours later, DNA was extracted, and DNA adducts were quantified (Tan et al., 1994). In this study DNA damage was less at night (when endogenous melatonin levels were high) than during the day (when endogenous melatonin levels were low), suggesting that the higher nocturnal melatonin concentrations probably reduced the oxidative damage to DNA. This idea was supported by the observation that surgical pinealectomy, which reduces circulating melatonin levels, further increased DNA adduct formation (Tan et al., 1994). Finally, as in their first study the administration of melatonin exogenously greatly reduced damage to DNA that was a consequence of safrole injection. These results confirmed that melatonin is indeed an important aspect of the antioxidative defense system and that it is so at physiological, as well as pharmacological, concentrations.

Another indirect test of melatonin's ability to protect against oxidative stress has been published. In this case, rats were treated with alloxan alone or in conjunction with melatonin (Chen et al., 1994). Alloxan damages cell membranes, especially those of the ß-cells of the pancreatic islets, because it generates free radicals (Malaise, 1982). Because of this action, alloxan also reduces the activity of the calcium pump in cardiomyocyte membranes. However, when melatonin was given concurrently with the exposure of rats to alloxan, it prevented the depression in the activity of the calcium pump in a dose-dependent manner (Fig. 4) (Chen et al., 1994). These findings, like those that preceded them, support the idea that melatonin is an important oxygen radical scavenger.

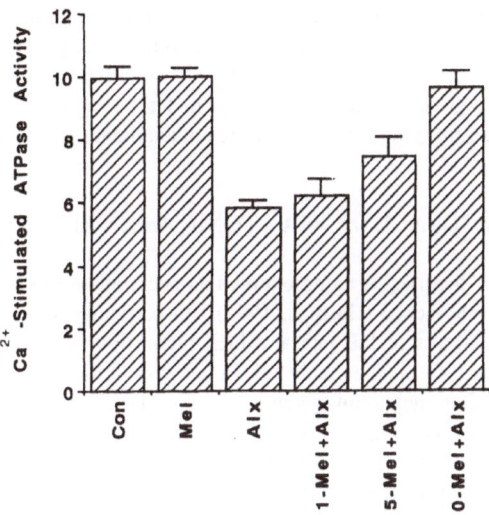

Figure 4. Effect of melatonin (either a 1, 5 or 10 mg/kg dose) on the depression of Ca^{2+}-ATPase activity in the heart of rats treated with alloxan (Alx). Melatonin reversed the depression in enzyme activity in a dose-dependent manner.

The protection afforded cellular macromolecules by melatonin was so marked that we suspected that melatonin may operate by more than one means to limit damage, i.e., melatonin was suspected of doing something in addition to merely scavenging toxic oxygen radicals. On this basis, we examined the effects of melatonin on an important antioxidative enzyme, GPx. In the brain GPx activity represents a major aspect of the antioxidative defense system since neural tissue in generally deficient in other defenses against oxidative attack. Within 30 min after melatonin administration rat brain GPx activity was found to increase 3-fold (L.R. Barlow-Walden, B. Poeggler, and R.J. Reiter, unpublished); this activity remained elevated for at least 180 min after the melatonin was injected. Considering, as already noted, that the brain relies heavily on GPx activity for its protection against oxygen free radicals, this finding suggests that the protection afforded the brain by melatonin may be very significant. The findings are also consistent with the high uptake of melatonin in the brain, particularly into the cell nuclei (Menendez-Pelaez and Reiter, 1993; Menendez-Pelaez et al., 1993).

In another test of melatonin's ability to overcome free radical damage *in vivo*, newborn rats were given buthionine sulfoximine (BSO) to inhibit the production of GSH; as a consequence of this treatment the new born animals, by the time they are 2 weeks of age, develop cataracts (Calvin et al., 1986). GSH normally prevents cataracts from developing because of its free radical scavenging ability. When BSO-treated rats were given melatonin for the first 2 weeks of life, cataract formation was essentially totally prevented (Abe et al., 1994). Thus melatonin readily substituted for GSH in preventing cataract formation, likely, because of its free radical scavening ability.

Not all the studies espousing melatonin as an antioxidant have come from the same laboratory. Using mice, Pierrefiche et al (1993) observed that melatonin, in a dose-dependent manner, lowered blood glucose levels in animals treated with alloxan to destroy their insulin producing cells in the pancreas. They also found that melatonin protected, to a small degree, lipid peroxidation in neural cell membranes induced by thiobarbituric acid.

Finally, in a publication by Blinkenstaff and colleagues (1994) melatonin was found to increase the survival of mice subjected to ionizing radiation. Ionizing radiation induces tissue damage and causes death because it induces the production of oxygen free radicals. In the same paper, Blinkenstaff et al (1994) cite the work of another individual who claimed an even greater protective effect of melatonin against irradiation; this was a personal communication so the actual data were not available.

CLINICAL SIGNIFICANCE OF MELATONIN AS AN ANTIOXIDANT

One of the major theories of aging claims that accumulated free radical damage is a causative factor of senescence (Harman, 1992). In particular, damage to DNA may be highly significant in terms of the rate at which organisms age (Halliwell and Aruoma 1991; Rao and Loeb, 1992). The studies of Tan et al (1993b, 1994), summarized above, show convincingly that melatonin potently protects DNA from oxidative damage by oxygen free radicals. An implication of these findings, and one already suggested (Poeggler et al., 1993; Reiter et al., 1993, 1994b), is that the hormone melatonin, which is greatly reduced during aging (Reiter, 1992), may have significant anti-aging protential.

Besides aging, a variety of age-related diseases have been at least tentatively linked to free radical damage (Harman, 1993; Oberley and Oberley, 1993; Strong et al., 1993). In particular, cancer initiation, which involves as an early step, DNA damage, could possibly be reduced if melatonin remained amply available throughout life (Reiter et al., 1994a). Also, neurodegenerative disorders are widely believed to involve free radical damage to at least selected areas of the brain (Strong et al., 1993). This being the case, the potential benefits of melatonin in deferring neurodegenerative changes seems obvious (Poeggeler et al., 1993; Reiter et al., 1993, 1994a, 1994b, 1994c; Reiter, 1994). This possibility is even more likely considering the ability with which melatonin is taken up by the brain (Menendez-Pelaez et al., 1993) and the rapidity with which melatonin stimulates neural GPx activity, an enzyme esepcially important in the anitoxidative defense system of the central nervous system.

INTERACTION OF MELATONIN WITH CALMODULIN

Besides its direct interaction intracellularly with free radicals (Poeggeler et al., 1993; Reiter et al., 1994b) and possibly with nuclear receptors (Acuña-Castroviejo et al., 1993, 1994), melatonin may have another important function which does not rely on the previously described membrane receptors for the hormone (Dubcovich, 1993; Stankov et al., 1993). Thus, it was initially described that, after its entrance into cells, melatonin interacted with calmodulin (CaM) (Benitz-King et al., 1991). Subsequent studies in fact revealed that melatonin actually binds to CaM (Benitz-King et al., 1990, 1993); indeed, in a series of experiments using a variety of different measurements this group characterized the kinetic constants of melatonin binding to CaM. CaM, of course, modulates a variety of intracellular metabolic pathways. Two cell lines, i.e., MDCK and N1E-115, were used to define CaM mediated melatonin actions. Using these *in vitro* models, it was shown that melatonin's effects on tubulin polymerization and its cytoskeletal effects are mediated by melatonin's antagonism to Ca^{2+}- calmodulin; these effects are observed in the presence of low concentrations ($10^{-9}M$) of melatonin. At melatonin concentrations of $10^{-5}M$ nonspecific binding of melatonin to tubulin occurs thereby overcoming the specific melatonin antagonism to Ca^{2+}-calmodulin. Interestingly, both melatonin and CaM are phylogenetically well preserved molecules and their interaction could represent a major mechanism for the regulation of cell physiology.

CONCLUDING REMARKS

Although the actions of melatonin that are mediated via membrane receptors on specialized cells are undoubtedly important, it is becoming increasingly apparent that melatonin's intracellular actions, which do not rely on membrane receptors, may be equally or more important. Certainly, the direct action of melatonin within cells greatly strengthens the argument that melatonin is a critically important hormone in all organisms. It also helps to explain the multiple functions of melatonin in a variety of organ systems that have heretofore gone unexplained (Reiter, 1991b). Considering its virtual absence of toxicity, its ease of administration, its ready absorbance, and its seemingly widely beneficial effects, it is anticipated that melatonin will find its way into clinical medicine in the rather near future.

ACKNOWLEDGMENT

Work by the author was supported by grants from the NSF.

REFERENCES

Abe, M., Reiter, R.J., Orhii, P.B., Hara, M., and Poeggeler, B., 1994, Inhibitory effect of melatonin on cataract formation in newborn rats: Evidence for an antioxidative role for melatonin. *J. Pineal Res.*, in press.

Acuña-Castroviejo, D., Pablos, M.I., Menendez-Pelaez, A., and Reiter, R.J., 1993, Melatonin receptors in purified cell neclei of liver. *Res. Commun. Chem. Pathol. Pharmacol.* 82:253.

Acuña-Castroviejo, D., Reiter, R.J., Menendez-Pelaez, A., Pablos, M.I. and Burgos, A., 1994, Characterization of high affinity melatonin binding sites in purified cell nuclei of rat liver. *J. Pineal Res.* 16:100.

Benitz-King, G., and Anton-Tay, F., 1993, Calmodulin mediates melatonin cytoskeletal effects. *Experientia* 49:635.

Benitz-King, G., Huerto-Delgadillo, L., and Anton-Tay, F., 1990, Melatonin effects on the cytoskeletal organization of MDCK and neuroblastoma N1E-115 cells. *J. Pineal Res.* 9:209.

Benitz-King, G., Huerto-Delgadillo, L., and Anton-Tay, 1991, Melatonin modifies calmodulin cell levels in MDCK and N1E-115 cell lines and inhibits phosphodiesterase activity *in vitro. Brain Res.* 557:289

Benitz-King, G., Huerto-Delgadillo, L., and Anton-Tay, F., 1993, Binding of 3H-melatonin to calmodulin. *Life Sci.* 53:201.

Bernstein, H., and Gensler, H.L., 1993, DNA damage and aging, *in*: "Free Radicals in Aging", B.P. Yu, ed., CRC Press, Boca Raton.

Blinkenstaff, R.T., Brandstadler, S.M., Reddy, S., and Witt, R., 1994, Protective radioprotective agents. 1. Homologs of melatonin. *J. Pharmaceut. Sci.* 83:216.

Borg, D.C., 1993, Oxygen free radicals and tissue injury; *in*: "Oxygen Free Radicals in Tissue Injury", M. Tarr and F. Samson, eds., Birkhäuser, Boston.

Calvin, H.I., Medvedovsky, C., and Worgul, B.V., 1986, Near-total glutathione depletion and age-specific cataracts induced by buthionine sulfoximine in mice. *Science* 223:553.

Chen, L.D., Tan, D.X., Reiter, R.J., Yaga, K., Poeggeler, B., Kumar, P., Manchester, L.C., and Chambers, J.P., 1993, In vivo and in vitro effects of the pineal gland and melatonin on $[Ca^{2+} + Mg^{2+}]$-dependent ATPase in cardiac sarcolemma. *J. Pineal Res*, 14:178.

Chen, L.D., Kumar, P., Reiter, R.J., Tan, D.X., Manchester, L.C., Chambers, J.P., Poeggeler, B., and Saarela, S, 1994, Melatonin prevents the suppression of cardiac Ca^{2+}- stimulated ATPase activity induced by alloxan. *Am. J. Physiol.*, 267:E57.

Dubocovich, M.L., 1993, Melatonin receptors in retina brain and pituitary, *in* "Advances in Pineal Research, Vol. 6", A. Foldes and R.J. Reiter, eds., John Libbey, London.

Finkelstein, E., Rosen, G.M., and Rauchman, E.J., 1980, Spin trapping. Kinetics of the reaction of superoxide and hydroxyl radicals with nitrones. *J. Am. Chem. Soc.* 102:4994.

Floyd, R.A., 1990, Role of oxygen free radicals in carcinogenesis and brain ischemia. *FASEB J.* 4:2587.

Floyd, R.A., 1993a, Basic free radical biochemistry, *in:* "Free Radicals in Aging", B.P. Yu, ed., CRC Press, Boca Raton.

Floyd, R.A., 1993b, In vivo detection of oxygen free radical species, *in:* "Oxygen Free Radicals in Tissue Damage," M. Tarr and F. Samson, eds., Birkhäuser, Boston.

Floyd, R.A., Lewis, C.A., and Wong, R.J., 1984, High pressure liquid chromatography-electrochemical detection of oxygen free radicals. *in:* "Methods of Enzymology, Vol. 105", L. Packer, ed., Academic Press, San Diego.

Fraga, C.G., Shigenaga, M.K., Park, J.W., Degan, P., and Ames, B.N., 1990, Oxidative damage to DNA during aging: 8-hydroxy-2'-deoxyguanosine in rat organ DNA and urine. *Proc. Natl. Acad. Sci. USA*, 87:4533,

Fridovich, I., 1978, The biology of oxygen radicals. *Science* 201:875

Fridovich, I., and Freeman, B., 1986, Antioxidant defenses in the lung. *Ann. Rev. Physiol.* 48:693.

Girotti, A.W., 1985, Mechanisms of lipid peroxidation. *Free Rad. Biol. Med.* 1:87.

Girotti, A.W., 1990, Photodynamic lipid peroxidation in biological systerms. *Photochem. Photobiol.* 51:497.

Halliwell, B., and Aruoma, O.I., 1991, DNA damage by oxygen-derived species. *FEBS Lett* 281:9.

Harman, D., 1992, Free radical theory of aging. *Mutat. Res.* 275:257.

Harman, D., 1993, Free radicals and age-related diseases, *in:* "Free Radicals in Aging", B.P. Yu, ed., CRC Press, Boca Raton.

Kaneko, M., Beamish, R.E., and Dhalla, N.S., 1989, Depression of heart sarcolemma Ca^{2+}-pumping ATPase activity by oxygen free radicals. *Am. J. Physiol.* 256:H368.

Malaise, W.J., 1982, Alloxan toxicity to the pancreatic ß-cell: a new hypothesis. *Biochem. Pharmacol.* 25:1085.

Meister, A., 1993, Approaches to the therapy of glutathione deficiency, *in:* "Free Radicals: From Basic Science to Medicine," G. Poli, E. Albano, and M.V. Dianzani, eds., Birkhäuser, Boston.

Menendez-Pelaez, A., and Reiter, R.J., 1993, Distribution of melatonin in mammalian tissues: the relative importance of nuclear versus cytosolic localization. *J. Pineal Res.* 15:59.

Menendez-Pelaez, A., Poeggeler, B., Reiter, R.J., Barlow-Walden, L., Pablos, M.I., and Tan, D.X., 1993, Nuclear localization of melatonin in different mammalian tissues: immunocytochemical and radioimmunossay evidence. *J. Cell. Biochem.* 53:373.

Morgan, J.P. 1993, The pars tuberalis as a target tissue for melatonin action, *in*: "Advances in Pineal Research, Vol. 6," A. Foldes and R.J. Reiter, eds., John Libbey, London.

Oberley, T.D., and Oberley, L.W., 1993, Oxygen radicals and cancer, *in*: "Free Radicals in Aging," B.P. Yu, ed., CRC Press, Boca Raton.

Pierrefiche, G., Topall, G., Courboin, G., Henriet, I., and Laborit, H., 1993, Antioxidant activity of melatonin in mice. *Res. Commun. Chem. Pathol. Pharmacol.* 80:211.

Poeggeler, B., Reiter, R.J., Tan, D.X., Chen, L.D., and Manchester, L.C., 1993, Melatonin, hydroxyl radical-mediated oxidative damage, and aging: a hypothesis *J. Pineal Res.* 14:151.

Pryor, W.A., 1993, Oxidative stress status measurements in humans and their use in clinical trials, in: "Active Oxygens, Lipid Peroxides and Antioxidants," Yagi, L., ed., CRC Press, Boca Raton.

Rao, K.S., and Loeb, L. A., 1992, DNA damage and repair in the brain: relationship to aging. *Mutat. Res.* 275, 317.

Reddy, M.V., and Randerath, K., 1986, Nuclease P1-mediated enhancement of sensitivity of ^{32}P-postlabeling test for structurally diverse DNA adducts. *Carcinogenesis* 7:1543.

Reiter, R.J., 1991a, Melatonin: cell biology of its synthesis and of its physiological interactions. *Endocrine Rev.* 12:151.

Reiter, R.J., 1991b, Melatonin: that ubiquitously acting pineal hormone. *News Physiol. Sci.* 6:223.

Reiter, R.J., 1992, The aging pineal gland and its physiological consequences. *BioEssays* 14:169.

Reiter, R.J., 1994, Pineal function during aging: attenuation of the melatonin rhythm and its neurobiological consequences. *Acta Neurobiol. Exp.*, 54 (Suppl.):31..

Reiter, R.J., Poeggeler, B., Tan, D.X., Chen, L.D., Manchester, L.C., and Guerrero, J.M., 1993, Antioxidant capacity of melatonin: a novel action not requiring a receptor. *Neuroendocrinol. Lett.* 15:103.

Reiter, R.J., Tan, D.X., Poeggeler, B., Chen, L.D., and Menendez-Pelaez, A., 1994a, Melatonin, free radicals and cancer initiation. *in*: "Advances in Pineal Research, Vol. 7", G.J.M. Maestroni, A. Conti, and R.J. Reiter, eds., John Libbey, London.

Reiter, R.J., Tan, D.X., Poeggler, B., Menendez-Pelaez, A., Chen, L.D., and Saarela, S., 1994b, Melatonin as a free radical scavenger: implications for aging and age-related diseases. *Ann. N.Y. Acad. Sci.*, 719:1.

Reiter, R.J., Poeggeler, B., Chen, L.D., Abe, M., Hara, M., Orhii, P.B. Atia, A.M., and Barlow-Walden, L.R., 1994c, Melatonin as a free radical scavenger: theoretical implications for neurodegenerative disorders in the aged. *Acta Gerontol.*, in press.

Shida, C.S., Castrucci, A.M., and Lamy-Freund, M.T., 1994, High melatonin solubility in aqueous medium. *J. Pineal Res.*, 16:198.

Stadtman, E.R., 1988, Protein modification with aging, *J. Gerontol.* 43: B112.

Stadtman, E.R., 1990, Metal iron catalyzed oxidation of proteins: biochemical mechanisms and biological consequences. *Free Radical Biol. Med.* 9:315.

Stankov, B., Fraschini, F., and Reiter, R.J., 1991, Melatonin binding sites in the central nervous system. *Brain Res. Rev.* 16:245.

Stankov B., Fraschini, F., and Reiter, R.J., 1993, The melatonin receptor: distribution, biochemistry and pharmacology, *in*: "Melatonin", H.S. Yu and R.J. Reiter, eds., CRC Press, Boca Raton.

Strong, R., Mattamal, M.B., and Andorn, A.C., 1993, Free radicals, the aging brain, and age-related neurodegenerative disorders, *in*: "Free Radicals in Aging", B.P. Yu, ed., CRC Press, Boca Raton.

Tan, D.X., Chen, L.D., Poeggeler, B., Manchester, L.C., and Reiter, R.J. 1993a, Melatonin: a potent, endogenous hydroxyl radical scavenger. *Endocrine J.* 1:57.

Tan, D.X., Poeggeler, B., Reiter, R.J., Chen, L.D., Chen, S., Manchester, L.C., and Barlow-Walden, L.R. 1993b. The pineal hormone melatonin inhibits DNA adduct formation induced by the chemical carcinogen safrole in vivo. *Cancer Lett.* 70:65.

Tan, D.X.., Reiter, R.J., Chen, L.D., Poeggeler, B., Manchester, L.C., and Barlow-Walden, L.R., 1994, Both physiological and pharmacological levels of melatonin reduce DNA adduct formation induced by the carcinogen safrole. *Carcinogenesis* 15:215.

Yatham, L.N., McHale, P.A., and Kinsella, A., 1988, Down's syndrome and its association with Alzheimer's disease. *Acta Psychiatr. Scand.* 77:38.

Yu, B.P., 1993, Oxidative damage by free radicals and lipid peroxidatin in aging, *in*:"Free Radicals and Aging", B.P. Yu, ed., CRC Press, Boca Raton.

PHYSIOLOGIAL EFFECTS AND BIOLOGICAL ACTIVITY OF MELATONIN

Paul Pévet, Bruno Pitrosky, Mireille Masson-Pévet, Raymond Kirsch, Bernard Canguilhem and Berthe Vivien-Roels

Neurobiologie des fonctions rythmiques et saisonnières, CNRS-URA 1332, Université Louis Pasteur, Strasbourg, FRANCE

INTRODUCTION

In recent years, it has become clear that the pineal gland is essential for the regulation of photoperiodic response in mammals and that melatonin (MEL), one of the hormones produced by this gland, is directly involved in this phenomenom (reviews Reiter 1987a, Pévet 1988, Goldman and Nelson 1993, Malpaux et al. 1993). Synthesis of Mel occurs during the dark period of the light/dark cycle and is tighly controlled by the photoperiod (review Arendt, 1985). Natural or experimental changes in photoperiod modify the temporal patterns of melatonin secretion. In most mammalian species studied to date the duration of the nocturnal peak of MEL secretion is proportional to the length of the dark phase (reviews Goldman 1983 ; Hoffmann 1985 ; Pévet 1988). These data suggest that changes in the pattern of MEL production is the key in the transmission of photoperiodic information to the central nervous system.

The major question arising out of this effect of MEL concerns its precise site and mechanism of action.

The effect of MEL is believed to be mediated by membrane bound receptors. The MEL receptors that mediate seasonal changes in reproductive physiology are thought to be those that are located on the pars tuberalis (Skene et al., 1993 ; Masson-Pévet and Gauer, 1994 ; Lincoln, 1994). This suggestion is strengthened by the observation that, in seasonal breeders, the density of MEL receptors in this structure, but not in other areas such as the

The Pineal Gland and Its Hormones
Edited by F. Fraschini *et al.*, Plenum Press, New York, 1995

suprachiasmatic nucleus (SCN) of the hypothalamus varies with season (Skene et al. 1993 ; Gauer et al. 1993, 1994a, Masson-Pévet and Gauer 1994). Although MEL receptors represent the substrate for the effects of MEL, their presence alone does not adequately explain how the brain interprets the nocturnal exposure to MEL and responds appropriately to the photoperiod. The rhythmic secretory pattern of MEL may also play some direct role. Several features of the MEL rhythm vary with photoperiod, including the phase of the 24h light/dark cycle exposed to MEL (phase coincidence), the amplitude of the MEL rhythm and the overall duration of MEL secretion. Acting via MEL-receptors one or more of these parameters have the potential to be directly involved.

The intention of this mini-review is to compare the relative importance of these three characteristics of the nocturnal MEL secretion pattern, in relation to the photoperiodic response.

CHANGES IN AMPLITUDE AND PHOTOPERIODIC RESPONSE

It has often been asserted that contrary to duration, the amplitude of the nocturnal MEL peak in the pineal and in plasma does not vary significantly when animals are transferred from one photoperiod to another (Hoffmann, 1981). Reviewing these results, it appears that this is only true in standard conditions, that is in animals facilities where all environmental parameters, except photoperiod, remains constant. However, when animals are kept under more or less natural environmental conditions not only the duration of nocturnal MEL secretion but also its amplitude varies with season (Steinlechner et al. 1987 ; Mc Connell, 1986 ; Arendt 1979 ; Vivien-Roels et al. 1992). In relation to the photoperiodic response, what could the significance of these changes in amplitude be ? Animals living in their natural habitat are not only subjected to photoperiodic changes. There is always a variation in environmental factors, e.g. temperature and combined types of action have been noted. For example, in the Syrian hamster, low temperature which by itself has no effect on gonadal activity, accelerates gonadal atrophy induced by short photoperiod (Pévet et al. 1989). Changes in MEL amplitude may result from an effect of neuropeptides on MEL synthesis, these compounds acting directly or indirectly by an effect on noradrenergic neurotransmission (review in Pévet 1991, Pévet et al. 1994). These changes might thus convey some environmental information (Pévet 1979, 1985, 1987 ; Vivien-Roels and Pévet 1983) in the same way that the duration of MEL secretion indicates daylength (see below).

If this hypothesis is true, changes in amplitude, although of probable physiological importance, would have nothing directly to do with transduction of the photoperiodic message. However, a direct relationship between photoperiodic changes and changes in the amplitude of the MEL peak cannot be completely excluded. For example, in the European hamster when kept under controlled experimental conditions where only photoperiod varied, changes in the duration as well as in the amplitude of MEL secretion were observed (fig. 1). Further studies are needed to understand the precise physiological role of changes in the amplitude of the nocturnal MEL peak.

CHANGES IN DURATION AND PHOTOPERIODIC RESPONSE

In most of the mammalian species studied to date, the duration of the nocturnal peak of MEL secretion is positively correlated (up to a point) to the length of the dark period. These observations led Hoffmann, in 1981, to suggest that it is the duration of the nocturnal peak of MEL secretion which was the critical parameter in the transmission of photoperiodic information (referred to now as the duration hypothesis). Administration of MEL to pinealectomized animals by methods designed to produce a daily MEL peak corresponding to either a long or a short duration have validated Hoffmann's concept. Carter and Goldman (1983b), for example, demonstrated that 12h MEL infusion, like short photoperiod, inhibited testicular development in juvenile Siberian hamster whereas 4 h or 6 h infusions failed to exhibit any inhibitory effect. This experimental approach has also been used later in studies of different photoperiodic species (Siberian and Syrian hamster, white-footed mice, mink, sheep) and in all situations, infusion of MEL induced the species specific gonadal response characteristic of either long (LP) or short (SP) photoperiod, depending upon the duration of the infusion (figure 2) (Bittman and Karsch 1984 ; Dowell and Lynch 1987 ; Bartness and Goldman 1988 ; Karch et al. 1988, Maywood et al. 1990 ; Bonnefond et al. 1990 ; Pitrosky et al 1991).

All these data have clearly established that signal duration is an important parameter for eliciting seasonally appropriate reproductive responses but it is not necessarily the only critical feature of the MEL signal.

The pattern of MEL secretion in Syrian hamster kept respectively under LP and SP is shown in Figure 3. The presence of a light-entrained circadian phase of sensitivity to MEL starting approximately 9 to 10 h after lights off might also explain the results. Under SP, due to the increased duration of MEL secretion, the presence of MEL would coincide with the sensitive phase

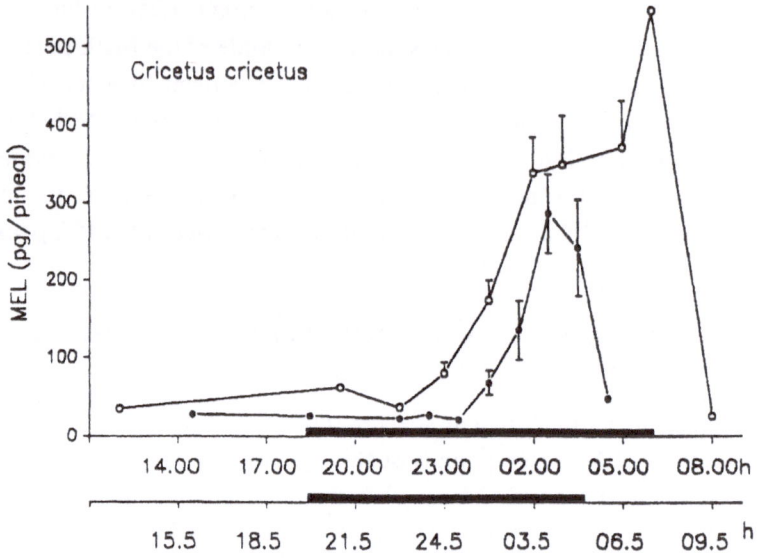

Figure 1 : Pineal melatonin content of European hamster kept under either LP(o) or SP (o) (B. Vivien-Roels et al., 1993, unpublished data).

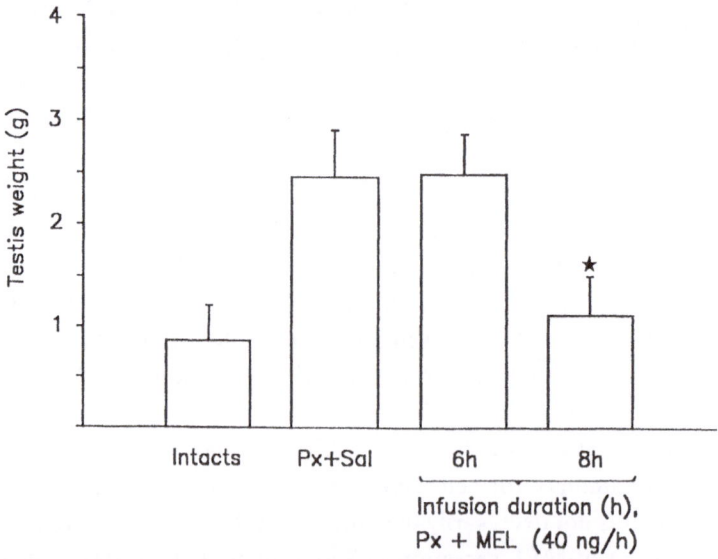

Figure 2 : Testis weight of intact and pinealectomized hamsters infused with saline (Px + sal) or melatonin 40 mg/h (Px + Mel) for 6 or 8h per day *P<0.05 when compared to saline-infused group. (from Pitrosky et al., 1991).

and so would trigger the appropriate response. This hypothesis of circadian-based variation in brain sensitivity to MEL, referred to as the coincidence hypothesis, has been put forward primarily by Stetson and Watson-Whitmyre (1984), who with others (review Reiter 1987b) observed that in intact Syrian hamsters kept under LP, the response to exogenous melatonin was dependent of the light/dark cycle. When injected late in the light phase, MEL was as effective as SP in inducing gonadal atrophy whereas MEL injections at other times (except for a brief interval before lights on) were ineffective (review Stetson and Watson-Whitmyre, 1984). The interpretation was that MEL administered in the afternoon advanced the circadian sensitive phase so that it was then coincident with the endogenous MEL peak (review Reiter 1987b).

Programmed daily infusions of MEL allow not only the duration of the MEL signal but also its phase relationships with the light/dark cycle as well as its frequency, to be varied systematically. Using this protocole it has therefore been possible to test the relative importance of these different parameters as well as the "coincidence hypothesis". In all the species studied (more than six), it has been observed that the gonadal response to MEL infusion does not depend upon the stage within the 24h light/dark cycle at which the signal is encountered. The effect of both the long and short duration infusions were the same despite the time of day at which they were administered (Carter and Golman 1983a ; Bartness and Goldman 1988, Wayne et al. 1988 ; Bonnefond et al 1990). The data suggested that the timing of the MEL infusion relative to the light/dark cycle was unimportant in determining the effect of MEL on the reproductive system and argue against the coincidence hypothesis. This conclusion was reinforced by the observation that, when the MEL signals were delivered at periodicities of less than 24h and, thus, during the course of the experiment scan across the light/dark cycle, a SP-like response still occurs (Darrow and Goldman 1986, Maywood et al. 1990).

Further evidence was also consistent with the idea that photoperiodic time measurement is not dependent upon a circadian rhythm of sensitivity to MEL in its target tissues (review Bartness et al., 1993). Interestingly in this context, is the fact that the experiment which led the coincidence hypothesis can also be interpreted using the duration hypothesis. Stetson and Watson-Whitmyre (1984) concluded that the gonadal atrophy observed in LP exposed Syrian hamsters following late afternoon injections of MEL was the consequence of a MEL induced advance of the sensitivity rhythm so that it was coincident with the endogenous MEL peak. We have observed that afternoon MEL injections given to Syrian hamsters does not interfere with the endogenous pineal MEL surge (Pitrosky et al. unpublished data). The injections, however, did cause a rapid rise in circulating MEL concentration.

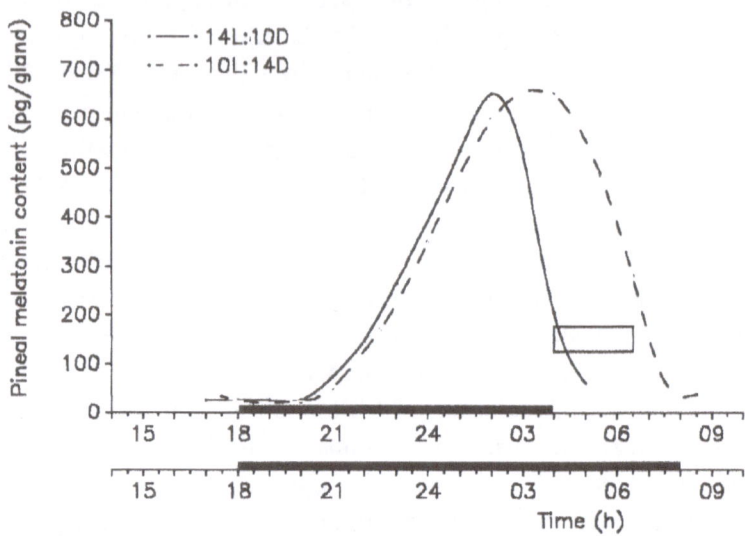

Figure 3 : Schematic representation of pineal melatonin content of Syrian hamster kept under either LP (-) or SP (---). ▭= supposed sensitivity period to melatonin. (modified from Pitrosky, 1994).

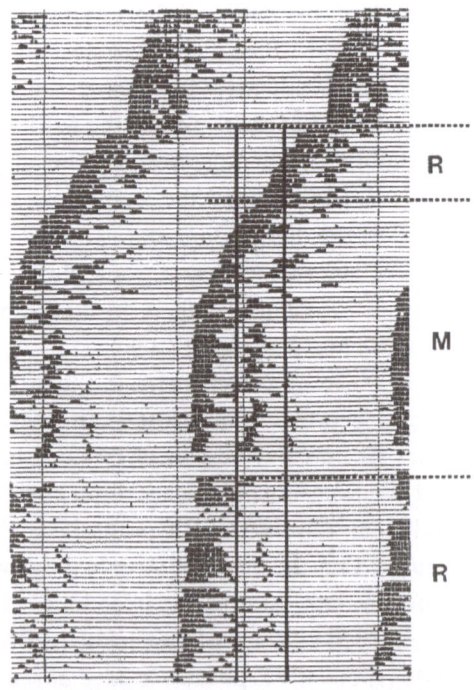

Figure 4 : Double plotted running wheel activity record of a Syrian hamster free-running with a period smaller than 24 hours. An entrainment of the activity is observed when the beginning of melatonin infusion coincides with the second part of the activity period. This effect ceases when melatonin is replaced by Ringer (from Kirsch et al., 1993).

According to the duration hypothesis, the exogenous afternoon MEL surge will summate with the endogenous MEL peak to prolong the duration of elevated MEL and as consequence, the neuroendocrine system is inhibited.

As noted by Bartness et al. (1993), all these findings by themselves do not refute completely the possibility of a circadian rhythm of sensitivity to MEL. In numerous species MEL receptors have been described in the SCN, the circadian clock. It thus cannot be excluded that the MEL signal itself drives such a rhythm via these SCN-receptors. However, lesioning of the SCN did not compromise the ability of the animals, at least the Syrian hamster, to exhibit an appropriate photoperiodic response to MEL infusion (Maywood et al. 1990). These authors thus concluded (review Bartness et al., 1993) that the duration signal was not dependent upon the circadian system and that the duration is the critical parameter of the nocturnal MEL profile for the photoperiodic control of reproduction as well as for other seasonal adaptative responses.

In our opinion, a note of caution should be expressed. The photoperiodic action is critically dependant upon the duration of MEL exposure each day. This aspect of the Mel signal has been studied, as reported above, with respect to a possible action of melatonin on the clock. It has, however, never been studied with respect to the idea that MEL is also a direct component of the circadian system.

MELATONIN SIGNAL AND THE CIRCADIAN SYSTEM IN THE INDUCTION OF THE PHOTOPERIODIC RESPONSE

The rhythm of MEL secretion is a circadian one which like many other circadian rhythms is directly generated by the SCN. This implies that it persists in the absence of light/dark cycles (period slightly different from 24 hours) and is entrained to a 24 h period by the light/dark cycle. The MEL rhythm is a important efferent hormonal signal from the clock. The periodic secretion of MEL might thus be used as an internal circadian zeitgeber or better (in the terminology defined by Moore-Ede et al, 1982) as a circadian mediator to any system than can "read" the message and so might be directly involved in the control of some circadian rhythms. In short, the daily MEL signal might be used by the SCN to impose a circadian rhythmicity to MEL-target structures, MEL directly driving these rhythms.

Clear data on the involvement of MEL in circadian rhythmicity have been obtained in non mammalian vertebrates (from review see Cassone, 1990, Armstrong and Redman, 1993). In mammals experimental evidence is still fragmentary. Injection of MEL at the same time every day entrains the circadian locomotor activity of male and female rats free-running in constant conditions (Redman et al., 1983). Contrary to rats, Syrian hamsters are not entrained by daily injections of MEL (Hastings et al. 1992). However recently it has been possible to demonstrate that when administered by infusion in a way which duplicates its endogenous plasma nocturnal peak, MEL is able to entrain the free running locomotor activity rhythm in pinealectomized Syrian (fig. 4) and Siberian hamsters. The observed effects correspond to a true entrainment since termination of the MEL infusion reestablished the free running rhythm (Kirsch et al. 1993). It has also been shown that the circadian rhythm of MEL receptor density within the pars tuberalis is directly induced by the SCN-dependent circadian rhythm of plasma MEL concentrations (Gauer et al., 1994b).

All these findings strongly support the idea that in mammals, like in non-mammalian vertebrates, MEL is also involved in the control of the expression of circadian rhythmicity. The question of major importance concerning this effect of MEL is its precise site and mechanism of action.

The SCN, the clock underlying circadian rhythmicity in mammals, contains MEL receptors (references in Masson-Pévet et al. 1994, Stankov et al. 1993). Its metabolism (Cassone 1990, 1991) as well as its electrical activity (Mc Arthur et al. 1991) are known to be directly affected by exogenous MEL. Moreover daily injections of MEL entrain free-running locomotor activity in intact rats, but not in SCN-lesioned animals (Cassone et al. 1986). Entrainment takes place when the onset of activity coincides with the time of injection, MEL inducing a phase advance (Redman et al. 1983). Based on these observations, especially on the phase advancing properties of exogenous MEL, most authors have concluded that MEL affects circadian rhythms by an action on the clock itself.

The presence of MEL receptors within the SCN as well as the in vitro data (see above) strongly supports the idea that MEL is involved in the regulation of SCN activity. Very probably MEL via the SCN-MEL receptors is important to prevent internal desynchronization (the neuroendocrine loop model proposed by Cassone and Lu 1994 ; Cassone and Menaker 1984). Exogenously administered MEL will thus have an effect on clock activity but this does not mean that it will systematically shift the clock activity rhythm and so affect circadian organization. At least, if exogenously administered MEL shifts the clock rhythm, it should act at the same time and in the same direction on all SCN-dependent circadian rhythms. To our knowledge no experimental study has specifically examined this point but results obtained in humans are interesting to consider. Exogenously administered MEL is used to manipulate human circadian rhythms (e.g. endogenous MEL production,

Arendt 1989, Lewy et al. 1992, Zaidan et al, 1994) and so to treat pathologically or socially induced disturbances of biological rhythms (disturbed sleep in jet-lag, shift workers or blind peoples ; review in Arendt, 1989). In a blind patient, Folkard et al., 1990, observed that exogenously administered MEL entrained the free running sleep-wake cycle rhythm but not that of rectal temperature or urinary cortisol. This observation argues against a MEL entrainment of the circadian oscillator and suggests that the daily administration of MEL imposed its rhythm to the sleep-wake cycle by acting on another unidentified system.

In nocturnal animals kept under constant conditions, the free-running period of locomotor activity is slighly longer than 24 hours. Consequently the onset of activity is delayed for a few minutes every day. In order to entrain this rhythm to a 24h period, MEL thus needs to induce an advance in the onset of activity every day. Based only on these results, is it correct to speak of a phase advance effect of MEL (a terminology which implies a direct effect of the hormone on the clock) ? A few animals in each nocturnal species always exhibit a free running period shorter than 24 h (fig. 4). Using these animals, Kirsch et al (1993) have demonstrated that daily infusions of MEL were also able to entrain this rhythm to a 24 h period (fig. 4). To achieve this, however, MEL has to induce a daily phase delay. Thinking in terms of a direct effect of MEL on the clock, it is thus necessary to accept that MEL administered at the same circadian time is able to induce, either a phase advance or a phase delay depending upon the animals. This is difficult to conceive. Again these observations suggests that MEL imposes its rhythm by acting on another yet unidentified structure.

How could such a system operate ? Different possibilities exist. "Trigger circuits" initiating specific patterns of hormone secretion or behaviours (e.g. locomotor activity) that possess a circadian component do not appear to be directly innervated by SCN neurons. Main target areas of SCN efferents are the paraventricular nuclei of the hypothalamus (PVN) and of the thalamus (PVT) (Watts and Swanson 1987, Watts et al. 1987 ; Kalsbeek et al. 1992, 1993). The most remarkable feature of the PVT is its prominent projection to the nucleus accumbens, olfactory tubercule, central amygdala and prefontal cortex (Berendse and Groenewegen 1990, Phillipson and Griffiths 1985). These projections provide the SCN with an almost direct input to the mesolimbo-cortical complex considered to be the main center in the central nervous system for the initiation and maintenance of (loco)motor behaviour. MEL binding sites have been observed in the PVT of the Siberian hamster, Syrian hamster and rat (review Masson-Pévet et al., 1994). MEL, acting on

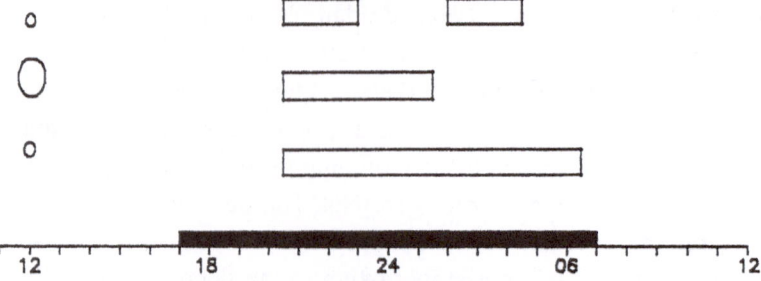

Figure 5 : Schematic representation of effects of timed melatonin infusions on the sexual activity of pinealectomized syrian hamsters (Modified from Pitrosky, 1994).

the PVT, to modulate the circadian message sent by the SCN could explain the above results as well as the negative results obtained with MEL injections in SCN-lesioned animals.

MEL binding sites/receptors have not only been identified in the SCN, the pars tuberalis or the PVT but also in a large number of brain and peripheral structures (more than 90 identified areas to date, reviews Masson-Pévet et al. 1994, Stankov and Fraschini 1993, Brown et al. 1994). Is it possible that on some of these structures MEL does not modulate the SCN-derived signal but directly imposes the circadian rhythm ? Such situation has been described in the pars tuberalis. The circadian rhythm of MEL receptor density in this structure disappeared after pinealectomy (Gauer et al. 1994b). MEL was shown to drive this circadian rhythm independently of the SCN or of the light/dark cycle. The rhythm, however, was circadian and entrained by light/dark cycle because MEL itself was dependent upon the clock.

As MEL receptors, within the pars tuberalis are thought to be involved in the photoperiodic response, such a concept implies that the photoperiodic response depends upon a circadian phase of sensitivity to MEL (as represented in fig. 3). We have tested this hypothesis in the Syrian hamster and positive correlations have been obtained. When MEL is infused in pinealectomized hamsters for two times 2.5h period, separated by a 3h non-infusion, gonadal atrophy similar to that observed following 10 h infusion is obtained whilst continuous 5h infusion was ineffective (fig. 5). Gonadal atrophy did not occur when the two 2.5h infusion periods were separated by a 5.5h interval (Pitrosky, 1994, in prep.). Measurement of plasma MEL revealed that MEL levels reached physiological values during the infusions and fell to daytime values 1h after the end of the infusion period. Two consecutive but separate MEL peaks were present in the plasma. These findings indicate that a period of sensitivity to MEL exists and that this is important for the action of MEL in transducing photoperiodic information.

CONCLUSION

With respect to the photoperiodic response, the relative importance of the phase versus duration of MEL signal remains an open question. Based on our experimental data, obtained during the last few years and presented above, we suggest that a circadian rhythm of sensitivity to MEL exists. This rhythm would be directly driven by MEL and would be circadian only because the MEL signal itself is dependent on the clock (MEL signal = circadian mediator). In physiological conditions, MEL synthesized and released at the beginning of the night would drive the rhythm of sensitivity. A photoperiodic response would only be obtained when the increase in the duration of MEL secretion due to the increasing night length is long enough to allow high levels of circulating MEL to coincide with the MEL-driven MEL sensitivity period. This hypothesis, which is of a "coincidence" type, also explains experimental data obtained by others in favor of the duration hypothesis. Indeed if, as suggested, the MEL signal itself drives directly the rhythm of sensitivity to MEL (thus independently of clock activity and the light/dark), daily infusions of MEL during daytime would be as effective as night time infusions precisely because the daytime infusion would shift the phase of MEL sensitivity into the daytime. The only condition would be that the infusion time was long enough to entrain and, at the same time, to cover this sensitivity period (approximately a minimum of 7-8 hours in the Syrian hamster). This was precisely the situation in all of the experimental procedures used by the different authors : a short photoperiod type of response was always obtained with an infusion time of 8 or more hours (review Bartness et al. 1993). Similarly, when the infusion time was at least 8 hours, this hypothesis also explains the possibility of the alternate infusion paradigm and of an infusion periodicity close to but less than 24 hours to obtain gonadal atrophy.

Lesion of the SCN blocks the response to SP because it prevents all circadian organizations and/or secretion of nocturnal MEL. Ablation of the SCN does not compromise the ability of, for example, the Syrian hamster to exhibit an appropriate photoperiodic response to MEL because daily MEL infusion, even in SCN lesioned animals, would reestablish the circadian/temporal organization of the MEL sensitivity rhythm. Atrophy will be obtained, if the infusion time is long enough to cover the MEL sensitive period timed by the MEL infusion itself. This interpretation permits also to explain the photoperiodic control of gonadal activity in animals species (such as the mink or the ferret) where no MEL receptors have been detected within the SCN.

Clearly this "hypothesis" (coincidence type hypothesis) which is compatible with the changes in duration of the MEL signal observed in physiological conditions does not answer all the questions. It probably oversimplifies the mechanisms. This hypothesis is, however, a working, testable hypothesis which in future can be rejected, confirmed or extended.

Acknowledgements

The authors wish to thank Dr. D. Skene for critical comments and Miss F. Murro for typing the manuscript.

REFERENCES

Arendt, J., 1989, Melatonin and the pineal gland, in : "Biological rhythms in clinical Practice", J. Arendt, J. Minors and J.M. Waterhouse, eds., Wright, London, pp. 184-206.

Arendt, J., 1985, Mammalian pineal rhythms. Pineal Res. Rev. 3 : 161-214.

Arendt, J., 1979, Radioimmunoassayable melatonin : circulating patterns in man and sheep, in : "The Pineal Gland of Vertebrates including Man", J. Ariëns-Kappers and P. Pévet, eds., Progress in Brain Res., vol. 52, Elsevier/North Holland, Amsterdam, pp. 243-258.

Armstrong, S.M. and Redman, J.R., 1993, Melatonin and circadian rhythmicity, in : "Melatonin : Biosynthesis, Physiological effects and clinical applications", H.S. Yu and R.J. Reiter, eds., CRC Press, Boca Raton, USA, pp.187-224.

Bartness, T.J. and Goldman, B.D., 1988, Peak duration of serum melatonin and short-day responses in adult Siberian hamsters. Am. J. Physiol. 255 : R812-R822.

Bartness, T.J., Powers, J., Hastings, M.H., Bittman, E.L. and Goldman, B.D., 1993, The timed infusion paradigm for melatonin delivery. What has it taught us about the melatonin signal, its reception and the photoperiodic control of seasonal responses ? J. Pineal Res. 15 : 161-190.

Berendse, H.W. and Groenewegen, H.J., 1990, Organization of the thalamostriatal projections in the rat, with special emphasis on the ventral striatum. J. Comp. Neurol. 299 : 187-228.

Bittman, E.L., Delosey, R.J. and Karsch, F.J., 1983, Pineal melatonin secretion drives the reproductive response to day-length in the ewe. Endocrinology 113 : 2276-2283.

Bittman, E.L. and Karsch, F.J., 1984, Nightly duration of pineal melatonin secretion determines the reproductive response to inhibitory day length in the ewe. Biol. Reprod. 30 : 585-593.

Bonnefond, C., Martinet, L. and Monnerie, R., 1990, Effects of timed melatonin infusions and lesions of the suprachiasmatic nuclei on prolactin and progesterone secretions in pregnant and pseudopregnant mink (Mustela vison). J. Neuroendocrinol. 2 : 583-592.

Brown, G.M., Pévet, P. and Pang, S.F., 1994, Melatonin binding sites in endocrine and immune systems. Biological Signals 3 : 2.

Carter, D.S. and Goldman, B.D., 1983a, Progonadal role of the pineal in the Djungarian hamster (Phodopus sungorus sungorus) : mediation by melatonin. Endocrinology 113 : 1268-1273.

Carter, D.S. and Goldman, B.D., 1983b, Antigonadal effects of timed melatonin infusion in pinealectomized male Djungarian hamsters (Photopus sungorus sungorus) : duration is the critical paramater. Endocrinology 113 : 1261-1267.

Cassone, V.M., 1990, Effects of melatonin on vertebrate circadian systems. T.I.N.S. 13 : 457-463.

Cassone, V.M., 1991, Melatonin and suprachiasmatic nucleus function, in : "Suprachiasmatic Nucleus : the Minds's Clock", D.C. Klein, R.Y. Moore and R.M. Reppert, eds., Oxford University Press, pp. 309-323.

Cassone, V.M., Chesworth, M.J. and Armstrong, S., 1986, Entrainment of rat circadian rhythms by daily injection of melatonin depends upon the hypothalamic suprachiasmatic nucleus. Physiol. Behav., 36, 111-112.

Cassone, V.M. and J. Lu, 1994, The pineal gland and avian circadian organization : the neuroendocrine loop, "Advance in Pineal Research 8", M. Moller and P. Pévet, eds., John Libbey, London (in press).

Cassone, V.M. and Menaker, M., 1984, Is the avian circadian system a neuroendocrine loop ? J. Exp. Zool., 232 : 539-549.

Darrow, J.M. and Goldman, B.D., 1986, Circadian regulation of pineal melatonin and reproduction in the Djungarian hamster. J. Biol. Rhythms, 1 : 39-54.

Dowell, S.F. and Lynch, G.R., 1987, Duration of the melatonin pulse in the hypothalamus controls testicular function in pinealectomized mice (Peromyscus leucopus). Biol. Reprod., 36 : 1095-1101.

Folkard, S., Arendt, J., Aldhous, M. and Kennett, H., 1990, Melatonin stabilises sleep onset time in a blind man without entrainment of cortisol or temperature rhythms. Neurosci. Lett., 113 : 193-198.

Gauer, F., Masson-Pévet, M., Saboureau, M., George, D. and Pévet, P., 1993, Differential seasonal regulation of melatonin receptor density in the pars tuberalis and the suprachiasmatic nuclei : a study in the hedgehog. J. Neuroendocrinol., 5 : 685-690.

Gauer, F., Masson-Pévet, M. and Pévet, P., 1994a, Seasonal regulation of melatonin receptors in rodents pars tuberalis : correlation to reproductive state. J. Neural transmission, 96 : 187-195.

Gauer, F., Masson-Pévet, M., Stehle, J. and Pévet, P., 1994b, Daily variations in melatonin receptor density of rat pars tuberalis and suprachiasmatic nuclei are distinctly regulated. Brain Res., 641 : 92-98.

Goldman, B.D., 1983, The physiology of melatonin in mammals, in : "Pineal Research Review", vol. 1, Reiter R.J., ed., alan R. Liss Inc., New York, pp.145-182.

Goldman, B.D. and Nelson, R.J., 1993, Melatonin and seasonality in mammals, in : "Melatonin, Biosynthesis, Physiological effects and clinical applications", H.S. Yu and R.J. Reiter, eds., CRC Press, Boca Raton, USA, pp. 225-252.

Hastings, M.H., Mead, S.M., Vindlacheruvu, R.R., Ebling, F.J.P., Maywood E.S. and Groose, J., 1992, Non-phototic phase shifting of the circadian activity rhythm of Syrian hamsters : The relative potency of arousal and melatonin. Brain Res, 591 : 20-26.

Hastings, M.H., Walker, A.P. and Herbert, J., 1985, Effect of asymmetrical reduction of photoperiod on pineal melatonin, locomotor activity and gonadal condition of male Syrian hamsters. J. Endocrinol., 114 : 221-229.

Hoffmann, K., 1981, Photoperiod function of the mammalian pineal organ, in : "The pineal organ : photobniology-biochronometry-endocrinology", A. Oksche and P. Pévet, eds., Amsterdam, Elsevier North-Holland, pp. 123-138.

Hoffmann, K., 1985, Interaction between photoperiod, pineal and seasonal adaptation in mammals, in : "The Pineal Gland : Current State of Pineal Research", B. Mess, C. Ruzsas, L. Tima and P. Pévet, eds., Akademiai Kiado, Budapest, pp. 211-227.

Kalsbeek, A., Buijs, R.M., Van Heerikuize, J.J., Arts, M. and Van der Woude, T.P., 1992, Vasopressin-containing neurons of the suprachiasmatic nuclei inhibit corticosterone release. Brain Res., 580 : 62-67.

Kalsbeek, A., Teclemarian-Mesbah, R., and Pévet, P., 1993, Efferent projections of the suprachiasmatic nucleus in the golden hamster (Mesocricetus auratus). A study using anterograde training specific lesions and immunocytochemistry as an approach. J. Comp. Neurol. 332, 293-314.

Karsch, F.J., Malpaux, B., Wayne, N.L. and Robinson, J.E., 1988, Characteristics of the melatonin signal that provide the photoperiodic code for timing seasonal reproduction in the ewe. Reprod. Nutr. Dev., 28 : 459-472.

Kirsch, R., Belgnaoui, S., Gourmelen, S. and Pévet, P., 1993, Daily melatonin infusion entrains free-running activity in Syrian and Siberian hamster, in : "Light and

Biological Rhythms in Man", L. Wetterberg, ed., Werner-Gren international Series, vol. 63, Pergamon Press, pp. 107-120.

Lewy, A.J., Ahmed, S., Latham Jackson, J.M. and Sack, R.L., 1992, Melatonin shifts human circadian rhythms according to a phase-response curve. Chronobiol. Int., 9 : 380-392.

Lincoln, G.A., 1994, Effects of placing micro-implants of melatonin in the pars distalis and the lateral septum of forebrain on the secretion of FSH and prolactin and testicular size rams. J. Endocrinol., 142 : 267-276.

Malpaux, B., Chemineau, P. and Pelletier, J., 1993, Melatonin and reproduction in sheep and goats, in : "Melatonin Biosynthesis, physiological effects and clinical application, H.S. Yu, and R.J. Reiter, eds., CRC Press, Boca Raton, USA, pp. 253-287.

Masson-Pévet, M. and Gauer, F., 1994, Seasonality and melatonin receptor in the pars tuberalis in some long day breeders. Biological Signals, 3 : 63-70.

Masson-Pévet, M., George, D., Kalsbeek, A., Saboureau, M., Lakhdar-Ghazal, N. and P. Pévet, 1994, An attempt to correlate brain areas containing melatonin-binding sites with rhytmic functions : a study in five hibernator species. Cell Tissue Res. (in press).

Maywood, E.S., Buttery, R.C., Vance, G.H.S., Herbert, J. and Hastings, M.H., 1990, Gonadal response of the male Syrian hamster to programmed infusions of melatonin are sensitive to signal duration and frequency but not to signal phase nor to lesions of the suprachiasmatic nuclei. Biol. Reprod., 43 : 174-182.

McArthur, A.J., Gillette, M.U. and R.A. Prosser, 1991, Melatonin directly resets the rat suprachiasmtic clock in vitro. Brain Res., 565 : 158-161.

McConnell, S.J., 1986, Seasonal changes in the circadian plasma melatonin profile of the tammar, Macropus eugenii. J. Pineal Res., 3 : 119-125.

Moore-Ede, M.C., Sulzman, F.M. and Fuller, C.A., 1982, The clocks that time us. Harward Univ. Press., Cambridge, USA.

Pévet, P., 1979. Secretory processes in the mammalian pinealocyte under natural and experimental conditions, in : "The pineal gland of Vertebrates including Man", J. Ariens-Kappers and P. Pévet, eds., Progress Brain Res., vol. 52, Elsevier/North-Holland Biomedical Press, Amsterdam, pp. 149-194.

Pévet, P., 1985, 5-methoxyindoles, pineal and seasonal reproduction. A new approach, in : "The Pineal Gland : Current state of pineal research", B. Mess, Cz. Ruszas, L. Tima and P. Pévet, eds., Akademia Kiado, Budapest and Elsevier Science Publishers, Amsterdam, pp. 163-186.

Pévet, P., 1987, Environmental control of the annual reproductive cycle in mammals, in : "Comparative Physiology of environmental adaptations, vol. 3, P. Pévet, ed., Karger, Basel, pp. 82-100.

Pévet, P., 1988, The role of the pineal gland in the photoperiodic control of reproduction in different hamster species. Reprod. Nutr. Develop., 28 : 443-458.

Pévet, P., 1991, Physiological role of neuropeptides in the mammalian pineal gland. Advances in Pineal Research : 6, Foldes A. and Reiter, eds., John Libbey, London, pp. 275-281.

Pévet, P., Simonneaux, V. and Vivien-Roels, B., 1994, Physiological role of neuropeptides in the mammalian pineal gland : a working hypothesis, in : "Advances in pineal research" 8, M. Moller and P. Pévet, eds., John Libbey, London.

Pévet, P., Vivien-Roels, B. and Masson-Pévet, M., 1989, Low temperature in the golden hamster accelerates the gonadal atrophy induced by short photoperiod but does not affect the daily pattern of melatonin secretion. J. Neural Transm., 76, 119-128.

Phillipson, O.T. and Griffiths, A.C., 1985, The topographic order of inputs to nucleus accumbens in the rat. Neurosciences, 16 : 275-296.

Pitrosky, B., 1994, Mélatonine et photopériodisme : importance des différents paramètres du profil de sécrétion de la mélatonine chez le hamster syrien (Mesocricetus auratus). Thèse de Doctorat, Université L. Pasteur, Strasbourg.

Pitrosky, B., Masson-Pévet, M., Kirsch, R., Vivien-Roels, B. and Canguilhem, B. and Pévet, P., 1991, Effect of different doses and durations of melatonin infusions on plasma melatonin concentrations in pinealectomized syrian hamsters : consequences in pinealectomized syrian hamsters : consequences at the level of sexual activity. J. Pineal Res., 11 : 149-155.

Redman, J.R., Armstrong, S.M. and Ng, K.T., 1983, Free running activity rhythms in the rat entrainment by melatonin. Science, 219 : 1089-1091.

Reiter, R.J., 1987, The melatonin message : duration versus coincidence hypotheses. Life Sci., 40 : 2119-2131.

Reiter, R.J., 1980, The pineal and its hormone in the control of reproduction in mammals. Endocrine Reviews, 1 : 109-131.

Reiter, R.J., 1987a, Photoperiod, pineal and reproduction in mammals including man, in : "Comparative physiology of environmental adaptation part 3, adaptations to climatic changes", P. Pévet, ed., Karger, Basel, pp.71-81.

Skene, D.J., Masson-Pévet, M. and Pévet, P., 1993, Seasonal changes in melatonin binding sites in the pars tuberalis of male European hamster and the effects of testosterone manipulation. Endocrinology, 132, 1682-1686.

Stankov, B. and Fraschini, F., 1993, High affinity melatonin binding sites in the vertebrates brain. Neuroendocrinol. Lett., 15 : 149-164.

Steinlechner, S., Buchberger, A. and Heldmaier, G., 1987, Circadian rhythms of pineal N-acetyltransferase activity in the Djungarian hamster, Phodopus sungorus, in response to seasonal changes of natural photoperiod. J. Comp. Physiol. A, 160 : 593-597.

Stetson, M.H., Watson-Whitmyre, M., 1984, The physiology of the pineal and its hormone melatonin in annual reproduction in rodents. In : "The Pineal Gland", R.J. Reiter, ed., Raven Press, New York, pp. 109-153.

Vivien-Roels, B. and Pévet, P., 1983, The pineal gland and the synchronization of reproductive cycles with variations of the environmental climatic conditions, with special reference to temperature, in : "Pineal Research Review", R.J. Reiter, ed., A.R. Liss, Inc., New York, pp. 92-143.

Vivien-Roels, B.,Pévet, P., Masson-Pévet, M. and Canguilhem 1992, Seasonal variation in the daily rhythm of pineal gland and/or circulating melatonin and 5-methoxytryptophol concentrations in the European hamster, Cricetus cricetus. Gen. Comp. Endocrinol., 86, 239-247.

Watts, A.G and Swanson, L.N., 1987, Efferent projections of the suprachiasmatic nucleus II. Studies using retrograde transport of fluorescent dyes and simultaneous peptide immunohistochemistry in the rat. J. Comp. Neurol., 258 : 230-252.

Watts, A.G., Swanson, L.W. and Sanchez-Watts, 1987, Efferent projections of the suprachiasmatic nucleus I. Studies using anterograde transport of Phaseolus vulgaris Leuco-agglutinin in the rat. J. Comp. Neurol. 258, 204-229.

Wayne, N.L., Malpaux, B., and Karsch, F.J., 1988, How does melatonin code for day length in the ewe. Duration of nocturnal melatonin release or coincidence of melatonin with a light-entrained sensitive period. Biol. Reprod., 39 : 66-75.

Zaidan, R., Geoffriau, M., Brun, J., Taillard, J., Bureau, C., Chazot, G. and Claustrat, B., 1994, Melatonin is able to influence its secretion in humans : description of a phase-response curve. Neuroendocrinology, 60 : 105-112.

BRAIN MELATONIN RECEPTORS: DISTRIBUTION AND PHYSIOLOGICAL SIGNIFICANCE

Bojidar Stankov and Franco Fraschini

Chair of Chemotherapy, Department of Pharmacology
University of Milan, via Vanvitelli, 32, 20129 Milan, Italy

INTRODUCTION

High affinity melatonin binding sites, having in most cases properties of a functional receptor, express a species-specific distribution pattern in the brain. In lower vertebrates binding is widespread, located in areas related to perception and decoding of the visual and auditory information. In lower mammals (*Marsupialia, Lagomorpha*) the distribution pattern is somewhat different; still widespread receptor allocation can be observed. In laboratory rodents, binding is limited to discrete brain regions. Primate brains show a trend for most restricted distribution. Though some species represent conspicuous exceptions, few areas have been most frequently mentioned as potential melatonin targets: the pars tuberalis (PT) and the zona tuberalis (ZT) of the pituitary gland, the preoptic hypothalamic area (POA) and the hypothalamic suprachiasmatic nuclei (SCN). In most ruminants, the hippocampal formation, the cerebral and cerebellar cortices express significant levels of binding, while in most rodents binding is absent in the cerebral cortex, but strong in the thalamus. In most species, where binding is strong in the POA, the SCN are apparently not labeled. In all examined areas the binding affinity of the putative receptor is in the low picomolar range and the binding site is linked to a guanine nucleotide binding protein, probably of the Gi class, regulating the adenylyl cyclase and the cellular cAMP levels. An exception is the hippocampal formation, where the binding characteristics are not influenced by coincubation with guanine nucleotides (R. Nonno et al., 1994, unpublished data).

In the recent years a number of papers describing the melatonin receptor distribution pattern and in some cases the existence of signal-transduction pathway were published (for recent reviews see Stankov et al., 1993; Morgan et al., 1994). It is generally accepted that most of the melatonin effects are mediated through a picomolar affinity binding site expressing the kinetic and pharmacological properties of a functional receptor and linked to at last one signal-transduction pathway, regulating the adenylyl cyclase and cAMP. In some cases two binding forms (high- and lower affinity) reportedly coexist (for reviews see Dubocovich, 1988; Morgan et al., 1994). As far as the high affinity binding site is linked to a regulatory G-protein in the first step of the signal-transduction mechanism (Morgan et al., 1989a, b; Rivkees et al., 1989a, 1990; Laitinen and Saavedra 1990b; Stankov et al., 1991d; 1992), some of the data describing low affinity binding may have been a result of measuring the low affinity state of the same receptor class, undergoing affinity shifts during the process of coupling and uncoupling to its G-protein. The recently cloned *Xenopus* melanophores melatonin receptor also expressed picomolar binding affinity (Ebisawa et al., 1994).

This paper will overview experiments performed under conditions allowing detection of high affinity melatonin binding in the vertebrate brain, using 2-[^{125}I]iodomelatonin as a labeled ligand. Data regarding the melatonin receptors in the retina (Dubocovich, 1985; Dubocovich and Takahashi, 1987; Laitinen and Saavedra, 1990a) and the melatonin binding sites in the arterial vasculature (Viswanathan et al., 1990; Seltzer et al., 1992; Stankov et al., 1993a) will not be included here for reason of space, and because the topic was covered by other speakers at this conference.

The initial reports demonstrating high affinity 2-[^{125}I]iodomelatonin binding in the vertebrate brain were performed on the laboratory rodents and pointed out to a discrete distribution pattern (Vanecek et al., 1987). Binding was identified in the SCN and the median eminence, but later experiments, performed in different laboratories (Williams and Morgan, 1988; Weaver et al., 1989) uncovered that the binding located in the median eminence area is in fact limited to the pars tuberalis of the pituitary gland: a tiny layer of cells, in close position to the external borders of the median eminence. Subsequently, binding in PT was detected in all mammalian species studied (see below).

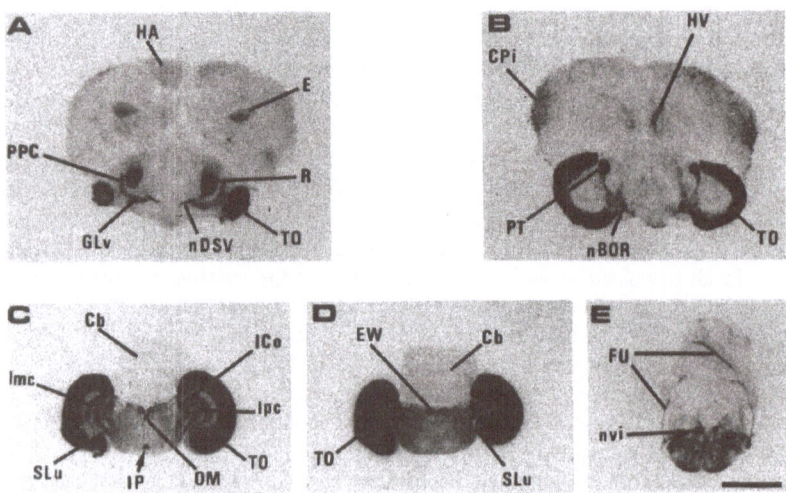

Figure 1. Distribution of the specific 2[125I]iodomelatonin binding in coronal sections of the quail brain. The nonspecific binding was very low, homogeneous and equal to the background. Cb, cerebellum; CPi, Cortex piriformis; E, Ectostriatum; EW, Edinger-Westphal nucleus; FU, fasciculus uncinatus (Russell); GLv, nucleus geniculatus lateralis, pars ventralis; HA, hyperstriatum accessorium; HV, hyperstriatum ventrale; ICo, nucleus intercollicularis; Imc, nucleus isthmi, pars magnocellularis; Ipc, nucleus isthmi, pars parvocellularis; IP, nucleus interpeduncularis; nVI, nucleus of the sixth cranial nerve; nBOR, nucleus opticus basalis; nDSV, nucleus decussatio supraoptica ventralis; OM, nucleus nervi oculomotorii; PPC, nucleus principals precommisuralis; PT, nucleus pretectalis; R, nucleus rotundus; SLu, nucleus semilunaris; TO, tectum opticum. Bar = 5 mm. (Modified from Cozzi et al., 1993).

NON-MAMMALIAN SPECIES

The earliest studies of non-mammalian species (Rivkees et al., 1989a) were confirmed and extended in different lower vertebrates. High density binding was detected in a number of areas, related to the visual pathways and to some extent to the auditory and olfactory pathways (Aggelopoulos and Demaine, 1990; Ekström and Vanecek, 1992; Wiechmann and Wiechmann, 1992, 1993). In the avian brain (Rivkees et al., 1989b; Siuciak et al., 1991; Cozzi et al., 1992) binding was also widely distributed, and taking as an example the Japanese quail (*Coturnix japonica*), high affinity melatonin binding sites can be found not only in the major targets of direct retinal input, such as the optic tectum, nucleus of the optic basal rout and ventrolateral geniculate nucleus, but also in areas, representing terminals in the visual pathways (nucleus rotundus, ectostriatum, the thalamo-hyperstriatal pathways). Binding is also found in the piriform cortex, the oculomotorius and the associated Edinger-Westphal nucleus and in the nuclei of the third, fourth and sixth cranial nerves, all involved in the control of eye movements. The ventral nuclei of the supraoptic decussation are intensely labeled (Figure 1). In several avian species these nuclei represent a specific retino-recipient region, which has been recognized as a homologue of the mammalian suprachiasmatic nuclei (Cassone, 1988; Cassone and Moore, 1988). Apparently, there is a typical avian melatonin receptor distribution pattern, the areas related to sight being critically involved. While binding in the cerebellum is strong in the some predatory species (*e.g.* the buzzard, *Buteo buteo*), it is very low or absent in the *Galliformes* (V. Cesaretti, R. Nonno and B. Stankov, unpublished data).

MARSUPIALS AND LAGOMORPHS

The lower mammalian species have not been a subject of profound investigation to date. In the brain of a marsupial (*Dendrolagus benettianus*) binding was also widespread, similar to what found in most of the non-mammalian brains (Paterson et al., 1990).

In the only lagomorph species studied (*Oryctolagus cuniculus*), a number of brain areas also expressed significant levels of high affinity melatonin binding sites. Binding was strong in many of the phylogenetically old cortical areas, such as the hippocampus and indusium griseum.

Binding, however, was intense in a great part of the neocortex, as well. While the frontal cortex presented diffuse, rather weak binding, the cingulate gyrus, the temporal and occipital lobes were heavily labeled. Certain formations, such as the dorsolateral thalamic nuclei, the superior colliculus and the amygdala expressed well discernible binding. The hypothalamus was diffusely labeled, especially in its anterior part, heavily labeled SCN being clearly visible on the background of diffuse binding. The pars tuberalis and the choroid plexus of the third ventricle expressed high levels of binding (Stankov et al., 1991d).

RODENTS

A common feature in all laboratory rodents brains studied to date is the limited distribution of the high affinity melatonin binding sites, compared to the lower vertebrates, including the lower mammals. An apparent exception is a New World rodent, the white-footed mouse (*Peromyscus leucopus*), where binding in the brain was widespread in the basal forebrain, the amygdala and the thalamus, as well as in a number of neocortical areas (Weaver et al., 1990). The laboratory bred Norwegian brown rat (*Rattus norvegicus*) showed a restricted receptor distribution, similar to the albino rat. In the brain of the guinea pig (*Cavia porcellus*) the paraventricular and paratenial thalamic nuclei were among the most intensely labeled structures (Stankov and Fraschini, 1993).

In the rest of the rodents studied binding was always detected in discrete brain regions. Three anatomical sites have been most frequently mentioned as potential melatonin targets: the PT, the SCN, and the paraventricular thalamic nuclei (Weaver et al., 1989; Laitinen and Saavedra, 1990b; Siuciak et al., 1990) Recent data pointed out to a more expansive presence of melatonin receptors in the thalamic nuclear complex, comprising the anterodorsal and the reticular nuclei, as well as in the entire anterior hypothalamus (Lindroos et al., 1992). These nuclei of the thalamic complex are connected with cortical and subcortical limbic structures and melatonin effects on learning, memory and alertness and sedation may be may be explained by influence on the limbic system.

Another putative site of melatonin action in the rodent brain is the area postrema (Laitinen et al., 1990), but binding was detected exclusively in the rat.

UNGULATES

Of the large Order of *Artiodactyla*, only one species, the domestic sheep (*Ovis aries*) has been a subject of profound investigation. Initial studies have demonstrated that binding of 2-[^{125}I]iodomelatonin was confined to the hypophyseal pars tuberalis (deReviers et al., 1989; Morgan et al., 1989c). Later research, however, verified that binding was more widespread in the brain, and a number of archi- and neocortical areas expressed high density binding sites. Pars distalis was diffusely labeled. (Stankov et al., 1990, 1991c; Bittman and Weaver, 1990; Helliwell and Williams, 1992). The SCN were seemingly devoid of binding, but the surrounding area expressed low to intermediate levels. Similar were the findings in the goat (*Capra hircus*) (Deveson et al., 1992), the red deer (*Cervus elaphus*) (Helliwell et al., 1991) and the bovine (*Bos taurus*) (Nonno et al., 1993) brains. The presence of high-density melatonin receptor population in the bovine PT and ZT is intriguing (Figure 2), because bovines are not photoperiodic in terms of their reproductive competence.

It should be pointed out that the apparent melatonin receptor distribution in the sheep hypothalamus and POA (Helliwell and Williams, 1992) corresponds very well to the location of the neuronal bodies and axons containing immunoreactive GnRH (Caldani et al., 1988). A notable new finding is the specific binding in the pars intermedia of the bovine pituitary (Nonno et al., 1993). In the domestic pig (*Sus scrofa scrofa*) brain, both, pars tuberalis and the suprachiasmatic nuclei were clearly labeled. The difference from the other *Artiodactyla* species is that the rest of the hypothalamus and the preoptic area were not labeled (Stankov, unpublished data).

Thus, there is a possibility that the binding in the SCN in a number of species may have remained undetectable, because of a relatively high level of binding in the surrounding area. A good example is the rabbit, where the binding in the SCN is barely visible on the background of high-density melatonin receptors in the area above the optic chiasm (Stankov et al., 1991. On the other hand, if the ruminants' SCN definitely are not a direct target for melatonin action, this does not preclude indirect effects of melatonin through interaction with nuclei in the preoptic area and mediobasal hypothalamus. Indeed, SCN lesions in sheep lead to an extension of the breeding activity well in the anoestrus period, but the same effect is achieved by lesions in the medial preoptic area (MPOA) or anterior deafferentation of the SCN from the MPOA. Moreover, melatonin infusions in the medial basal hypothalamus of sheep bearing MPOA lesions, and experiencing extended estrous cycles in the anoestrus period, blocked the release of LH and ovulation (Domanski et al., 1975, 1980). These results would support the idea that SCN receives and transfers information from the MPOA or even from extrahypothalamic formations.

Of the three Recent families contained in Order *Perissodactyla* the horse (*Equus caballus*) has been a subject of investigation. A number of a areas were examined by *in vitro* ligand-receptor binding and autoradiography. Putative melatonin receptors, however were identified in a limited number of them, the pars tuberalis region expressing high density of high affinity binding sites (Stankov et al., 1991c). Autoradiography demonstrated high density binding in both the horse and donkey (*Equus asinus*) pars tuberalis and diffuse binding in zona tuberalis (R. Nonno et al., unpublished data).

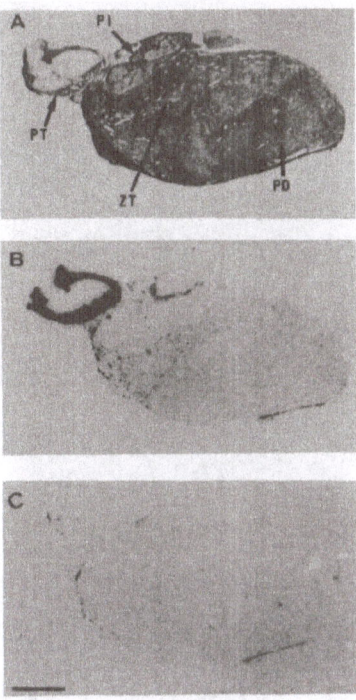

Figure 2. Melatonin receptors in the bovine pituitary gland. A, prasagittal 20 μm section stained with OFG, showing the morphology: PT, pars tuberalis; PD, pars distalis; PI, pars inremedia; ZT, zona tuberalis. B, specific binding (autoradiograph). C, nonspecific binding binding determined in presence of excess of unlabeled 2-iodomelatonin. Note the strong binding in PT, the diffuse binding in ZT and PI. Bar = 3 mm. (R. Nonno and B. Stankov, unpublished data).

CARNIVORES

In two of the carnivore species studied so far (*Mustela putorius furo* and *Spilogale putorius latifrons*) autoradiography detected binding in the PT and pars distalis only (Weaver and Reppert, 1990; Duncan and Mead, 1992). The domestic dog (*Canis familiaris*), however, expressed high-affinity GTP-regulated binding in the olfactory bulbs, pars tuberalis and pars distalis, as well as in the suprachiasmatic nuclei (Figure 3). High-resolution autoradiography demonstrated that the binding in pars distalis was located exclusively over the clusters of neutrophils and basophils that enter pars distalis proper from pars tuberalis, to form the *zona tuberalis* of the adenopituitary. The eosinophil cell population was virtually devoid of binding. Pars intermedia was clearly labeled (Stankov et al., 1994). Similar were the findings in the bovine pituitary (R. Nonno et al., unpublished).

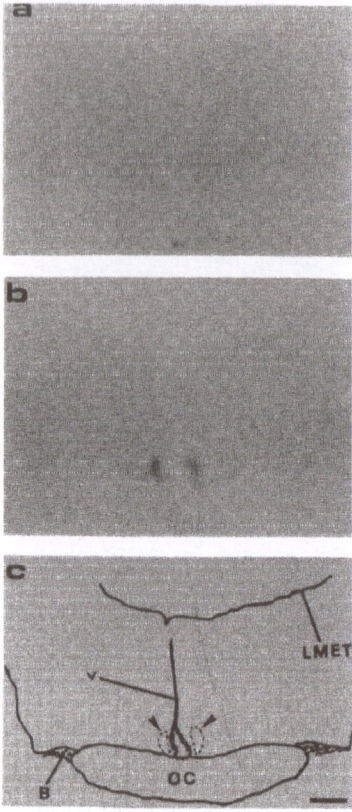

Figure 3. Autoradiographs generated from two consecutive coronal sections through the anterior hypothalamus of the dog. The specific binding (b) is confined to the suprachiasmatic nuclei (arrowheads, the apparent borders are depicted with a dashed line in the drawing, c) in contrast to the extremely low nonspecific binding (a). B, pial blood vessels; OC, optic chiasm; LMET, lamina medullaris externa thalami; v, third ventricle. Bar = 1.5 mm. (From Stankov et al., 1994).

PRIMATES

The first studies of 2-[^{125}I]iodomelatonin binding in the primate brain were limited to the human hypothalamus (Reppert et al., 1988). The SCN were labeled in both the fetus and the adult (Reppert et al., 1988; Fauteck and Stankov, unpublished data). Initially, no binding was reported in the PT (Reppert et al., 1988). In the rest of the primates studied (Stehle et al., 1991; Stankov et al., 1992b), binding was intense in PT and present in the SCN. In both, the baboon (*Papio ursinus*) and the vervet monkey (*Cecopithecus aethiops*), pars distalis was diffusely labeled. The intermediate lobe appeared labeled as well, similar to the findings in other species as discussed above (Stankov et al., 1993a). It should be pointed out that while the Rhesus monkey is clearly photoperiodic in terms of reproductive competence, neither *Papio ursinus* nor *Cecopithecus aethiops* are. Therefore, taken also the findings in the bovine, the presence of melatonin receptors in PT should not necessarily imply photoperiod-dependent melatonin effects on the reproductive activity. Very recent studies demonstrated that the human pars tuberalis and pars distalis also express detectable levels of specific, high-affinity binding, although in a limited number of the examined subjects (Weaver et al., 1993; B. Stankov, unpublished data).

Another interesting site of melatonin action in the human appears to be the molecular layer of the cerebellum (Fauteck et al., 1993). Binding in this area was found in a limited number of other vertebrates' brains, such as the Atlantic salmon (Ekström and Vanecek, 1992), the sheep (Helliwell and Williams, 1992) the horse and donkey, (R. Nonno and B. Stankov, unpublished data). This site might have been overlooked in most cases, because the binding density in most species is low, and longer exposure times (that might interfere with the evaluation of the results in other areas) are necessary.

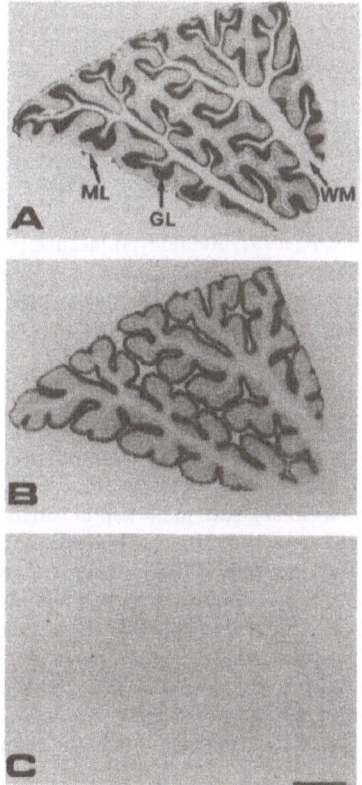

Figure 4. Putative melatonin receptors in the human cerebellar cortex. A, cresyl violet-stained section, used to generate the autoradiographic image in (B). C, nonspecific binding. GL, granule layer; ML, molecular layer; WM, white matter. Bar = 3 mm. (From Fauteck et al., 1993).

In conclusion, the distribution of the high affinity melatonin binding sites in the vertebrate brain presents unusual characteristics: in the lower vertebrates binding is associated mainly with the pathways mainly involved in the processing of visual and auditory information, and the distribution is wide throughout; similarly, widespread distribution is seen in some lower mammals. In higher mammals, high affinity binding sites show discrete distribution, and in all species studied the PT and ZT of the pituitary gland has been found a potential melatonin target, with possible exception of the human, where binding in PT and ZT is not readily detectable in all samples examined.

Some discrepancies observed between the mammalian species of the same Order might also be due to methodological problems. The limited number of comparative studies still does not allow for reliable conclusions from phylogenetic point of view.

REFERENCES

Aggelopoulos, N. and Demaine, C., 1990, Quantitative receptor autoradiography of melatonin receptors in three non-mammalian vertebrates: the green anole lizard *(Anolis carolinensis)*, the edible frog *(Rana esculenta)* and the rainbow trout *(Salmo gairdneri)*. In: V^{th} *Colloquium of the European Pineal Study Group*, 2-7 September, Guildford, U.K., abstr. 66.

Bittman, E.L. and Weaver, D.R., 1990, The distribution of melatonin binding sites in neuroendocrine tissues of the ewe. *Biol. Reprod.* 43: 986.

Caldani, M., Batailler, M., Thiéry, J.C. and Dubois, M.P, 1988, LHRH immunoreactive structures in the sheep brain. *Histochemistry* 89:129.

Cassone, V.M., 1988, Circadian variation of $[^{14}C]$2-deoxyglucose uptake within the suprachiasmatic nucleus of the house sparrow, *Passer domesticus. Brain Res.* 459: 178.

Cassone, V.M. and Moore, R.Y., 1987, Retinohypothalamic projection and suprachiasmatic nucleus of the house sparrow, *Passer domesticus. J. Comp. Neurol.* 266: 171.

Carlson, L.L., Weaver, D.R. and Reppert, S.M., 1989, Melatonin signal transduction in hamster brain: inhibition of adenylyl cyclase by a pertussis toxin-sensitive G protein. *Endocrinology* 125: 2670.

Cozzi, B., Stankov, B., Panzica, C.V., Capsoni, S., Aste, N., Lucini, V., Panzica, G.C. and Fraschini, F., 1993, Distribution and characterization of melatonin receptors in the brain of the Japanese quail, *(Coturnix japonica). Neurosci. Lett.* 150: 149.

Deveson, S., Howarth, J., Arendt, J. and Forsyth I.A., 1992, In vitro autoradiographical localization of melatonin binding sites in the caprine brain. *J. Pineal Res.* 13: 6-12.

Domanski, E., Przekop, F., Skubiszewski, B. and Wolinska, E., 1975, The effect and site of action of indolamines on the hypothalamic centres involved in the control of LH-release and ovulation in sheep. *Neuroendocrinology* 17: 265.

Domanski, E., Przekop, F. and Polkowska, J., 1980, Hypothalamic centres involved in the control of gonadotrophin secretion. *J. Reprod. Fert.* 58: 493.

Dubocovich, M.L. and Takahashi, J.S., 1987, Use of 2-$[^{125}I]$iodomelatonin to characterize melatonin binding sites in chicken retina. *Proc. Natl. Acad. Sci USA.* 84: 3916.

Dubocovich, M.L., 1985, Characterization of a retinal melatonin receptor. *J. Pharmacol. Exp. Ther.* 234: 395.

Dubocovich, M.L., 1988, Pharmacology and function of melatonin receptors. *FASEB J.* 2: 2765.

Duncan, M.J., Takahashi, J.S. and Dubocovich, M.L., 1989, Characteristics and autoradiographic localization of 2$[^{125}I]$iodomelatonin binding sites in Djungarian hamster brain. *Endocrinology* 125: 1011.

Duncan, M.J. and Mead, R.A.,1992, Autoradiographic localization of binding sites for 2-$[^{125}I]$iodomelatonin in the pars tuberalis of the western spotted skunk *(Spilogale putorius latifrons). Brain Res.* 569: 152 .

Ebisawa, T., Karne, S., Lerner, M.R. and Reppert, S.M., 1994, Expression cloning of a high-affinity melatonin receptor from *Xenopus* dermal melanophores, *Proc. Natnl. Acad. Sci., USA.* 91: 6133.

Ekström, P. and Vanecek, J., 1992, Localization of 2-[^{125}I]iodomelatonin binding sites in the brain of the Atlantic salmon, *Salmo salar. Neuroendocrinology,* 55: 529.

Fauteck, J.-D., Lerchl, A. Bergman, M., Wittkowski W., and Stankov, B., High-affinity melatonin binding sites in the molecular layer of the human cerebellar cortex. Advances in Pineal Research, Vol. 7, 1994 (in press).

Helliwell, R., Adam, C.L., Hannah, L.T., Kyle, C.E and Williams, L.M., 1991, The distribution of central 2-[^{125}I]-iodomelatonin binding sites in the adult and foetal red deer (*Cervus elaphus*). *J. Reprod. Fert.* Abstr. Series 8, Abstr. 117.

Helliwell, R. and Williams, L.M., 1992, Melatonin binding sites in the ovine brain and pituitary: Characterization during the oestrous cycle. *J. Neuroendocrinol.* 4: 287.

Laitinen, J.T. and Saavedra, J.M., 1990a, The chick melatonin receptor revisited: localization and modulation of agonist binding with guanine nucleotides. *Brain Res.* 528: 349.

Laitinen, J.T. and Saavedra, J.M., 1990b, Characterization of melatonin receptors in the rat suprachiasmatic nuclei: modulation of affinity with cations and guanine nucleotides, *Endocrinology* 126: 2110.

Laitinen, J.T., Flugge, G. and Saavedra, J.M., 1990, Characterization of melatonin receptors in the rat area postrema: modulation of affinity with cations and guanine nucleotides. *Neuroendocrinology* 51: 619.

Lindroos, O.F.C., Leinonen, L.M. and Laakso, M-L., 1992, Melatonin binding to the anteroventral and anterodorsal thalamic nuclei in the rat. *Neurosci. Lett.* 143: 219.

Moore-Ede, M.C., Sulzman, F.M. and Fuller, C.A., 1982, *The clocks that time us,* . Cambridge-London: Harvard University Press pp. 1 - 448.

Morgan, P.J., Lawson, W., Davidson, G. and Howell, H.E., 1989a, Guanine nucleotides regulate the affinity of melatonin receptors on the ovine pars tuberalis. *Neuroendocrinology* 50: 359.

Morgan, P.J., Lawson, W., Davidson, G. and Howell, H.E., 1989b, Melatonin inhibits cyclic AMP production in cultured ovine pars tuberalis cells. *J. Mol. Endocr.* 3: R5-R8.

Morgan, P.J., Williams, L.M., Davidson, G., Lawson, W. and Howell H.E., 1989c, Melatonin receptors on ovine pars tuberalis: characterization and autoradiographical localization. *J. Neuroendocrinol.* 1: 1-4.

Morgan, P.J., Barret, P., Howell, H.E. and Helliwell R., 1994, Melatonin receptors: localization, molecular pharmacology and physiological significance. *Neurochem. Int.* 24:101.

Nonno, R., Capsoni, S., Lucini, V., Fraschini F. and Stankov, B., 1993, Distribution of the melatonin receptor in the bovine brain and pituitary. In: *VIth Colloquium of the European Pineal Society,* 23-27, Copenhagen, Denmark, abstr D3.

Paterson, A.M., Chong, N., Sugden, D., Brinklow, B.R. and Loudon A.S.I., 1990, Studies of the role of melatonin in determining the transition to seasonal reproductive quiescence and preliminary observations on melatonin binding sites in the brain of a marsupial, the Bennett's wallaby. In: *Vth Colloquium of the European Pineal Study Group,* 2-7 September, Guildford, U.K., abstr. 60.

Reppert, S.M., Weaver, D.R., Rivkees, S.A. and Stopa, E.G., 1988, Putative melatonin receptors in human biological clock. *Science* 242: 78.

deReviers, M-M., Ravault, J-P., Tillet, Y. and Pelletier, J., 1989, Melatonin binding sites in the sheep pars tuberalis. *Neurosci. Lett.* 100: 89.

Rivkees, S.A., Carlson, L.L. and Reppert, S.M., 1989a, Guanine nucleotide-binding protein regulation of melatonin receptors in lizard brain. *Proc. Natl. Acad. Sci.* 86: 3882.

Rivkees, S.A., Cassone, V.M., Weaver, D.R. and Reppert, S.M., 1989b, Melatonin receptors in chicken brain: characterization and location. *Endocrinology* 125: 363.

Rivkees, S.A., Conron, R.W., Jr and Reppert, S.M., 1990, Solubilization and purification of melatonin receptors from lizard brain. *Endocrinology* 127: 1206.

Seltzer, A., Viswanathan, M. and Saavedra, J.M., 1992, Melatonin-binding sites in brain and caudal arteries of the female rat during the estrous cycle and after estrogen administration. *Endocrinology* 130: 1896.

Siuciak, J. A., Fang J-M. and Dubocovich M.L., 1990, Autoradiographic localization of 2-[^{125}I]-iodomelatonin binding sites in the brains of C3H/HeN and C57BL/6J strains of mice. *Eur. J. Pharmacol.* 180: 387.

Siuciak, J.A, Krause, D.N. and Dubocovich M.L., 1991, Quantitative pharmacological analysis of 2-[^{125}I]iodomelatonin binding sites in discrete areas of the chicken brain. *J Neurosci.* 11: 2855.

Stankov, B. and Reiter R.J., 1990, Melatonin receptors: current status, facts and hypotheses. *Life Sci* 46: 971.

Stankov, B., Cozzi, B., Lucini, V., Fumagalli, P., Scaglione, F. and Fraschini, F., 1990, Characterization and mapping of melatonin receptors in the brain of three mammalian species: rabbit, horse and sheep. In: *Vth Colloquium of the European Pineal Study Group*, 2-7 September, Guildford, U.K., abstr. 61.

Stankov, B., Fraschini, F. and Reiter, R.J., 1991a, Melatonin binding sites in the central nervous system. *Brain Res. Rev.* 16: 245 .

Stankov, B., Lucini, V., Scaglione, F., Cozzi, B., Righi, M., Canti, G., Demartini, G. and Fraschini, F., 1991b, 2-[^{125}I]iodomelatonin binding in normal and neoplastic tissues. In: *Role of Melatonin and Pineal Peptides in Neuroimmunomodulation*, eds. F. Fraschini & R.J. Reiter, Plenum Press, New York-London pp. 117-125.

Stankov, B., Cozzi, B., Lucini, V., Fumagalli, P., Scaglione, F. and Fraschini F., 1991c, Characterization and mapping of melatonin receptors in the brain of three mammalian species: rabbit, horse and sheep. *Neuroendocrinology* 53: 214.

Stankov, B., Cozzi, B., Lucini, V., Capsoni, S., Fauteck, J., Fumagalli, P. and Fraschini, F., 1991d, Localization and characterization of melatonin binding sites in the brain of the rabbit *(Oryctolagus cuniculus)* by autoradiography and *in vitro* ligand-receptor binding. *Neurosci. Lett.* 133: 68.

Stankov, B., Biella, G., Panara, C., Capsoni, S., Lucini, V., Fauteck, J., Cozzi, B. and Fraschini F., 1992, Melatonin signal transduction and mechanism of action in the central nervous system: Using the rabbit cortex as a model. *Endocrinology* 130: 2152.

Stankov, B., Capsoni, S., Lucini, V., Fauteck, J., Gatti, S., Gridelli, B., Biella, G., Cozzi, B. and Fraschini F., 1993a, Autoradiographic localization of putative melatonin receptors in the brains of two Old World primates: *Cercopithecus aethiops* and *Papio ursinus*. *Neuroscience* 52, 459.

Stankov, B., Fraschini, F. and Reiter R.J., 1993b, The melatonin receptor: distribution, biochemistry and pharmacology. In: *Melatonin, Biological Effects and Clinical Application*, eds. H.S. Yu & R.J.Reiter, CRC Press, Boca Raton, pp. 155-186.

Stankov, B., Lucini, V., Capsoni, S. and Fraschini, F., 1994, A carnivore species (*Canis familiaris*) expresses circadian melatonin rhythm in the peripheral blood and melatonin receptors in the brain. *Europ. J. Endocrinol* 131: 191.

Stankov, B. and Fraschini, F., 1993, High affinity melatonin binding sites in the vertebrate brain. *Neuroendocrinol. Lett.* 15: 149.

Stehle, J.H., Weaver, D.R. and Reppert, S.M., 1991, Autoradiographic localization of melatonin receptors in the rhesus monkey hypothalamus. In: *Society for Neuroscience Abstracts* 17, abstr. 365.9 .

Vanecek, J., Pavlik, A. and Illnerova, H., 1987, Hypothalamic melatonin receptor sites revealed by autoradiography. *Brain Res.* 435: 59.

Viswanathan, M., Laitinen, J.T. and Saavedra, J.M., 1990, Expression of melatonin receptors in arteries involved in thermoregulation. *Proc. Natl. Acad. Sci. USA* 87: 6200.

Weaver, D.R. and Reppert, S.M., 1990, Melatonin receptors are present in the ferret pars tuberalis and pars distalis, but not in brain. *Endocrinology* 127: 2607.

Weaver, D.R., Rivkees, S.A. and Reppert S.M., 1989, Localization and characterization of melatonin receptors in rodent brain by in vitro autoradiography. *J. Neurosci.* 9: 2581.

Weaver, D.R., Carlson, L.L. and Reppert, S.M., 1990, Melatonin receptors and signal transduction in melatonin-sensitive and melatonin-insensitive populations of white-footed mice *(Peromyscus Leucopus)*. *Brain Res.* 506: 353.

Weaver, D.R., Stehele, J., Stopa, E.G. and Reppert, S.M., 1993, Melatonin receptors in human hypothalamus and pituitary: implication for circadian and reproductive responses to melatonin. *J. Clin. Endoc. Metab.* 76: 295.

Wiechmann, A.F. and W-Wiechmann, C.R., 1992, Asymmetric distribution of melatonin receptors in the brain of the lizard *Anolis carolinensis. Brain Res.* 593: 281.

Wiechmann, A.F. and W-Wiechmann,C.R., 1993, Distribution of melatonin receptors in the brain of the frog *Rana pipiens* as revealed by in vitro autoradiography. *Neuroscience* 52: 469.

Williams, L.M. and Morgan P.J., 1988, Demonstration of melatonin-binding sites on the pars tuberalis of the rat. *J. Endocr.* 119: R1-R3.

LOCALIZATION AND PHYSIOLOGICAL ROLE OF MELATONIN RECEPTORS IN THE VISUAL AND CIRCADIAN SYSTEMS

Margarita L. Dubocovich[1], Diana N. Krause[2], Stanca Iacob[1], Susan Benloucif[1], and Monica I. Masana[1]

[1]Department of Molecular Pharmacology and Biological Chemistry
Northwestern University Medical School
Chicago, IL 60611, U.S.A.

[2]Department of Pharmacology
College of Medicine
University of California, Irvine
Irvine, CA 92717, U.S.A.

INTRODUCTION

The production of melatonin is driven by internal circadian clocks and modified by the environmental light/dark cycle (Krause and Dubocovich, 1990). It now appears that melatonin in turn regulates neural structures involved in visual and circadian processing. Melatonin receptors are located in the retina where they can mediate dark adaptive processes in the eye, inhibit retinal dopaminergic systems, and influence the transmission of light signals to the brain (Blazynski and Dubocovich, 1991; Dubocovich, 1985, 1988a,b; Dubocovich and Takahashi, 1987; Laitinen and Saavedra, 1990a). Melatonin receptors also are found in the suprachiasmatic nucleus (SCN) of the hypothalamus, an area which receives direct retinal input and is thought to contain circadian pacemakers (Reppert et al., 1988; Laitinen and Saavedra, 1990b; Rivkees et al., 1989; Siuciak et al., 1990). The melatonin receptors in the SCN are likely to be involved in melatonin's ability to synchronize and reset circadian rhythms (Benloucif, S. and Dubocovich M.L., 1994a; Warren et al., 1993). Other retinal target nuclei such as the superior colliculus/optic tectum also contain putative melatonin receptors which may function to modulate visual processing (Krause et al., 1992; Siuciak et al., 1991; Rivkees et al., 1989; Beresford et al., 1994; Stankov and Fraschini, 1993). In some species, melatonin receptors are observed all along specific visual pathways. For example in avian brain, the tectofugal pathway, which is involved in brightness discriminations, and the accessory optic pathway, which mediates visual-ocular reflexes, appear to be regulated at every level by melatonin receptors (Siuciak et al., 1991; Rivkees et al., 1989; Krause et al., 1994a). The cellular effects of melatonin receptors in the visual

The Pineal Gland and Its Hormones
Edited by F. Fraschini *et al.*, Plenum Press, New York, 1995

system are not well understood, but presynaptic regulation and inhibition of cyclic AMP are two functions which have been demonstrated (Dubocovich, 1985, 1988a; Iuvone and Gan, 1994; Krause et al., 1992, 1994a,b). Recent evidence suggests that the expression of ML-1 melatonin receptors is regulated by the level of activity in certain visual pathways (Krause et al., 1994b; Dubocovich et al., 1994). This chapter will discuss localization, function and regulation of melatonin receptors in the mammalian and avian visual system (retina, optic tract nerve terminals, central visual pathways), and the mammalian circadian system (SCN).

MELATONIN RECEPTORS IN THE RETINA

Melatonin secreted from retinal photoreceptors during the dark period is thought to mediate dark adaptive processes in the eye (Cahill et al., 1991; Wiechmann, 1986). Melatonin modulates retinal dopaminergic activity (Dubocovich, 1983; 1988b; Dubocovich et al., 1985). In the guinea pig eye, melatonin induces pigment aggregation in the retinal pigment epithelium and choroid (Pang and Yew, 1979), while in *Xenopus* laevis retina, it causes cone elongation, thus mimicking the effect of darkness (Pierce and Besharse, 1985). Melatonin also activates outer segment disc shedding and phagocytosis in photoreceptors (Besharse and Dunis, 1983; Ongino et al., 1983).

The characterization of melatonin receptors in the retina provides further evidence for a local action of melatonin synthesized within this tissue. In both rabbit and chicken retina, activation of melatonin heteroreceptors inhibits the calcium-dependent release of dopamine (Dubocovich, 1985; 1988a,b; Dubocovich and Takahashi, 1987). These receptors exhibit the pharmacological characteristics that define the ML-1 melatonin receptor subtype (Dubocovich, 1988a). Recently it has been reported that activation ML-1 like melatonin receptors inhibit cyclic AMP accumulation in cells cultured from chick retina (Iuvone and Gan, 1994). Picomolar concentrations of melatonin also inhibit cyclic AMP formation in hamster retina (Faillace et al., 1994).

2-[^{125}I]-Iodomelatonin has been used to identify high affinity binding sites in the retina which have the same pharmacological profile as the functional ML-1 receptors (Dubocovich and Takahashi, 1987; Dubocovich, 1988a). 2-[^{125}I]-Iodomelatonin binding sites have been localized in retina from albino and pigmented rabbits, chickens and frogs, primarily to the inner plexiform layer where the dopamine amacrine cells are found (Blazynski and Dubocovich, 1991; Laitinen and Saavedra, 1990a; Weichmann and Wirsig-Wiechmann, 1991). Retina from albino mice exhibit specific 2-[^{125}I]-iodomelatonin binding in both the inner plexiform and the outer and inner segment layers (Blazynski and Dubocovich, 1991). 2-[^{125}I]-Iodomelatonin binding sites were also reported for the retinal pigmental epithelium of mammalian and avian species (Blazynski and Dubocovich, 1991; Laitinen and Saavedra, 1990a). The latter sites may modulate electrical properties of the pigment epithelial cells (Nao-I et al., 1989).

Melatonin levels in the retina show a diurnal rhythm with high concentrations at night and low concentrations during the day (Dubocovich et al., 1985b; Cahill and Besharse, 1993). The activity of dopaminergic neurons in the retina also follows a diurnal cycle with higher levels during the light period. Although the dopaminergic amacrine cells are activated by light, the mechanisms by which these cells are stimulated are not known (Iuvone, 1986; Iuvone et al., 1978; Wirz-Justice et al., 1984; Parkinson and Rando, 1983a,b). The potent inhibitory effect of melatonin on the retinal release of dopamine has led to the suggestion that changes in illumination may modulate the activity of dopaminergic amacrine cells indirectly through changes in retinal melatonin levels (Figure 1; Bubenik et al., 1978; Binkley et al., 1979; Dubocovich, 1983). The effect of melatonin on dopamine

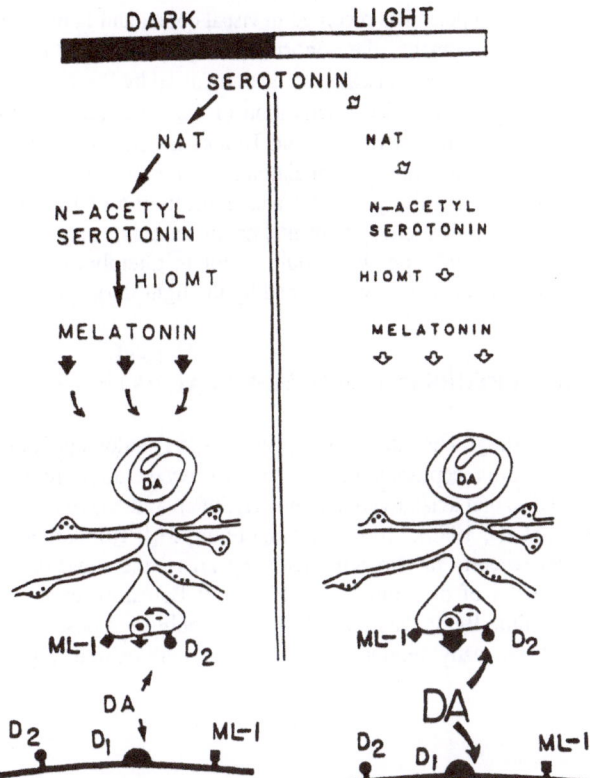

Figure 1. Schematic representation of dopamine (DA)-containing amacrine neurons during the light and dark period. In the vertebrate retina, melatonin is synthesized following a diurnal rhythm with peak levels during the dark period. In contrast, the activity of DA-containing neurons is higher during the light period. Presynaptic melatonin (ML-1) heteroreceptors and D2 dopamine autoreceptors (D2) inhibit DA release. Similar receptors are located on retinal cells postsynaptic to dopamine neurons and modulate a variety of retinal functions. Presynaptic and postsynaptic D1 and D2 dopamine and melatonin receptors are not necessarily located in the same synapse. NAT: N-acetyltransferase; HIOMT: hydroxyindole-O-methyltransferase (Adapted from Dubocovich, 1988b).

release *in vivo* has been studied indirectly by correlating retinal melatonin levels with specific [^{3}H]-spiperone binding to D2 dopamine receptors in retina from rabbits kept under different lighting regimes. Light-induced down-regulation of dopamine binding sites can be reversed by treatments that increase melatonin levels, such as constant dark and melatonin administration (Dubocovich et al., 1985). These studies suggest that the decrease in melatonin levels in constant light disinhibits dopamine retinal neurons *in vivo*, thus leading to elevated dopamine release and subsequent D2 dopamine receptor down-regulation. In a different series of experiments using rabbits implanted with melatonin capsules, melatonin did not inhibit the calcium-dependent release of [^{3}H]-dopamine from the retina *in vitro* (Dubocovich, 1987). This evidence suggests that prolonged increases in the circulating levels of melatonin decrease the sensitivity of presynaptic melatonin receptors and result in increased dopaminergic activity (Dubocovich, 1987).

Biochemical and electrophysiological studies suggest that the retinal dopaminergic system may be involved in transmitting photoperiodic information to the brain. McCullogh and colleagues (1980) have reported that apomorphine, administered *in vivo*, increases the metabolic activity of the rat superior colliculus, probably through activation of dopamine receptors in the retina. Decreases in retinal dopaminergic activity increase the latency of early peaks in flash-evoked potentials recorded in visual cortex and lateral geniculate nucleus (Dyer et al., 1981). In humans, the abnormal delays of visually evoked potentials in untreated patients with Parkinson's disease can be normalized by levodopa/carbidopa therapy (Bodis-Wollmer et al., 1982). Direct activation of D2 dopamine receptors in rat retina enhances the sensitivity to light (Morgan and Brooks-Eidelberg, 1986). It follows that interactions between the dopamine and melatonin systems in the retina may directly or indirectly control retinal sensitivity to light and transmission of light information to the brain. Diurnal or pathological changes in melatonin secretion may lead to alterations in retinal dopaminergic transmission that would result in alterations in the relay of light information to brain centers normally entrained by the light/dark cycle.

MELATONIN RECEPTORS IN CENTRAL VISUAL PATHWAYS

Receptor autoradiographic studies of specific 2-[^{125}I]-iodomelatonin binding indicate the presence of melatonin receptors in brain areas that receive direct retinal input, e.g., the SCN of the mammalian and avian hypothalamus (Laitinen and Saavedra, 1990a; Rivkees et al., 1989; Siuciak et al., 1990). In mammalian brain, 2-[^{125}I]-iodomelatonin binding has been demonstrated recently in two retinorecipient areas, the superior colliculus and the lateral geniculate nuclei of the guinea pig (Figure 2; Beresford et al., 1994) and rabbit (Stankov and Fraschini, 1993; Dubocovich M.L., unpublished observations). Avian and amphibian brains have many retinorecipient regions that exhibit 2-[^{125}I]-iodomelatonin

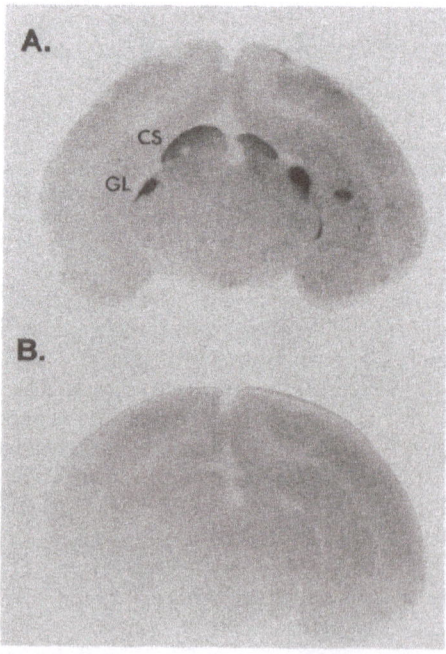

Figure 2. Localization of 2-[^{125}I]-Iodomelatonin (90 pM) binding sites to the superior colliculi (CS) and lateral geniculate nuclei (GL) of guinea pig brain. Autoradiograms of (A) total binding and (B) non-specific binding (in the presence of 1 μM melatonin).

specific binding (Rivkees et al., 1989; Siuciak et al., 1991, Wiechmann and Wirsig-Wiechmann, 1993; Cassone and Brooks, 1991). As shown in coronal sections of the chicken brain (Figure 4), binding is observed in the optic tectum, lateroventral and dorsolateral geniculate nuclei, tectal grey, nucleus of the basal optic root (nBOR) and the dorsolateral anterior thalamus (Siuciak et al., 1991). These findings suggest that melatonin modulates the processing of visual information at the first level of analysis in the brain.

Avian and amphibian studies suggest that melatonin can also modulate the processing of visual information all along central visual pathways (Rivkees et al., 1989; Siuciak et al., 1991; Cassone and Brooks, 1991; Wiechmann and Wirsig-Wiechmann, 1993). In the chicken brain, some of the highest densities of specific 2-[^{125}I]-iodomelatonin binding are found along the major route of visual input, the tectofugal pathway (retina-tectum-nucleus rotundus-ectostriatum-periectostriatum) (Figure 3; Rivkees et al., 1989, Siuciak et al., 1991). Binding is also observed in nuclei of the visual SCN-Edinger-Westphal (EW) circuit

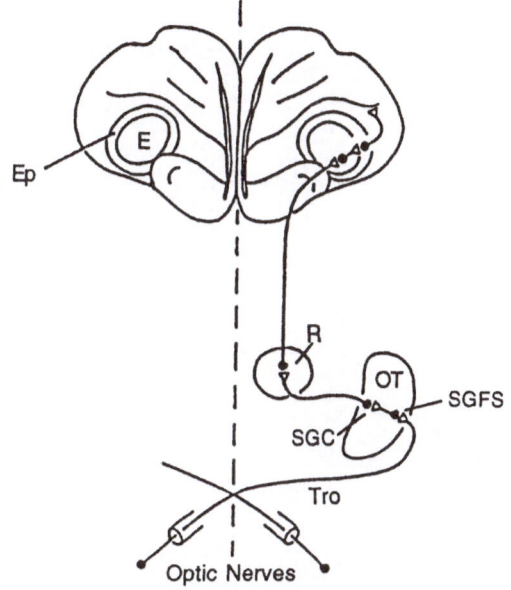

Figure 3. Schematic diagram of the tectofugal pathway. Abbreviations: Ep (periectostriatal belt), E (ectostriatum), R (nucleus rotundus), OT (optic tectum), SCG (stratum griseum centrale), SGFS (stratum griseum et fibrosum superficiale) and TrO (optic tract). (Modified from Karten, 1969)

and the accessory optic pathway (retina-nBOR-oculomotor nuclei-eye) (Rivkees et al., 1989; Siuciak et al., 1991), two systems involved in visual-ocular reflexes (Brecha et al., 1980; Reiner et al., 1983). Lower levels of specific 2-[^{125}I]-iodomelatonin binding are associated with the thalamofugal visual pathway (retina-thalamus-hyperstriatum) (Siuciak et al., 1991). Overall, the 2-[^{125}I]-iodomelatonin binding pattern in avian brain suggests that melatonin regulates incoming signals of light and visual perception (Cassone and Brooks, 1991), circadian rhythmicity (Ebihara and Kawamura, 1981; Oshima et al., 1989) and light-mediated ocular reflexes. These central binding sites show the pharmacological characteristics of ML-1 melatonin receptors (Siuciak et al., 1991).

MELATONIN RECEPTORS ON OPTIC NERVE AXON TERMINALS

Unlike the retina where presynaptic inhibition is a well established function of melatonin receptors, little is known regarding the cellular actions of melatonin in the CNS. Interestingly, specific 2-[^{125}I]-iodomelatonin binding is detected in the avian and amphibian optic tract and several other relevant fiber tracts associated with visual pathways, e.g. the tectothalamic tract, ventral supraoptic decussation and oculomotor nerve (Siuciak et al., 1991; Wiechmann and Wirsig-Wiechmann, 1993). These observations led to the hypothesis that melatonin receptor protein is transported in certain nerve fibers to axon terminal locations in the brain where melatonin acts to regulate presynaptic function (Krause et al., 1992, 1994a,b).

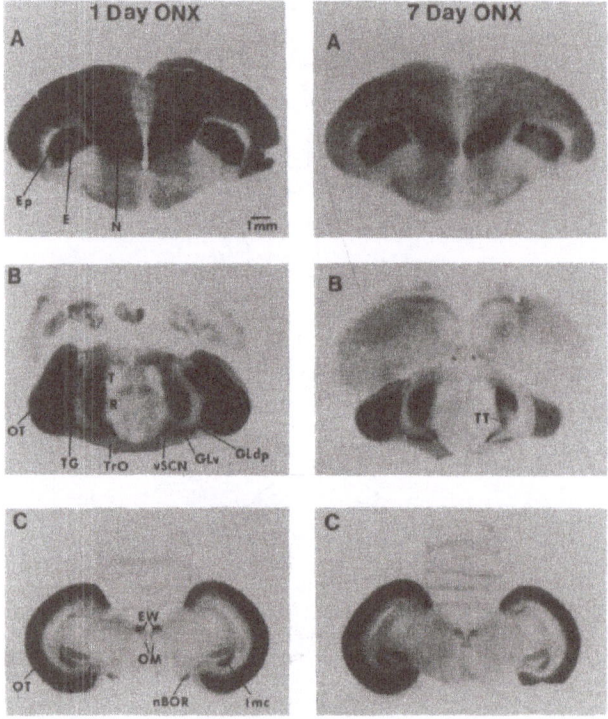

Figure 4: Autoradiograms of 2[^{125}I]-iodomelatonin binding in visual areas of chick brain. Coronal sections were cut at the level of the telencephalon (A), diencephalon (B) and mesencephalon (C) from chicks surviving either 1 or 7 days after transection of the left optic nerve (ONX). Lesioned brain areas are located on the right side of the sections and control areas on the left side. Differences in binding between the two sides are apparent only at 7 days after surgery. Non-specific binding (in the presence of 3 μM melatonin) is similar to background levels (not shown) (from Krause et al., 1994a). Relevant areas exhibiting 2-[^{125}I]-iodomelatonin (75 pM) binding are abbreviated as in Figure 3, except for: N (neostriatum), TG (tectal gray), T (nucleus triangularis), Gldp (dorsolateral geniculate nucleus), GLv (lateroventral geniculate nucleus), vSCN (visual suprachiasmatic nucleus), EW (nucleus of Edinger-Westphal), OM (oculomotor nucleus), Imc (nucleus isthmi, pars magnocellularis), nBOR (nucleus of the basal optic root and TT (tectothalamic tract) (Adapted from Krause et al., 1994).

Unilateral transection of the chick optic nerve was performed to test the hypothesis that melatonin receptors reside on optic nerve terminals. Quantitative autoradiography was used to measure 2-[^{125}I]-iodomelatonin binding in selected brain areas at various survival times (Figure 4). Because the majority of fibers in the chick optic tract cross in the optic chiasm (Ehrlich and Mark, 1984; Hunt and Brecha, 1984), the retinorecipient regions ipsilateral to the lesion provided a useful control for the corresponding contralateral, denervated nuclei. One week after surgery, a significant decrease in 2-[^{125}I]-iodomelatonin binding was found in those layers of the optic tectum that receive retinal ganglion cell input (Krause et al., 1992; Figure 4,5A). This finding correlates with the known time course of degeneration of retinotectal terminals (Ehrlich and Mark, 1984) and is compatible with a presynaptic location of melatonin receptors. By two weeks after transection, 2-[^{125}I]-iodomelatonin binding was decreased by 91% in the central portion of the transected optic tract which is consistent with the interruption of axonal transport of melatonin receptors in optic fiber axons (Krause et al., 1992, 1994a).

Other retinal targets also exhibited substantial decreases in 2-[^{125}I]-iodomelatonin binding following optic nerve transection (Figure 4; Krause et al., 1994a). These areas included the contralateral nucleus of the basal optic root (nBOR; 89% decrease), the lateroventral geniculate nucleus (58%), dorsolateral geniculate nucleus (60%) and tectal gray (58%). In contrast, optic nerve transection did not alter 2-[^{125}I]-iodomelatonin binding in the visual SCN and dorsolateral anterior thalamus, which are primary level nuclei in the circadian/oculomotor and thalamofugal pathways, respectively. In the latter areas, melatonin receptors may be located on the postsynaptic cells or non-retinal afferents.

Recently, electrophysiological effects of melatonin have been examined on retinotectal transmission in a brain slice preparation of the chick. Melatonin suppressed tectal field potentials evoked by stimulation of retinal afferents in a manner consistent with a presynaptic site of action for melatonin (Krause et al., 1994b). It is proposed that elevated levels of melatonin at night regulate the flow of visual information from the retina to the brain by inhibiting the release of optic nerve transmitters in a manner similar to the presynaptic action of melatonin in the retina (Dubocovich, 1985). The optic tectum is normally activated by photic stimulation and the firing frequency is thought to be an important component of the visual message (LaVail and Cowan, 1971; Cassone and Brooks, 1991). Presynaptic melatonin receptors could modify the frequency of transmission or act as a filter suppressing low levels of retinal input and thus sharpening the signal of a light-dark transition.

ACTIVITY-DEPENDENT REGULATION OF MELATONIN RECEPTORS

An unexpected result of the chick optic nerve transection studies was that 2-[^{125}I]-iodomelatonin binding was also decreased in certain secondary (nucleus rotundus, isthmi nuclei) and tertiary level (ectostriatum) visual nuclei, in particular those along the prominent tectofugal pathway (Figures 4,5A; Krause et al., 1994a). The ectostriatum is an integrative visual center analogous to the mammalian visual cortex where similar transsynaptic decreases in GABA receptor binding have been observed following eye enucleation (Hendry et al., 1990). 2-[^{125}I]-Iodomelatonin binding was also decreased in the nucleus of Edinger-Weshphal (EW) but not in the oculomotor nuclei (OM), indicating selective effects of the optic nerve transection in chick oculomotor pathways. In birds, the EW are secondary level oculomotor nuclei involved in pupillary light reflexes and visual control of choroidal blood flow (Reiner et al., 1983).

The transsynaptic changes in 2-[^{125}I]-iodomelatonin binding most likely reflect functional receptor plasticity since nerve terminal degeneration is not observed in secondary

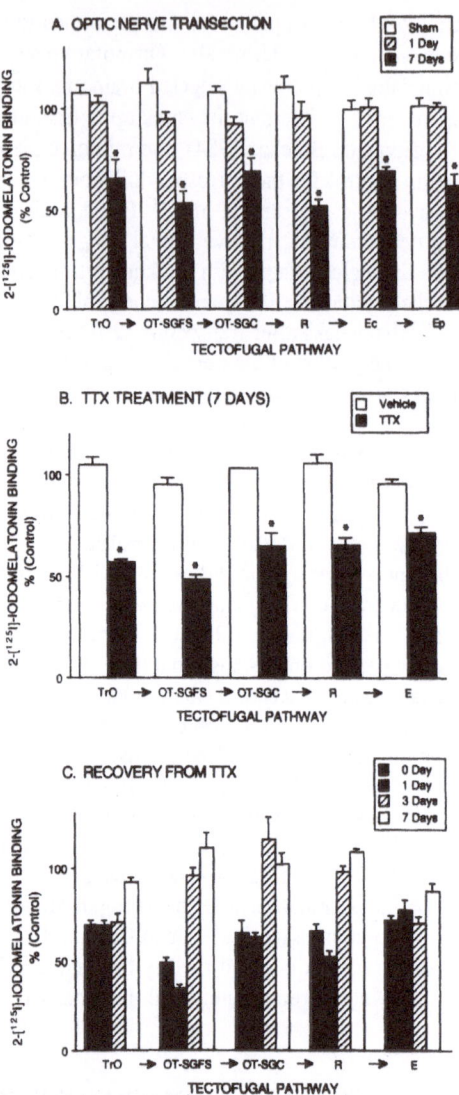

Figure 5. 2-[^{125}I]-Iodomelatonin binding in the tectofugal visual pathway of the chick brain: (A) Effect of unilateral optic nerve transection (B) Effect of intravitreal treatment with tetrodotoxin (TTX) for 7 days and (C) recovery from TTX treatment. Binding data from brain regions contralateral to ONX (lesioned) or the treated eye are expressed as % of binding in the corresponding ipsilateral control regions. Values represent the mean ± S.E.M. of data taken from 3-5 animals. * p < 0.05 as compared to sham. (Data taken in part from Dubocovich et al., 1994; Krause et al., 1992, 1994a)

and tertiary level visual areas at one week after lesion of the optic nerve (Erhlich and Mark, 1984). The binding decreases suggest that specific functional changes in melatonin receptivity are occurring in response to altered activity in a particular pathway. It is possible that transsynaptic degeneration and/or modulation also may have contributed to the reduction in 2-[^{125}I]-iodomelatonin binding observed in retinorecipient areas. While degeneration of postsynaptic cells is unlikely (Reperant and Angaut, 1977), physical or functional loss of synaptic input may alter the metabolism, physiology or receptor expression of postsynaptic neurons.

The optic nerve transection studies suggest that the expression of certain melatonin receptors in chick brain is regulated by the level of visual input. To examine this hypothesis, unilateral intraocular injections of tetrodotoxin (TTX) were used to suppress the electrical activity of chick retinal ganglion cells and consequently the input of visual information to the brain for one week. As a result, specific 2-[^{125}I]-iodomelatonin binding was significantly reduced in a number of contralateral primary, secondary, and tertiary visual areas (Dubocovich et al., 1994; Krause et al., 1994b). Binding was decreased throughout the tectofugal pathway, i.e., in the optic nerve, optic tectum, nucleus rotundus and ectostriatum (Figure 5B). Binding in these areas recovered to control levels by 3-7 days following cessation of the TTX treatment (Figure 5C). The time course of recovery correlated with the return of the pupillary reflex which was suppressed by TTX. These data further support the hypothesis that expression of ML-1 melatonin receptors in certain chick visual pathways is selectively modulated by the level of visual input.

It has been suggested that melatonin acts in a number of visual areas of the avian brain to depress neuronal activity triggered by visual stimulation (Cassone and Brooks, 1991). When the visual input is chronically reduced, less modulation by melatonin may be required. The visual system of the chick brain appears to be a good model for the study of activity-dependent regulation and functional plasticity of central melatonin receptors.

EFFECT OF LIGHT AND MELATONIN RECEPTOR ACTIVATION IN THE MOUSE CIRCADIAN SYSTEM

In mammals, the suprachiasmatic nucleus (SCN) plays an important role as the pacemaker which maintains circadian rhythms (Moore and Klein, 1974; Moore and Eichler, 1972; Stephen and Zucker, 1972; Ibuka and Kawamura, 1975). Entrainment of circadian rhythms to a light/dark cycle occurs via activation of the retinohypothalamic tract projecting from retinal ganglion cells to the SCN (Moore, 1973). The presence of high affinity 2-[^{125}I]-iodomelatonin binding sites in the SCN (Dubocovich, 1988a; Suciak et al., 1990) strongly suggests that melatonin affects circadian activity through activation of specific receptors in this brain region. Modulation of circadian rhythms has been used to determine the influence of melatonin, and other neurotransmitters and neuromodulators such as NPY, acetylcholine, and GABA on the circadian clock (Turek, 1987, Lewy et al., 1992, Colwell et al., 1993; Miller, 1993; Benloucif and Dubocovich, 1994a). Although removal of endogenous melatonin by pinealectomy has no effect on circadian rhythms in constant dark (Cheung and McCormack, 1982; Quay, 1968), removal of the pineal alters SCN neuronal activity (Rusak and Yu, 1993) and free running activity rhythms measured under constant light (Cassone, 1992; Aguilar-Roblero and Vega-Gonzalez, 1993). The exogenous administration of melatonin induces phase dependent shifts in circadian rhythms at dusk (i.e., phase advances) and dawn (i.e., phase delays) both *in vivo* and *in vitro* (Mc Arthur et al., 1991; Lewy et al., 1992; Benloucif and Dubocovich, 1994a). Furthermore, melatonin entrains circadian rhythms and facilitates reentrainment to a new light/dark cycle (Benloucif and Dubocovich, 1994b; Warren et al., 1993).

Light pulses induce time- dependent shifts in circadian rhythms in animals maintained in the absence of entraining cues. Under these conditions the circadian rhythms of locomotor activity become free running. Plotting of these shifts as a function of circadian time (CT) results in a phase-response curve (DeCoursey, 1961; Turek, 1987). Figure 6 shows a phase response curve to a 300 lux pulse of light on the circadian rhythm of wheel running activity in C3H/HeN mice. Although C3H/HeN mice possess a mutant gene (rd) that causes selective degeneration of photoreceptors in the retina, the (rd) mutation does not

LIGHT-INDUCED PHASE SHIFTS IN CIRCADIAN LOCOMOTOR ACTIVITY

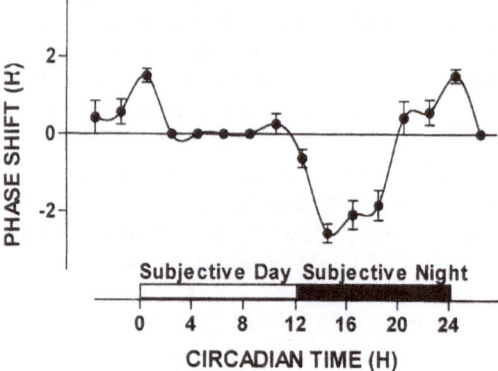

Figure 6. Phase-response curve to light pulses on the circadian rhythm of wheel running activity in the C3H/HeN mouse. C3H/HeN mice, entrained to a 12:12 light:dark cycle, were placed in constant darkness with activity monitored by running wheels. After a stable free running rhythm was obtained, a phase response curve was generated by applying a 15 min light pulse (300 lux) at various circadian times (CTs). Delays in the onset of the active period (CT 12: onset of activity) were observed when the light pulse occurred between CT 12 and CT 19, and advances in activity onset were observed when light was pulsed between CT 22 and CT 1.

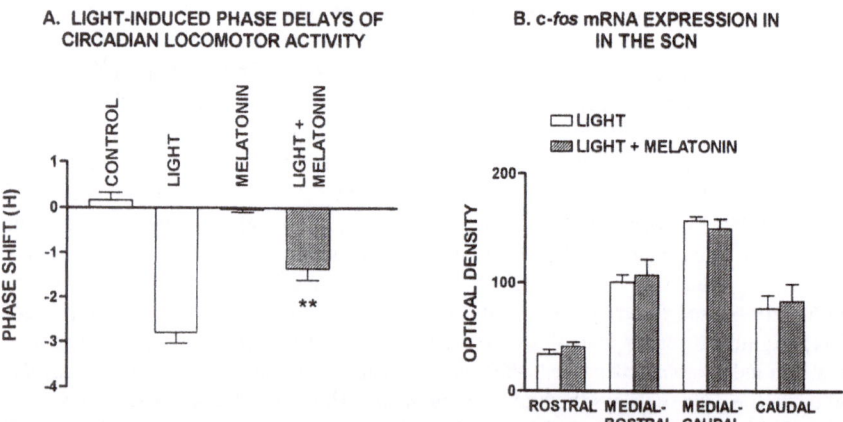

Figure 7. Effect of melatonin on light-induced (A) Phase delays in the circadian rhythm of wheel running activity and (B) *c-fos* mRNA expression at various levels of the SCN in the C3H/HeN mouse. C3H/HeN mice were placed in constant dark with activity monitored by running wheels. At circadian time (CT) 14, mice were injected with either melatonin (3 mg/kg s.c.) or vehicle (10% ethanol/saline) and then kept in the dark (control) or exposed to a 15 min light pulse (300 lux). (A) Melatonin by itself had no effect on the phase of the locomotor activity rhythm but inhibited the phase shift induced by light. (B) Light induced increases in the expression of *c-fos* mRNA in the SCN at CT 14, which is consistent with the period of behavioral sensitivity to light in the C3H/HeN mouse. Animals were sacrificed 15 min after the light pulse and adjacent brain sections (20μm) were cut throughout the SCN and processed for *in-situ* hybridization. Relative optical density (ROD) was measured from the autoradiograms.

appear to affect circadian photoreception (Foster et al., 1991; Colwell and Foster, 1992, Argamaso et al., 1993). The circadian responses of C3H/HeN are similar to those reported for other mammalian species, with phase delays in activity when light is applied in the early subjective night (CT12-CT19) and advances in phase when light is applied in the late subjective night (CT 22-CT1) (Daan and Pittendrigh, 1976; Benloucif and Dubocovich, 1994a). Is of interst to point out that the light-induced phase shifts occurred at times when melatonin administration alone does not shift circadian rhythms.

The question then arises as to whether melatonin could influence the sensitivity of the SCN to light during the early subjective night, i.e, when this hormone does not affect the phase of biological rhythms (Fig 7 A). Therefore, the effect of melatonin administration on light-induced phase shifts was tested at a time when light induced phase delays (CT 14). Mice, free running in constant dark, were injected with either melatonin (3 mg/kg s.c.) or vehicle (10% ethanol/saline) 30 min prior to a light pulse, delivered at CT 14. As shown in Figure 7A, melatonin alone did not shift the onset of activity, but significantly attenuated the phase delay induced by a pulse of light at circadian time 14. These results suggest that melatonin affects circadian rhythms by interfering with the signal arriving from the retina or with the output from the SCN to other components of the circadian system.

The induction of c-*fos* mRNA and other immediate early genes in the SCN is part of the response cascade associated with light-induced phase shifts. This response is dependent upon the phase of the circadian cycle in which the light pulse is administered (Rusak et al., 1990, 1992; Ebling et al., 1991; Aronin et al., 1990; Rea et al, 1989, 1992; Colwell and Foster, 1992; Sutin and Kilduff, 1992). Because we observed that melatonin modulates light-induced phase shifts in circadian locomotor activity in C3H/HeN mice (Figure 7A), we examined the effect of melatonin administration on light-induced expression of c-*fos* mRNA in the SCN. Animals kept in constant dark and not subjected to a light pulse show a very low level of c-fos mRNA in the SCN. Light pulses, (15 min, 300 lux) delivered at CT14 to C3H/HeN mouse induced the expression of c-*fos* mRNA. Figure 7B shows the gradient of c-*fos* mRNA expression across the SCN with maximal values observed in the medial region of the nucleus. The response, however, was not modified by melatonin (3 mg/kg, s.c.) which was administered five minutes prior to the light pulse. This result indicates that the inhibition of light-induced phase shifts by melatonin occurs through a mechanism which does not involve c-*fos*. A similar dissociation between the inhibition of light-induced phase shifts and c-*fos* expression was recently reported for nitric oxide synthase inhibitors (Weber et al., 1994).

CONCLUSION

The distribution of high affinity ML-I melatonin receptors in neural tissue is striking in that it is correlated with those areas that process light, visual and circadian information. Melatonin acts via these receptors in the retina, the first level for sensory perception of environmental light. In addition putative melatonin receptors are present in the pathway by which light influences circadian function (retina-SCN) as well as pathways which receive and interpret visual information (e.g. tectofugal pathway). We know that melatonin can alter biological rhythms, however, we are still a long way from understanding the precise roles that melatonin plays in the modulation of circadian and visual function. We are beginning to understand the cellular changes triggered by melatonin receptor activation, such as inhibition of transmitter release and cyclic AMP formation. Interestingly, the number of melatonin receptors can be regulated by neural activity. Continued focus on elucidating the actions and regulation of melatonin receptors in the visual and circadian systems will be important for understanding the physiological significance of melatonin and its role in various diseases.

ACKNOWLEDGMENTS

This work was supported in part by US Public Health Service Grant MH42922 and a Glaxo Grant to M.L. Dubocovich and NRSA fellowships T32-NS07140 and F 32-AG05608 to S. Benloucif. The authors thank Gina Sian for secretarial assistance.

REFERENCES

Aguilar-Roblero, R., and Vega-Gonzalez, A., 1993, Splitting of locomotor circadian rhythmicity in hamsters is facilitated by pinealectomy, *Brain Res.* 605: 229.

Argamaso, S.M., Knowlton, M.K., and Foster, R.G., 1993, Photoreception and circadian behavior in *rd* and *rds* mice, *Invest. Ophthalmol. Vis. Sci.* 34:1077.

Aronin, N., Sagar, S., Sharp, F., and Schwartz, W., 1990, Light regulates expression of a Fos-related protein in rat suprachiasmatic nuclei, *Proc. Natl. Acad. Sci.*, 87:5959.

Benloucif, S. and Dubocovich M.L., 1994a, Melatonin and light induce phase shifts of circadian rhythms in the mouse, *Biol. Rhythms*, in press.

Benloucif, S. and Dubocovich, M.L., 1994b, Individual differences in reentrainment correlate with specific 2-[^{125}I]-iodomelatonin binding in the suprachiasmatic nucleus, *Soc. Neurosci. Abstr.* 20:1438.

Beresford, I.J.M., Dubocovich, M.L., Andrews, J., Nicholls, R., Coughlan, J.C., Iacob, S., Hayes, A. G., and Hagen, R.M., 1994, Localisation and characterisation of 2-[^{125}I]-iodomelatonin binding sites in guinea pig brain, *Soc. Neurosci. Abstr.* 20:1440.

Besharse, J.C. and Dunis, D.A., 1983, Methoxyindoles and photoreceptor metabolism: Activation of rod shedding, *Science* 219:1341.

Binkley, S., Hryshchyshyn, M., and Reilly, L., 1979, N-Acetyltransferase activity responds to environmental lighting in the eye as well as in the pineal gland, *Nature* 281:479.

Blazynski, C. and Dubocovich, M.L., 1991, Localization of 2-[^{125}I]-iodomelatonin binding sites in mammalian retina, *J. Neurochem.* 56:1873.

Brecha, N., Karten, H.J., and Hunt, S.P., 1980, Projections of the nucleus of the basal optic root in the pigeon: an autoradiographic and horseradish peroxidase study, *J. Comp. Neurol.* 189:615.

Bodis Wollmer, I., Yahr, M.D., Mylin, L., and Thorton, J., 1982, Dopaminergic deficiency and delayed visual evoked potential in humans. *Ann. Neurol.* 91:237.

Bubenik, G.A., Purtill, R.A., Brown, G.M., and Grota, L.J., 1978, Melatonin in the retina and the Harderian gland. Ontogeny, diurnal variations and melatonin treatment. *Exp. Eye Res.* 27:323.

Cahill, G.M., Grace, M.S., and Besharse, J.C., 1991, Rhythmic regulation of retinal melatonin: metabolic pathways, neurochemical regulation and the ocular circadian clock, *Cell. Mol. Neurobiol.* 11:529.

Cassone, V.M., 1992, The pineal gland influences rat circadian activity rhythms in constant light, *J. Biol.Rhythms* 7:27.

Cassone, V.M. and Brooks, D.S., 1991, Sites of melatonin action in the brain of the house sparrow, *Passer domesticus, J. Exp. Zool.* 260:302.

Cheung, P.W. and McCormack, C.E., 1982, Failure of pinealectomy or melatonin to alter circadian activity rhythm of the rat, *Am. J. Physiol.* 242:R261.

Colwell, C.S. and Foster R.G., 1992, Photic regulation of Fos-like immunoreactivity in the suprachiasmatic nucleus of the mouse, *J. Comp. Neurol.* 324:135.

Colwell, C.S., Kaufman, C.M., and Menaker, M., 1993, Phase-shifting mechanisms in themammalian circadian system: New light on the carbachol paradox, *J. Neurosci.* 13:1454.

Daan, S. and Pittendrigh, C.S., 1976, A functional analysis of circadian pacemakers in nocturnal rodents, *J. Comp. Physiol.* 106:253.

DeCoursey, P.J., 1961, Effect of light on the circadian activity rhythm of the flying squirrel, *Glaucomys volans., Z. vergl. Physiol.* 44:331.

Dubocovich, M.L., 1983, Melatonin is a potent modulator of dopamine release in the retina, *Nature* 306:782.

Dubocovich, M.L., 1985, Characterization of a retinal melatonin receptor, *J. Pharmacol. Exp. Ther.* 234:395.

Dubocovich, M.L., 1987, Sensitivity of presynaptic D2 dopamine autoreceptors and melatonin receptors in rabbit retina following chronic treatment with melatonin, *Proc 6th Int. Catecholamine Symposium*, p. 24.

Dubocovich, M.L., 1988a, Pharmacology and function of melatonin receptors, *FASEB J.* 2:2765.

Dubocovich, M.L., 1988b, Role of melatonin in retina, *in: Progress in Retinal Research, vol. 8,* N.N. Osborne and G.J. Cheder, eds., Pergamon Press, Oxford, p. 129.

Dubocovich, M.L., Iacob, S., and Krause, D.N., 1994, Plasticity of ML-1 melatonin receptors in chick brain visual areas following intravitreal tetrodotoxin (TTX), *Soc. Neurosci. Abstr.* 20, 1168.

Dubocovich, M.L., Lucas, R.C., and Takahashi, J.S., 1985, Light-dependent regulation of dopamine receptors in mammalian retina, *Brain Res.* 335:321.

Dubocovich, M.L. and Takahashi, J.S., 1987, Use of 2-[^{125}I]-iodomelatonin to characterize melatonin binding sites in chicken retina, *Proc. Natl. Acad. Sci. USA,* 84:3916.

Dyer, R.S., Howell, W.F., and MacPhail, R.C., 1981, Dopamine depletion slows retinal transmission, *Exp. Neurol.* 71:326.

Ebihara, S. and Kawamura, H., 1981, The role of the pineal organ and the suprachiasmatic nucleus in the control of circadian locomotor rhythms in the java sparrow, *Padda oryzivora, J. Comp. Physiol.* 141:207.

Ebling, F.P.J., Maywood, E.S., Staley, K., Humby, T., Hancock, D.C., Waters, C.M., Evan, G.I., and Hastings, M. H., 1991, The role of N-methyl-D-aspartate-type glutamatergic neurotransmission in the photic induction of immediate-early gene expression in the suprachiasmatic nuclei of the Syrian hamster, *J. Neuroendocrinol.* 3:641.

Ehrlich, D. and Mark, R., 1984, An atlas of the primary visual projections in the brain of the chick *Gallus gallus, J. Comp. Neurol.* 223:592.

Faillace, M.P., Keller Sarmiento, M.I., Siri, L.N., and Rosenstein, R.E., 1994, Diurnal variations in cyclic AMP and melatonin content of golden hamster retina, *J. Neurochem.* 62:1995.

Foster, R.G., Provencio, I., Hudson, D., Fiske, S., De Grip, W., and Menaker, M., 1991, Circadian photoreception in the retinally degenerate mouse (*rd/rd*), *J. Comp. Physiol.A* 169:39.

Hendry, S.H.C., Fuchs, J., deBlas, A.L., and Jones, L.G., 1990, Distribution and plasticity of immunocytochemically localized GABA$_A$ receptors in adult monkey visual cortex, *J. Neurosci.* 10:2438.

Hunt, S.P. and Brecha, N., 1984, The avian optic tectum: a synthesis of morphology and biochemistry, *in: Comparative Neurology of the Optic Tectum,* H. Vanegas, ed., Plenum Press, New York, pp. 619.

Ibuka, N. and Kawamura, H., 1975, Loss of circadian rhythm in sleep-wakefulness cycle in the rat by suprachiasmatic nucleus lesions, *Brain Res.* 59:20.

Iuvone, P.M., 1986, Neurotransmitters and neuromodulators in retina: regulation, interaction, and cellular effects, *in: The Retina: A Model for Cell Biology Studies, vol. 2,* R. Adler and D. Farber, eds., Academic Press, London, pp. 1-72.

Iuvone, P.M., Galli, C.L., Garrison-Gund, C.K., and Neff, N.H., 1978, Light stimulates tyrosine hydroxylase activity and dopamine synthesis in retinal amacrine neurons, *Science* 202:901.

Iuvone, P.M. and Gan, J., 1994, Melatonin receptor-mediated inhibition of cyclic AMP accumulation in chick retinal cell cultures, *J. Neurochem.* 63:118.

Karten, H.J., 1969, The organization of the avian telencephalon and some speculations on the phylogeny of the amniotic telencephalon, *Ann. N.Y. Acad. Sci.* 167:164.

Krause, D.N. and Dubocovich, M.L., 1990, Regulatory sites in the melatonin system of mammals, *Trends Neurosci.* 13:464.

Krause, D.N., Siuciak, J.A., and Dubocovich, M.L., 1992, Optic nerve transection decreases 2-[^{125}I]-iodomelatonin binding in the chick optic tectum, *Brain Res.* 590:325.

Krause, D.N., Siuciak, J.A., and Dubocovich, M.L., 1994a, Unilateral optic nerve transection decreases 2-[^{125}I]-iodomelatonin binding in retinorecipient areas and visual pathways of chick brain, *Brain Res.* 654:63.

Krause, D.N., Dye, J.C., Iacob, S., Karten, H.J. and Dubocovich, M.L., 1994b, Melatonin receptors in the chick optic tectum: presynaptic effects and activity-dependent regulation, *Soc. Neurosci. Abstr.* 20:1168.

Laitinen, J.T. and Saavedra, J.M., 1990a, The chicken retinal melatonin receptor revisited: localization and modulation of agonist binding with guanine nucleotides, *Brain Res.* 528:349.

Laitinen, J.T. and Saavedra, J.M., 1990b, Characterization of melatonin receptors in the rat suprachiasmatic nuclei: modulation of affinity with cations and guanine nucleotides, *Endocrinol.* 126:2110.

LaVail, J.H. and Cowan, W.M., 1971, The development of the chick optic tectum. I. Normal morphology and cytoarchitectonic development, *Brain Res.* 28:391.

Lewy, A.J., Ahmed, S., Latham-Jackson, J.M., and Sack, R.L., 1992, Melatonin shifts human circadian rhythms according to a phase-response curve, *Chronobiol. Int.* 9:380.

McArthur, A.J., Gillette, M.U., and Prosser, R.A., 1991, Melatonin directly resets the rat suprachiasmatic circadian clock in vitro, *Brain Res.* 565:158.

McCullogh, J., Savaki, H.F., McCullough, M.D., and Sokoloff, L., 1980, Retinal-dependent activation by apomorphine of metabolic activity in the superficial layers of the superior colliculus, *Science* 207:313.

Miller, J.D., 1993, On the nature of the circadian clock in mammals, *Am. J. Physiol.* 264:R821.

Moore, R.Y., 1973, Retinohypothalamic projection in mammals: A comparative study, *Brain Res.* 49:201.

Moore, R.Y. and Eichler, V.B., 1972, Loss of a circadian adrenal corticosterone rhythm following suprachiasmatic lesions in the rat, *Brain Res.* 42:201.

Moore, R.Y. and Klein, D.C., 1974, Visual pathways and the central neural control of a circadian rhythm in pineal serotonin n-acetyltransferase activity, *Brain Res.* 71:17.

Morgan, W.W. and Brooks-Eidelberg, B.A., 1986, Intraocular injection of a selective D2 agonist enhances visual evoked output from the rat retina, *Soc. Neurosci. Abstr.* 12:624.

Nao-I, N., Nilsson, S.E.G., Gallemore, R.P., and Steinberg, R.H., 1989, Effects of melatonin on the chick retinal pigment epithelium: Membrane potentials and light-evoked responses, *Exp. Eye Res.* 49:573.

Ogino, N., Matsumura, M., Shirakawa, H. and Tsukahara, I., 1983, Phagocytic activity of culture pigment epithelial cells from chicken embryo: inhibition by melatonin and cyclic AMP, and its reversal by taurine and cyclic GMP, *Ophthalmic Res.* 15:72.

Oshima, I., Yamada, H., Goto, M., Sato, K., and Ebihara, S., 1989, Pineal and retinal melatonin is involved in the control of circadian locomotor activity and body temperature rhythms in the pigeon, *J. Comp. Physiol.* 166:217.

Pang, S.F. and Yew, D.T., 1979, Pigment aggregation by melatonin in the retinal pigment epithelium and choroid of guinea pigs, *Cavia Porcellus, Experientia* 35:231.

Parkinson, D. and Rando, R.R., 1983a, Effect of light on dopamine turnover and metabolism in rabbit retina, *Invest. Ophthalmol. Vis. Sci.* 24:384.

Parkinson, D. and Rando, R.R., 1983b, Effects of light on dopamine metabolism in the chick retina, *J. Neurochem.* 40:39.

Pierce, M.E. and Besharse, J.C., 1985, Circadian regulation of retinomotor movements. I. Interaction of melatonin and dopamine in the control of cone length. *J. Gen. Physiol.* 86:671.

Quay, W.B., 1968, Individualization and lack of pineal effect in the rat's circadian locomotor rhythm, *Physiol. Behav.* 3:109.

Rea, M.A., 1989, Light increases *fos*-related protein immunoreactivity in the rat suprachiasmatic nuclei, *Brain Res.Bull.* 23:577.

Rea, M.A., 1992, Different populations of cells in the suprachiasmatic nuclei express *c-fos* in association with light-induced phase delays and advances of the free-running activity rhythm in hamsters, *Brain Res.*579:107.

Reiner, A., Karten, H.J., Gamlin, P.D.R., and Erichsen, J.T., 1983, Parasympathetic ocular control: functional subdivisions and circuitry of the avian nucleus of Edinger-Westphal, *Trends Neurosci.* 6:140.

Reperant, J. and Angaut, P., 1977, The retinotectal projections in the pigeon. An experimental optical and electron microscope study, *Neuroscience,* 2:119.

Rivkees, S.A., Cassone, V.M., Weaver, D.R., and Reppert, S.M., 1989, Melatonin receptors in chick brain: characterization and localization, *Endocrinology* 125:363.

Rusak, B. and Yu, G.D., 1993, Regulation of melatonin-sensitivity and firing-rate rhythms of hamster suprachiasmatic nucleus neurons: Pinealectomy effects, *Brain Res.* 602:200.

Rusak, B., Robertson, H., Wisden, W., and Hunt, S., 1990, Light pulses that shift rhythms induce gene expression in the suprachiasmatic nucleus, *Science* 248:1237.

Rusak, B., McNaughton, L., Robertson, H.A., and Hunt, S.P., 1992, Circadian variation in photic regulation of immediate-early gene mRNAs in rat suprachiasmatic nucleus cells, *Mol. Brain. Res.* 14:124.

Siuciak, J.A., Fang, J.M., and Dubocovich, M.L., 1990, Autoradiographic localization of 2-[^{125}I]-iodomelatonin binding sites in the brains of C3H/HeN and C57BL/6J strains of mice, *Eur. J. Pharmacol.* 180:387.

Siuciak, J.A., Krause, D.N. and Dubocovich, M.L., 1991, Quantitative pharmacological analysis of 2-[^{125}I]-iodomelatonin binding sites in discrete areas of the chicken brain, *J. Neurosci.* 11:2855.

Stankov, B. and Fraschini, F., 1993, High affinity melatonin binding sites in the vertebrate brain, *Neuroendocrinol. Lett.* 15:149.

Stephen, R. and Zucker, I., 1972, Circadian rhythms in drinking behavior and locomotor activity of rats are eliminated by hypothalamic lesions, *Proc. Natl. Acad. Sci.* 69:1583.

Sutin, E.L. and Kilduff, T.S., 1992, Circadian and light-induced expression of immediate early gene mRNAs in the rat suprachiasmatic nucleus, *Mol. Brain Res.* 15:281.

Turek, F.W., 1987, Pharmacological probes of the mammalian circadian clock: use of the phase response curve approach, *Trends Pharmacol. Sci.* 8:212.

Warren, W.S., Hodges, D.B., and Cassone, V.M., 1993, Pinealectomized rats entrain and phase-shift to melatonin injections in a dose-dependent manner, *J. Biol.Rhythms* 8:233.

Weber, E.T., Gillette, M.W., and Rea, M.A., 1994, Nitric oxide synthase inhibitor blocks light-induced phase shifts the free-running activity rhythm in hamsters, *Soc. Neurosc.Abstr.* 20:162.

Wiechmann, A.F., 1986, Melatonin: parallels in pineal gland and retina, *Exp. Eye Res.* 42:507.

Wiechmann, A.F. and Wirsig-Wiechmann, C.R., 1993, Distribution of melatonin receptors in the brain of the frog *Rana pipiens* as revealed by in vitro autoradiography, *Neuroscience* 52:469.

Wirz-Justice, A., DaPrada, M., and Reme, C., 1984, Circadian rhythm in rat retinal dopamine, *Neurosci. Lett.* 45:21.

MELATONIN RECEPTORS IN BRAIN AND PERIPHERAL ARTERIES

Mohan Viswanathan[1], Simona Capsoni[2] and Juan M. Saavedra[1]

[1]Laboratory of Clinical Science
National Institute of Mental Health
Bethesda, MD 20892-1514, USA
[2]Department of Pharmacology
University of Milan
20129 Milano, Italy

INTRODUCTION

With the development of [^{125}I]2-iodoMelatonin as a ligand, (Vakkuri et al., 1984) melatonin receptors could be localized and characterized in small, restricted brain areas (Vanecek, 1988; Laitinen et al., 1989). Melatonin receptors were soon identified by autoradiography to be present also in cerebral and caudal arteries of the rat (Viswanathan et al., 1990). In this report, we analyze the localization, characterization and regulation of these arterial melatonin receptors. Our results indicate that arterial melatonin receptors could be physiologically active, and perhaps play a role in the regulation of cerebral blood flow and body temperature.

MATERIALS AND METHODS

2-[^{125}I]Iodomelatonin, specific activity 1800 to 2000 Ci/mmol, was purchased from Amersham (Arlington Heights, IL). Drugs and chemicals were obtained from Sigma Chemical Co. (St. Louis, MO) unless otherwise stated. 2-Iodomelatonin was obtained from Research Biochemicals (Natick, MA). Guanosine-5'-O-(3-thiotriphosphate) (GTPγS) and adenosine-5'-O-(3-thiotriphosphate) (ATPγS) were purchased from Boehringer Mannheim (Indianapolis, IN). The cAMP RIA kit (NEK-033) was obtained from New England Nuclear Corp. (Boston, MA), and pertussis toxin was purchased from ICN Biomedicals (Irvine, CA).

Sprague-Dawley rats, male (200-250 g), female and male, 150-200 g, and ovariectomized (OVX) female rats, 150-200 g, were purchased from Zivic Miller (Zelienople, PA). Wistar male rats, 9, 96, and 306 days old, were purchased from Charles River (Wilmington, MA). Young (4 weeks old) and adult (14 weeks old) male spontaneously hypertensive (SHR) and age-matched male normotensive controls, Wistar Kyoto (WKY) were obtained from Taconic Farms (Germantown, NY).

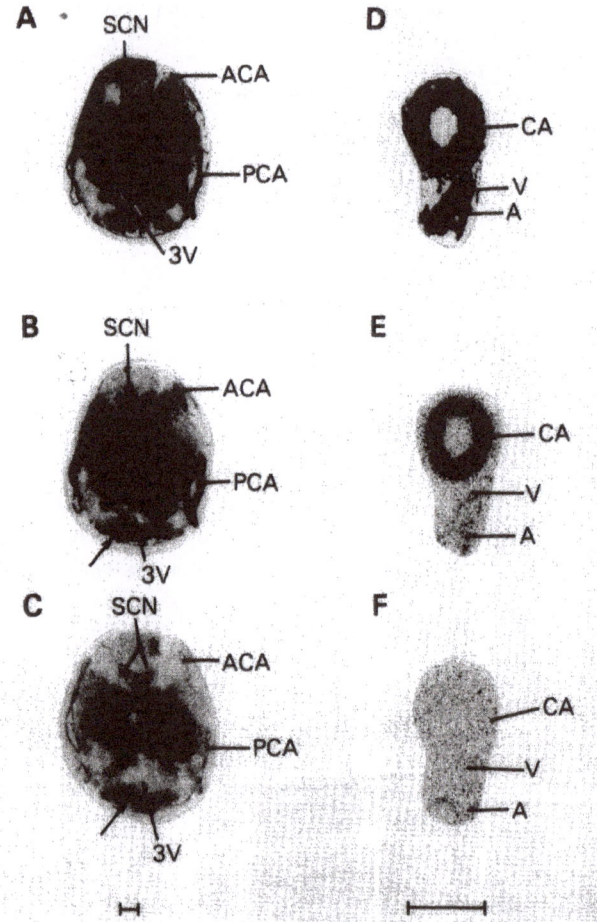

FIGURE 1. Autoradiography of [^{125}I]2-iodomelatonin binding to rat brain and caudal arteries. ACA: anterior cerebral artery; PCA: posterior communicating artery; 3V: third ventricle; SCN: suprachiasmatic nucleus; CA: caudal artery; V: vein; A: small artery. A and D: hematoxylin and eosin staining; B and E: total binding; C and F: non-specific binding, incubated as in total binding plus 1 μM unlabeled melatonin. (From Viswanathan et al., 1990).

Intact females were monitored for estrous cyclicity by taking daily vaginal smears between 1200-1300 h. Only females showing three consecutive 4-day cycles were used. OVX rats were randomly divided into two groups. One group was implanted with a 17β-estradiol pellet (OVX-E$_2$; Innovative Research of America, Toledo, OH; 0.05 mg/pellet, sc) for 1 week.

All rats were kept under a 12-h light, 12-h dark cycle, with lights on at 0630 h. Standard food and water were provided ad libitum. Animals were killed by decapitation between 3 and 6 h after light onset. All procedures for handling and killing the animals were approved by the NIMH Animal Care and Use Committee.

Brains with circle of Willis arteries were rapidly dissected, frozen in isopentane at -30 ˚C, and stored at -80˚ C. Caudal arteries were dissected immediately after death, frozen, and stored at -80˚C until used. Coronal sections, 16 μm thick, were cut at the level of the optic chiasm and containing sections of the anterior cerebral arteries, were cut in a cryostat at -20 ˚C. Caudal (tail) arteries from each rat were sectioned at various levels along the length of the vessel. Sections were mounted on chrome alum gelatin-coated slides, dried overnight in a desiccator at 4 ˚C, and kept at - 80˚C for no longer than 2 weeks.

For autoradiography, sections were incubated in the presence of 2-[^{125}I]Iodomelatonin as described elsewhere (Capsoni et al., 1994). After incubation, sections were exposed to [^{3}H]Hyperfilm (Amersham), developed as described (Capsoni et al., 1994) and values (fmol/mg protein) were obtained by using a computerized image processing and analysis program, Image v1.42 (written by W. Rasband, NIMH, Bethesda, MD).

The effects of drugs on cerebral artery cAMP were determined in the complete circle of Willis, as described (Capsoni et al., 1994), and the content of cAMP was determined using a RIA kit (New England Nuclear).

RESULTS

During the course of our studies on melatonin receptors in the rat suprachiasmatic nucleus, we detected 2-[^{125}I]Iodomelatonin binding in coronal cuts of the anterior cerebral arteries (Viswanathan et al.,1990). A screening of most major arteries in the rat revealed that the only arteries where melatonin binding was easily displaced were the cerebral arteries and the caudal artery (Viswanathan et al., 1990). All arteries belonging to the circle of Willis contained melatonin receptors. In the caudal artery, the receptors were located throughout the medial smooth muscle layer (Viswanathan et al., 1990) (Figure 1).

Pharmacological characterization of the cerebral artery melatonin binding revealed the presence of two binding sites: a high affinity site, with a Kd of 13 pM, and a low affinity site, with a Kd of 823 pM (Capsoni et al., 1994) (Figure 2A). Melatonin binding to the cerebral arteries was inhibited by simultaneous incubation with the GTP analog GTPγS, and the high affinity binding site could no longer be detected (Capsoni et al., 1994) (Figure 2B).

In the cerebral arteries, whereas melatonin alone did not influence the basal level of cAMP, melatonin was effective to inhibit the forskolin-stimulated cAMP accumulation, an effect blocked by preincubation with pertussis toxin (Capsoni et al., 1994) (Figure 3). Conversely, no effect of melatonin on forskolin-stimulated cAMP had been observed in the tail artery (Viswanathan et al., 1990).

The effect of melatonin in arterial contractility was studied in the isolated tail artery of the adult rat. Melatonin by itself did not affect the resting tension of the artery. However, melatonin significantly prolonged and potentiated the norepinephrine-induced contractions (Viswanathan et al., 1990).

Developmental studies revealed that age was an important factor for the regulation of melatonin receptors in the vasculature (Laitinen et al., 1992), and melatonin binding in both the cerebrovascular and caudal arteries is inversely correlated with age (Laitinen et al., 1992).

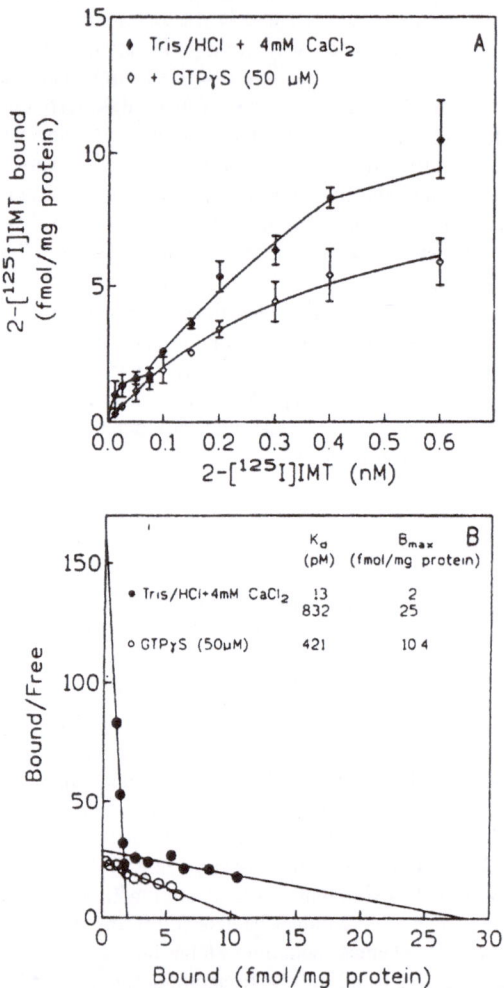

FIGURE 2. Effects of GTPγS on [¹²⁵I]2-iodomelatonin binding to rat cerebral arteries. A: Saturation curve. Each point is the mean ± S.E.M. of measures from eight different animals; B: Scatchard plot. The Figure represents the mean of 3 experiments.

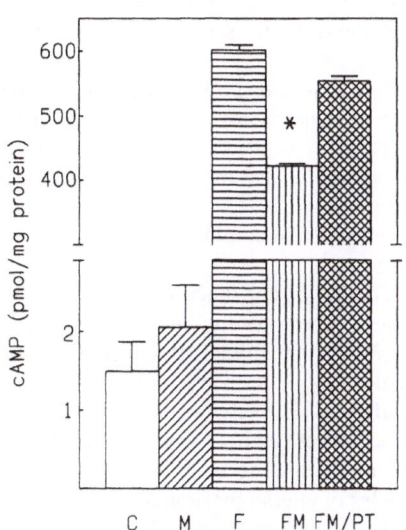

FIGURE 3. Inhibitory effect of melatonin on cAMP production in rat circle of Willis. C: control; M: melatonin (10 nM) alone; F: forskolin (10 μM) alone; FM: forskolin plus melatonin; FM/PT, forskolin plus melatonin, preincubation with pertussis toxin. (From Capsoni et al., 1994).

Reproductive hormones seem to be another important factor in the regulation of vascular melatonin receptors. Both in the caudal and cerebral arteries of female rats, the number of melatonin receptors is higher during diestrous than during estrous (Seltzer et al., 1992). In addition, in ovariectomized rats, estradiol replacement decreased the expression of melatonin receptors in the vasculature (Seltzer et al., 1992). Thus, it appears that in the female rat, estrogen downregulates the number of vascular melatonin receptors.

In caudal arteries from spontaneously hypertensive rats (SHR), the expression of melatonin receptors is lower with respect to their normotensive controls, the Wistar Kyoto rats (WKY). In brain arteries from SHR, the expression of the receptors is so low as to be undetectable in adult hypertensives (Viswanathan et al, 1992).

DISCUSSION

The present data indicate that the melatonin binding site present in the rat brain arteries of the circle of Willis may be considered a melatonin receptor, because of its high affinity, possible coupling to a pertussis-toxin sensitive G-protein, and inhibition of cAMP production.

Whether the melatonin receptor plays a significant role in the physiological regulation of the cerebral circulation still needs to be determined. However, there are several indications that this could probably be the case. First, the brain arterial melatonin receptor is developmentally regulated, and is also regulated by female reproductive hormones and in a model of genetic hypertension. Second, brain arterial melatonin receptors have been reported not only in the rat, but also in non-human primates (Stankov et al., 1993a, 1993b).

A similar case for a physiological function can be proposed for the caudal artery melatonin receptors. These receptors are also developmentally regulated, and their expression is much higher in younger animals (Laitinen et al., 1992). In vitro stimulation of caudal artery melatonin receptors in the rat enhances the contractile effect of norepinephrine, but this effect is relatively weak (Viswanathan et al., 1990). However, melatonin has a direct vasoconstrictive effect in caudal arteries of young rats (Evans et al., 1992) and this correlates with the high receptor expression in immature animals.

The two most likely roles of arterial melatonin receptor are the regulation of body and brain temperature and the modulation of cerebral blood flow. In the rat, the caudal artery is important for the regulation of body temperature. The blood flow through the circle of Willis is also known to contribute to regulate the brain temperature. The possible regulation by melatonin of blood flow in the caudal artery and the circle of Willis could explain the effects of the hormone in temperature regulation (Ralph et al., 1979; Cagnacci et al., 1992; Strassman et al., 1991). This hypothesis could explain the alterations in the regulation of body temperature present in genetically hypertensive rats, when the arterial melatonin receptors are very low (Wright et al., 1977; Yen et al., 1978).

So far, the evidence for a melatonin mediated regulation of cerebral blood flow is not complete. However, preliminary data indicate that intravenous administration of melatonin can alter the cerebral blood flow autoregulation in the rat (unpublished data). The expression of brain arterial melatonin receptors is higher in young rats, when the cerebral blood flow is highest. It has been reported that melatonin regulates a large number of brain functions, and this can be achieved by regulation of blood flow to selective brain areas. Melatonin effects are related to reproductive function. Female reproductive hormones, by regulating the expression of brain arterial melatonin receptors, could also modulate blood flow to selected brain areas, and by this mechanism affect reproductive physiology and behavior.

Further study of arterial melatonin receptors may help to further clarify the physiological effects of this hormone.

REFERENCES

Cagnacci, A., Elliot, J.A., and Yen, S.S.C., 1992, Melatonin: A major regulator of the circadian rhythm of core temperature in humans, J. Clin. Endocrinol. Metab. 75:447.

Capsoni, S., Viswanathan, M., De Oliveira, and Saavedra, J.M., 1994, Characterization of melatonin receptors and signal transduction system in rat arteries forming the circle of Willis, Endocrinology 135:373.

Evans, B.K., Mason, R. and Wilson, V.G., 1992, Evidence for direct vasoconstrictor activity of melatonin in 'pressurized' segments of isolated caudal artery from juvenile rats. Naunyn-Schmiedegerg's Arch. Pharmacol. 346:362.

Laitinen, J.T., Viswanathan, M., Vakkuri, O., and Saavedra, J.M, 1992, Differential regulation of the rat melatonin receptors: selective age-associated decline and lack of melatonin-induced changes, Endocrinology 130:2139.

Ralph, C.L., Firth, B.T., Gern, W.K., and Owens, D.W., 1979, The pineal complex and thermoregulation, Biol. Rev. 54:41.

Seltzer, A., Viswanathan, M., and Saavedra, J.M, 1992 Melatonin binding sites in brain and caudal arteries of the female rat during the estrous cycle and after estrogen administration, Endocrinology 130:1896.

Stankov, B., Capsoni, S., Lucini, V., Fautek, J., Gatti, S., Gridelli, B., Biella, G., Cozzi, B., and Fraschini, F., 1993b, Autoradiographic localization of putative melatonin receptors in the brains of two old world primates: Ceropithecus aethiops and Papio ursinus, Neuroscience 52:459.

Stankov, B. and Fraschini, F., 1993a, High affinity melatonin binding sites in the vertebrate brain, Neuroendocrinology 56:149.

Strassman, R.J., Qualls, C.R., Lisansky, E.J., and Peake, G.T., 1991, Elevated rectal temperature produced by all-night bright light is reversed by melatonin infusion in men, J. Appl. Physiol. 71:2178.

Vakuri, O., Lamsa, E., Rahkamaa, E., Ruotsalainen, H., and Leppaluotto, J., 1984, Iodinated melatonin: Preparation and characterization of the molecular structure by mass and ^{1}H NMR spectroscopy, Anal. Biochem. 142:284.

Viswanathan, M., Laitinen, J.T., and Saavedra, J.M, 1992 Differential regulation of melatonin receptors in spontaneously hypertensive rats, Neuroendocrinology 56:864.

Viswanathan, M., Laitinen, J., and Saavedra, J.M, 1990, Expression of melatonin receptors in arteries involved in thermoregulation, Proc. Nat. Acad. Sci. USA. 87:6200.

Wright, G., Iams, S., and Kecht, E., 1977, Resistance to heat stress in the spontaneously hypertensive rats, Can. J. Physiol. Pharmacol. 55:975.

Yen, T.T., Pearson, D.V., Powell, C.E., and Krischner, G.L., 1978, Thermal stress elevates the systolic blood pressure of spontaneously hypertensive rats, Life Sci. 22: 359.

MELATONIN RECEPTORS IN MODEL SYSTEMS

Peter J. Morgan and Perry Barrett

Molecular Neuroendocrinology Group
Rowett Research Institute
Greenburn Road
Bucksburn
Aberdeen AB2 9SB
Scotland, UK

INTRODUCTION

One of the most striking features to emerge from the many studies on the distribution of $2\text{-}^{125}\text{I}$-iodomelatonin binding sites is the wide range of tissues to which binding has been localised. (Morgan et al 1994a). Often these sites have been found to be species specific, and in some cases their expression is limited to a particular developmental period. It is self-evident from these data that the list of potential target sites vastly outnumber the range of biological responses for which we have robust and rigorous experimental data. Therefore what is the function of these numerous sites and what is the significance of their species-specific distribution and developmental appearance? A parsimonious interpretation might be that there is gross redundancy in melatonin receptor expression. Such an interpretation might seem reasonable given that following pinealectomy the concentration of plasma melatonin falls below undetectable levels (Lewy et al 1981; Reiter 1991), suggesting that the pineal is the main source of circulating melatonin. This being the case then all sites would theoretically receive the same limited endocrine signal. Alternatively not all sites may require circulating melatonin for their function. In the retina and the gut, for example, there is evidence that melatonin is synthesised locally (Vivien-Roels et al 1981; Zawilski and Iuvone 1989; Raikhlin et al 1975) and thus the function of melatonin is independent of the pineal. Likewise in the brain the possibility of localised synthesis within neural terminals cannot yet been excluded. Sensitive molecular biological techniques may yet show that hydroxy-0-methyltransferase, an enzyme important to the biosynthesis of melatonin, is expressed in the nerve terminals which synapse to areas of the brain that express melatonin receptors. Therefore we must be open to the possibility that melatonin may have important biological functions beyond the limitations of the pineal gland.

The Pineal Gland and Its Hormones
Edited by F. Fraschini *et al.*, Plenum Press, New York, 1995

Two recent advances will aid us in defining the true functions of melatonin at each of its potential target sites. The first is the cloning and expression of the first high affinity melatonin receptor from *Xenopus* melanophores (Ebisawa et al 1994), the second is the development of novel analogues of melatonin which act with high affinity at the melatonin receptor (Yous et al 1992; Howell et al 1994, Sugden 1994, Copinga et al 1993, Spadoni et al 1993). Both of these advances will help us to define melatonin receptor pharmacology, and will enable the development of new tools with which to probe the function of melatonin at selected target sites. However, neither can occur without a sound basic understanding of the pharmacological functioning of the melatonin receptor within native model systems.

THE PITUITARY AS A MODEL SYSTEM

Whilst most evidence for the photoperiodic and circadian effects of melatonin point to a site of action within the brain (Morgan et al 1994a), inevitably study of the cellular action of melatonin within neurones is particularly difficult. This is due to the intrinsic problem of culturing neuronal tissue, the heterogeneity of the tissue and the generally low yield of cells. The cells of the pituitary gland, on the other hand, are much easier to culture and they are also a target site for melatonin action (Morgan et al 1994a). The pituitary offers two model systems each of which have proved useful in the study of the mode of action of the melatonin receptor. The first is the pars distalis of the neonatal rat, and the second in the pars tuberalis of the sheep pituitary. Each tissue has its merits. For the rat neonatal pars distalis, it was established well in advance of the first demonstration of melatonin receptors on its cells, that melatonin would inhibit luteinising hormone releasing hormone-stimulated release of luteinising hormone from the gland (Martin and Klein 1976). Subsequently the signal transduction pathways which might mediate this response have been investigated extensively (Vanecek and Vollrath 1989; Vanecek and Vollrath 1990; Vanecek and Klein 1992a,b). In general these studies have revealed that, through the melatonin receptor, melatonin inhibits a number of different second messenger pathways, including cyclic AMP production (Vanecek and Vollrath 1989), calcium mobilisation (Vanecek and Klein 1992b), arachidonic acid release, diacylglycerol turnover (Vanecek and Vollrath 1990) and membrane hyperpolarisation (Vanecek and Klein 1992a). In each case melatonin alone has no or only a weak effect. Thus its main action appears to be to inhibit receptor-activated second messenger pathways, and these actions each involve pertussis toxin sensitive G-proteins (Vanecek and Vollrath 1989; Vanecek and Vollrath 1990; Vanecek and Klein 1992a,b). The biological effect of melatonin in these cells is clearly to inhibit the release of LH, but as melatonin receptors are present on the gonadotrophs solely during the neonatal period (Martin and Sattler 1979; Vanecek 1988) the physiological role of melatonin in this gland remains to be established. The main disadvantage of the neonatal rat pars distalis as a model tissue is not only the transient appearance of melatonin receptors, but also the small size of the gland.

In marked contrast the pars tuberalis of the sheep offers a gland of considerably larger size, which is easily dissociated by enzymatic dispersion and therefore provides more tissue for biochemical experiments. A number of lines of circumstantial evidence have indicated that the pars tuberalis may play an important role in the photoperiodic effects of melatonin (Morgan et al 1994a). Perhaps the strongest evidence for its role in sheep has recently been provided by some elegant experiments on Soay rams. In these studies the hypothalamus was surgically disconnected from the pituitary, maintaining the pars tuberalis intact. Subsequently it was shown that the rams expressed the expected changes in serum prolactin levels following manipulations of either photoperiod or administration of melatonin by subcutaneous implant (Lincoln and Clarke 1994). These experiments directly showed that the brain is not required to mediate the photoperiodic responses on prolactin in the sheep, and therefore strongly implicate the pars tuberalis in the response (Lincoln and Clarke 1994). Like the rat neonatal pars distalis the main effect of melatonin appears to be the inhibition of an activated second messenger pathway, as melatonin dose-dependently inhibits forskolin-stimulated cyclic AMP production (Morgan et al 1991b). As in the pars distalis melatonin alone has no effect on a number signal transduction pathways, including the accumulation of cyclic AMP (Morgan et al 1989a, 1991b), the moblisation of calcium, or the activities of phospholipases C, A2 or D (Morgan et al 1991a McNulty et al 1994). Nevertheless it was established that such second messenger activities do exist in pars tuberalis cells (Morgan et al 1989a, 1991a; McNulty et al 1994). The main disadvantage of the pars tuberalis as a model tissue is that the main secretory cell type of the pars tuberalis is of unknown phenotype, and therefore the natural agonist, which stimulates cyclic AMP, and potentially other pathways, in pars tuberalis cells is not yet known (Morgan 1991). Although we have been able to isolate a novel isoform of the α-MSH receptor which stimulates cyclic AMP from the ovine pars tuberalis. we do not yet know the physiological role of this receptor (Barrett et al 1994). Therefore to date it has been necessary to utilise the pharmacological tool, forskolin, to stimulate pars tuberalis cells to study the mode of action melatonin. This notwithstanding it has been possible to establish some important features of melatonin receptor pharmacology.

G-PROTEIN COUPLING AND RECEPTOR PHARMACOLOGY

It is now well established that the melatonin receptor present in the ovine pars tuberalis and a number of other tissues belongs to the G-protein-coupled class of receptor (Morgan et al 1994a). The initial evidence for this was that the binding of the radioligand $2\text{-}^{125}\text{I}$-iodomelatonin to ovine pars tuberalis membranes was modulated by guanine nucleotides, and in cultured sheep pars tuberalis cells, melatonin inhibited forskolin-stimulated accumulation of cyclic AMP (Morgan et al 1989a,b,) The cloning of a gene that encodes a protein, which when transiently expressed in COS-7 cells binds $2\text{-}^{125}\text{I}$-iodomelatonin with high affinity, and inhibits forskolin-stimulated cyclic AMP production when stably transfected into Chinese hamster ovary cells (Ebisawa et al 1994), confirms the existence of such a receptor. The function of the G-protein is threefold. Firstly it regulates the affinity state of the receptor, secondly it directs the transfer of information from a particular receptor to a specific effector enzyme, and thirdly it regulates the activity of the specific effector enzyme (Conklin and Bourne 1993). Thus it plays a critical role in defining the pathway of signal transduction and in determining the pharmacology of a receptor. On the basis of the diversity in the structure of the α-sub-unit of G-proteins, at least four classes of G-protein are now recognised to mediate signal transduction events. These have been classified as Gs, Gi, Gq and G11, and several members exist within each class (Simon et a 1991) (see Fig 1). Different G-proteins exist within different tissues, and the stochiometries of the G-proteins present also varies. Thus as the interaction between a particular receptor and a G-protein is a selective rather than an absolute process, clearly it is essential to know with which G-proteins the melatonin receptor interacts within a given tissue. This information becomes particularly important when transfection techniques are used to study the pharmacology of a receptor expressed in a mammalian cell line. Usually it is only the receptor cDNA that is transfected not that of its associated G-protein. As such transfection studies lead to the over-expression of the receptor in the membrane of the host cell, this can lead to interactions between the receptor and G-proteins which may not be representative of a coupling which occurs in the native cell (Milligan 1991). As a consequence the pharmacological profiles of a series of drugs may be different and full agonists may only display partial agonism. For example it has been suggested recently that 8-OH-DPAT acts as a full agonist at 5-HT1A receptors where Gi3 is the main modulator of the effector, yet where Gi2 predominates, the same ligand acts only as a partial agonist (Gettys et al 1994). Such differences may help to explain why, when mRNA isolated from ovine pars tuberalis and microinjected into *Xenopus* oocytes to express the melatonin receptor protein, the potency of the effect of melatonin was very much lower than that for the inhibition of cyclic AMP in ovine pars tuberalis cells (Fraser et al 1991; 1994). In Fig 2 the potency of the inhibition of forskolin-stimulated cyclic AMP levels in ovine pars tuberalis cells is compared with the potency of the dose-dependent inhibition of aluminium fluoride (AlF^-_4)-induced current oscillations in *Xenopus* oocytes after microinjection of ovine pars tuberalis mRNA. Whereas melatonin inhibits the cyclic AMP response with an IC_{50} of about 190pM, the inhibition of the oscillatory currents had an IC_{50} of between 0.1 and 1 mM. Clearly therefore, study of the pharmacology of a given receptor must achieve a balance between the power and convenience of molecular biological techniques and an understanding of the biology reality in the native cell.

G-PROTEIN COUPLING IN THE PARS TUBERALIS

Generally it has been found that for receptors which mediate the inhibition of cyclic AMP the G-protein involved is a member of the Gi class of G-protein, and characteristically such G-proteins are sensitive to the bacterial toxin, *Bordetella pertussis* (Simon et al 1991). Pertussis toxin inactivates the G-protein by catalysing the ADP-ribosylation of the α-sub-unit of the G-protein. The effects are to render the receptor in its low affinity conformation, and to block information transfer between the receptor and the effector enzyme (adenylate cyclase) (Sunyer et al 1989). Thus the effect of the receptor is blocked. Whilst pertussis toxin ADP-ribosylates another G-protein, Go, this G-protein seems to be primarily involved with receptor which regulate phospholipase C pathways (Simon et al 1991). Pertussis toxin has no effect on G-proteins which belong to the other three classes of G-proteins (Simon et al 1991). Therefore the effects of pertussis toxin are reasonably specific. In the hamster and sheep pars tuberalis it has been shown that pretreatment of the tissue with pertussis toxin blocks the ability of melatonin to inhibit forskolin-stimulated cyclic AMP (Morgan et al 1990; Carlson et al 1989). As in sheep pars tuberalis melatonin has no effect on either calcium mobilisation or the turnover of inositol phosphates (Morgan et al 1991b), it is reasonable to conclude that the G-protein involved is a Gi rather than a Go protein. However, in contrast to the situation in the hamster pars tuberalis, and in the rat neonatal pituitary, the effect of pertussis toxin in the sheep pars tuberalis was shown to be incomplete. This raised the possibility that the effect of the melatonin receptor might be mediated through a second and independent G-protein (Morgan et al 1990). Recent evidence suggests that the G-protein involved is a substrate for cholera toxin (Morgan unpublished). The best documented of the substrates for cholera toxin is the G-protein which mediates the stimulation of adenylate cyclase, Gs. Cloning of the ovine Gs has indicated that it is unlikely that a novel variant of Gs mediates the pertussis toxin insensitive component of the melatonin response (Barrett

G-protein class	Main effector to which associated	Tissue distribution	Reference
Gs			
Gαs	Stimulation of AC and Ca^{2+} channels	Widespread	30
Gαolf	Stimulation of AC	Moderate	11,43
Gi			
Gαi1	Inhibition of AC	Widespread	30
Gαi2	and stimulation	Widespread	
Gαi3	K$^+$ channels	Widespread	
GαoA	Activation of PLC	Widespread	30
GαoB	and stimulation of ion channels		
Gαt1	Stimulation of	Restricted	30
Gαt2	cGMP PDE		
Gαz	Unknown	Moderate	27
Gq			
Gα15	All of the Gq	Restricted	30
Gα16	class stimulate		
Gα14	PLC (β-isoform)		
Gα11			
Gαq		Widespread	
G12			
Gα12	Unknown	Widespread	27
Gα13			

Fig 1   The four classes of G-protein proposed by Simon et al (1991) based upon the sequence homology of the α-sub-units of the respective G-proteins, and the main effector coupling with which they have been associated.   AC=adenylate cyclase; PLC=phospholipase C; cGMP-PDE-=cyclic GMP dependent phosphodiesterase.

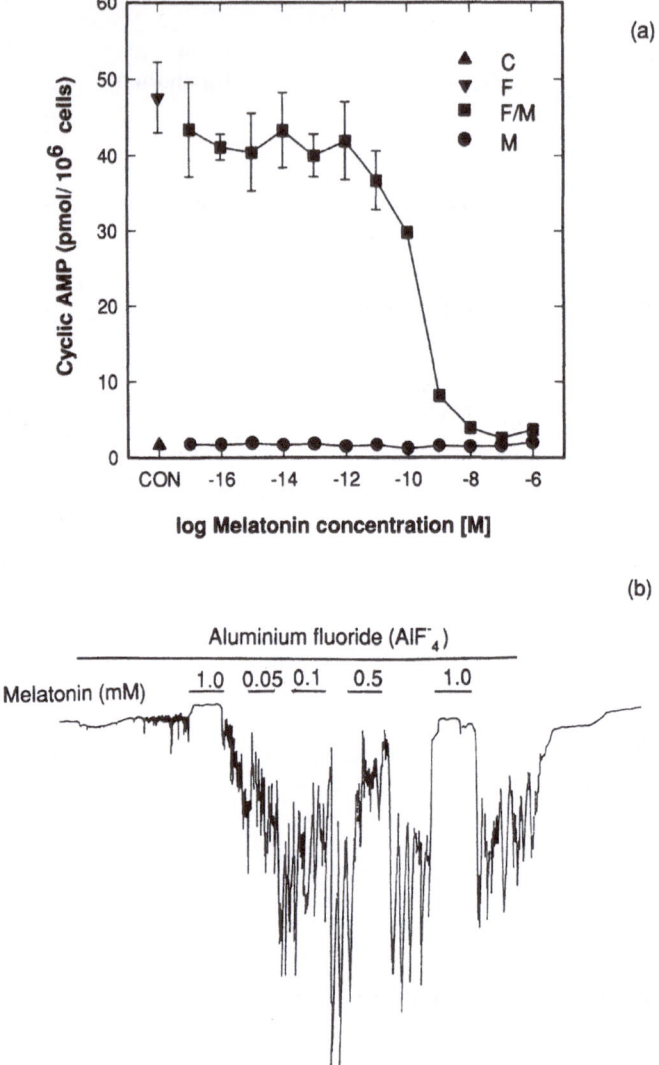

Fig 2 Contrast in the dose-response effects of melatonin upon the inhibition of forskolin-stimulated production of cyclic AMP in ovine pars tuberalis cells (a), and the inhibition of AlF_4^--induced current oscillations in *Xenopus* oocytes microinjected with mRNA from ovine pars tuberalis (b). Symbols in (a) indicate cells which were unstimulated (C), cells stimulated with 1μM forskolin (F) or stimulated with varying concentrations of melatonin alone (M) or in the presence of 1μM forskolin (F/M). Bars in (b) indicate the period of administration of drug. AlF_4^- was 0.1 mM $AlCl_3$ plus 10mM NaF

unpublished). Therefore we presently tentatively conclude that the pertussis toxin insensitive component may be a member of the Gq or G11 classes of G-protein or possibly a novel G-protein (Morgan unpublished). Given that members of the Gq class of G-protein have been mainly associated with the stimulation of phospholipase C (β-isoform), these G-proteins would seem the least likely of these possibilities. These results predict that either one melatonin receptor exists in the pars tuberalis which can act with two different G-proteins or that two different receptors exist each with its own G-protein. In either case differences in the pharmacology might result. Some preliminary evidence for this may be reflected in the complex competitive binding curves which some melatonin analogues have been shown to display (Howell et al 1994). These data indicate that for some analogues of melatonin, the competitive binding of these compounds against $2\text{-}^{125}\text{I}$-iodomelatonin cannot be adequately explained by a single site binding model. Instead the binding curves appear to be biphasic (Howell et al 1994)

The involvement of a cholera toxin sensitive G-protein with the melatonin receptor appears not to be limited to the pars tuberalis, as we have found that the binding of $2\text{-}^{125}\text{I}$-iodomelatonin to the membranes of chicken and lizard brains are sensitive to pre-treatment with cholera toxin (Morgan unpublished). Clearly therefore this establishes the general importance of a distinct pertussis toxin-insensitive G-protein to the signal transduction mechanism of the melatonin receptor, and to its pharmacology.

HETEROGENEOUS MELATONIN RECEPTOR/G-PROTEIN COMPLEXES

From the evidence for other receptors it would seem unlikely that melatonin receptors will exist in only one form. Most receptor exist as multiple homologues or sub-types, and these different forms can then interact differentially with different G-proteins. We have obtained primary evidence for the heterogeneity which potentially exists in the melatonin receptor/G-protein complex. Using the detergent digitonin we have been able to solubilise the melatonin receptor from a number of tissues, including the ovine pars tuberalis, the lizard brain, the chicken brain and the ovine hippocampus (Barrett et al 1994). When the melatonin receptor was solubilised, with $2\text{-}^{125}\text{I}$-iodomelatonin pre-bound, the receptor/G-protein complexes could be resolved by gradient non-denaturing polyacrylamide electrophoresis. Remarkably we found that the molecular masses of the complexes from different species were highly variable in size (Fig 3). The complex isolated from the ovine pars tuberalis was found to have the highest molecular mass of 525 kD. In contrast the complexes isolated from the other tissues were less discrete bands, ranging from 365 kD up to 525 kD. This suggests that the complexes are heterogeneous in composition, and illustrates that the receptors in the ovine pars tuberalis are clearly distinct from those in the neural tissues. It must be remembered that several factors contribute to the molecular mass estimates, including post-translational modifications such as glycosylation, and therefore the mass estimates will not correspond to the molecular mass predicted from the primary amino acid or cDNA sequence. Nevertheless, the fact that there is such variation in the molecular masses does reflect that the receptor/G-protein complexes are not structurally identical, and therefore provides the first indication of the heterogeneity which exists in the structure of the melatonin receptor

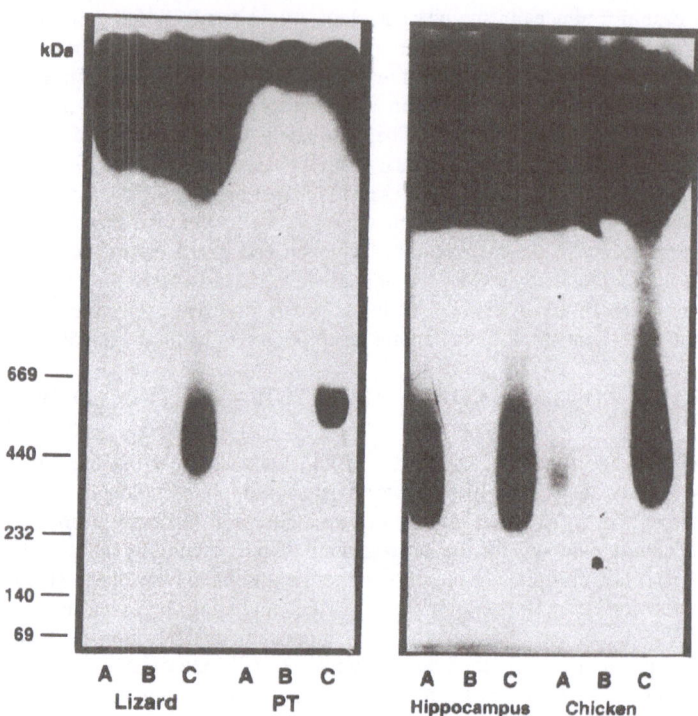

Fig 3 Electrophoretic separation of melatonin/G-protein complexes from different tissue sources on a 4-10% gradient polycarylamide native gel. For each tissue, membranes were incubated with 200 pM of ^{125}I-Mel (A) in the presence of 100 µM GTPγS, (B) in the presence of 1µM unlabelled melatonin and (C) alone (see Barrett et al 1994). Tissues were whole lizard brain, sheep pars tuberalis (PT) and hippocampus and whole chicken brain.

It was interesting to note that the isolated complexes were each sensitive to the modulatory effect of the guanine nucleotide GTPγS, with the exception of the ovine hippocampus (Barrett et al 1994). This observation which has been substantiated in homogenate binding assays (Morgan unpublished) might appear to indicate that the melatonin receptors in the ovine hippocampus are non-G-protein coupled. However, on the basis of its pharmacology using a number of novel melatonin analogues, this must be discounted, as the order of potency and affinities for the drugs were identical to that using the pars tuberalis (Morgan unpublished). Thus although the molecular mass of the receptor in the ovine pars tuberalis and the ovine hippocampus are clearly different, indicating large differences in structure, it would seem that the binding pocket for melatonin is well conserved. At present the reason why the hippocampal receptor is not sensitive to guanine nucleotides is not clear, but the observation is not in isolation as similar insensitivity to guanine nucleotides has also been observed in the dentate gyrus of the hippocampus, and the frontal cortex of the vervet monket and the baboon (Stankov et al 1993).

CONCLUSION

Our present view of how melatonin acts at the cellular level stems in the main from work on only a few model systems. In addition to the neonatal pituitary and the sheep pars tuberalis mentioned above, the amphibian melanophore and the vertebrate retina have also been important model systems for the study of melatonin receptor pharmacology. It is striking that from studies of these tissues the inhibition of cyclic AMP levels through the melatonin receptor is a common pathway (see Morgan et al 1994 for references). These results thereby reflect the widespread nature and importance of this receptor/signal transduction pathway to the diverse cellular functions of melatonin. In the pars tuberalis the inhibition of cyclic AMP by melatonin has been shown to cause parallel changes in protein kinase A activity, and phosphorylation of the cyclic AMP response element binding protein (CREB) (Hazlerigg et al 1991; McNulty et al 1994). Similarly it has been shown that protein synthesis and secretion in the pars tuberalis is regulated through cyclic AMP pathways, and that by inhibiting cyclic AMP production, melatonin blocks forskolin-stimulated translation and secretion of proteins (Morgan et al 1994b). In toto our results seem to indicate that a primary function of melatonin in the pars tuberalis is to terminate, or prevent the activation of, translational and transcriptional processes. Likewise from the studies of the signal transduction responses in other tissues, including the neonatal pars distalis, the amphibian melanophore and the vetebrate retina, it would seem that similar cellular effects of melatonin are likely to pertain. Nevertheless it is important not to overlook the fact that melatonin may have stimulatory effects through other signal transduction pathways not yet studied. Undoubtedly the cloning of the melatonin receptor, and the subsequent expression of its potentially different forms in mammalian cell lines will provide the future model systems which will resolve such issues.

Acknowledgements: The authors would like to acknowledge the financial support of SOAFD in this work.

REFERENCES

1 P. Barrett, A. MacLean and P.J. Morgan Evidence for multiple forms of melatonin receptor-G-protein complexes by solubilization and gel electrophoresis. *J Neuroendocrinol.* (In press) (1994).

2. L.L. Carlson, D.R. Weaver, and S.M. Reppert, Melatonin signal transduction in hamster brain:inhibition of adenylyl cyclase by pertussis toxin-sensitive G protein, *Endocrinol* 125:2670-2676 (1989).

3. B.R. Conklin and H.R. Bourne, Structural elements of Gα subunits that interact with G βγ, receptor and effectors, *Cell* 73:631-641 (1993).

4. S. Copinga, P.G. Tepper, C.J. Grol, A.S. Horn, and M.L. Dubocovich, 2-Amido-8-methoxytetralins - a series of non indolic melatonin-l like agents, *J. Med. Chem.* 36:2891-2898 (1993).

5. T. Ebisawa, S. Karne, M.R. Lerner, and S.M. Reppert, Expression cloning of a high-affinity melatonin receptor from *Xenopus* dermal melanophores, *Proc Nat Acad Sci USA* 91:6133-6137 (1994).

6. S.P. Fraser, P. Barrett, Moon C, P.J. Morgan, and M.B.A. Djamgoz Electrophysiological characterstics of AlF⁻4 induced response and melatonin receptor expression in *Xenopus* oocytes. Mol Neuropharmacol (in press).

7. S.P. Fraser, P. Barrett, M.B.A. Djamgoz, and P.J. Morgan, Melatonin receptor mRNA expression in Xenopus oocytes: inhibition of G-protein activated response, *Neurosci lett* 124:242-245 (1991).

8. T.W. Gettys, T.A. Fields, and J.R. Raymond, Selective activation of inhibitory G-protein alpha- subunits by partial agonists of the human 5-HT(1A) Receptor, *Biochemistry* 33:4283-4290 (1994).

9. D.G. Hazlerigg, P.J. Morgan, W. Lawson, and M.H. Hastings, Melatonin inhibits the activation of cyclic AMP-dependent kinase in cultured pars tuberalis cells from ovine pituitary, *J Neuroendocrinol* 3:597-603 (1991).

10. H.E. Howell, B. Guardiola, P. Renard adn P.J. Morgan Naphthalenic ligands reveal melatonin binding site heterogeneity. *Endocrine J* (in press) (1994).

11. D.T. Jones and R.R. Reed, Golf: An olfactory neuron specific-G-protein involved in odorant signal transduction, *Science* 244:790-795 (1989).

12. A.J. Lewy, M. Tetsuo, S.P. Markey, F.K. Goodwin, and I.J. Kopin, Pinealectomy abolishes plasma melatonin in rats, *J Clin Endocrinol Metab* 50:204-205 (1980).

13. G.A. Lincoln and I.J. Clarke, Photoperiodically-induced cycles in secretion of prolactin in hypothalamo-pituitary disconnected rams. Evidence for translation of the melatonin signal in the pituitary gland, *J Neuroendocrinol* 6:251-260 (1994).

14. J.E. Martin and D.C. Klein, Melatonin inhibition of the neonatal pituitary response to luteinizing-releasing factor, *Science* 191:301-302 (1976).

15. J.E. Martin and C. Sattler, Developmental loss of the acute inhibitory effect of melatonin on the in vitro pituitary luteinizing hormone and follicle-stimulating hormone responses to luteinizing hormone-releasing hormone, *Endocrinol* 105:1007-1012 (1979).

16. S. McNulty, P.J. Morgan, M. Thompson, G. Davidson, W. Lawson, and M.H. Hastings, Phospholipases and melatonin signal transduction in the ovine pars tuberalis, *Mol Cell Endocrinol* 99:73-79 (1994).

17. S. McNulty, S. Ross, P. Barrett, M.H. Hastings, and P.J. Morgan. Melatonin regulates the phosphorylation of CREB in ovine pars tuberalis. J Neuroendocrinol 6: (in press) (1994)

18. G. Milligan, Mechanisms of mutifunctional signalling G-protein-linked receptors, *TIPS* 14:239-244 (1993).

19. P.J. Morgan, The pars tuberalis as a target tissue for melatonin action, *Adv Pineal Res* 6:149-158 (1991).

20. P.J. Morgan, P. Barrett, H.E. Howell, and R. Helliwell, Melatonin receptors: localization, molecular pharmacology and physiological significance, *Neurochem Int* 24:101-146 (1994a).

21. P.J. Morgan, P. Barrett, G. Davidson, W. Lawson, and D. Hazlerigg, p72, a Marker Protein for Melatonin Action in Ovine Pars Tuberalis Cells - Its Regulation by Protein Kinase-A and Protein Kinase-C and Differential Secretion Relative to Prolactin, *Neuroendocrinol* 59:325-335 (1994b).

22. P.J. Morgan, G. Davidson, W. Lawson, and P. Barrett, Both pertussis toxin-sensitive and insensitive G-proteins link melatonin receptor to inhibition of adenylate cyclase in the ovine pars tuberalis, *J Neuroendocrinol* 2:773-776 (1990).

23. P.J. Morgan, M.H. Hastings, M. Thompson, P. Barrett, W. Lawson, and G. Davidson, Intracellular signalling in the ovine pars tuberalis: an investigation using aluminium fluoride and melatonin, *J Mol Endocrinol* 7:137-144 (1991a).

24. P.J. Morgan, W. Lawson, and G. Davidson, Interaction of forskolin and melatonin on cyclic AMP generation in pars tuberalis cells of ovine pituitary, *J Neuroendocrinol* 3:497-501 (1991b).

25. P.J. Morgan, W. Lawson, G. Davidson, and H.E. Howell, Melatonin inhibits cyclic AMP in cultured ovine pars tuberalis cells, *J Mol Endocrinol* 5:R3-R8 (1989a).

26. P.J. Morgan, W. Lawson, G. Davidson, and H.E. Howell, Guanine nucleotides regulate the affinity of melatonin receptors on the ovine pars tuberalis, *Neuroendocrinol* 50:359-362 (1989b).

27. S. Offermanns and G. Schultz, What are the functions of the pertussis toxin-insensitive G proteins G(12), G(13) and G(Z), *Mol Cell Endocrinol* 100:71-74 (1994).

28. N.T. Raikhlin, I.M. Kvetnoy, and V.N. Tolkachev, Melatonin may be synthesised in enterochromaffin cells, *Nature* 255:344-345 (1975).

298. R.J. Reiter, Pineal melatonin: cell biology of its synthesis and of its physiological interactions, *Endocrine Rev* 12:151-180 (1991).

30. M.I. Simon, M.P. Strathmann, and N. Gautum, Diversity of G-proteins in signal transduction, *Science* 252:802-808 (1991).

31. G. Spadoni, B. Stankov, A. Duranti, G. Biella, V. Lucini, A. Salvatori, and F. Fraschini, 2-substituted 5-methoxy-N-acyltryptamines - synthesis, binding affinity for the melatonin receptor, and evaluation of the biological activity, *J Med Chem* 36 4069-4074 (1993).

32. B. Stankov, S. Capsoni, V. Lucini, J. Fauteck, S. Gatti, B. Gridelli, G. Biella, B. Cozzi, and F. Fraschini, Autoradiographic localization of putative melatonin receptors in the brains of 2 old world primates - *Cercopithecus aethiops* and *Papio ursinus*, *Neuroscience* 52:459-468 (1993).

33. D. Sugden, N-Acyl-3-amino-5-methoxychromans - A new series of non- iIndolic melatonin analogues, *Eur J Pharmacol* 254:271-275 (1994).

34. T. Sunyer, B. Monastirsky, J. Codina, and L. Birnbaumer, Studies on nucleotide and receptor regulation of Gi proteins: effects of pertussis toxin, *Mol Endocrinol* 3:1115-1124 (1989).

35. J. Vanecek, The melatonin receptors in rat ontogenesis, *Neuroendocrinol* 48:201-203 (1988).

36. J. Vanecek and D.C. Klein, Sodium-dependent effects of melatonin on membrane potential of neonatal rat pituitary cells, *Endocrinol* 131:939-946 (1992a).

37. J. Vanecek and D.C. Klein, Melatonin inhibits gonadotropin-releasing hormone-induced elevation of intracellular Ca-$^{2+}$ in neonatal rat pituitary cells, *Endocrinol* 130:701-707 (1992b).

38. J. Vanecek and L. Vollrath, Melatonin inhibits cyclic AMP and cyclic GMP accumulation in the rat pituitary, *Brain Res* 505:157-159 (1989).

39. J. Vanecek and L. Vollrath, Melatonin modulates diacylglycerol and arachidonic acid metabolism in the anterior pituitary of immature rats, *Neurosci lett* 110:199-203 (1990).

40. B. Vivien-Roels, P. Pevet, M.P. Dubois, J. Arendt, and G.M. Brown, Immunohistochemical evidence for the presence of melatonin in the pineal gland, the retina, and the Harderian gland, *Cell Tissue Res* 217:105-115 (1981).

41. S. Yous, J. Andrieux, H.E. Howell, P.J. Morgan, P. Renard, B. Pfeiffer, D. Lesieur, and B. Guardiola-lemaitre, Novel naphthalenic ligands with high affinity for the melatonin receptor, *J Med Chem* 35:1484-1486 (1992).

42. J.B. Zawilska and P.M. Iuvone, Catecholamine receptors regulating serotonin N-acetyltransferase activity and melatonin content of chicken retina and pineal gland: D2-dopamine receptors in retina alpha-2 adrenergic receptors in pineal gland, *J Pharmacol Exp Ther* 250:86-92 (1989).

43. J.M. Zigman, G.T. Westermark, J. LaMendola, E. Boel and D.F. Steiner. Human Golfα: complementary deoxyribonucleic acid strucure and expression in pancreatic islets and other tissues outside the olfactory neuroepithelium and central nervous system. *Endocrinol* 133:2508-2514 (1993)

THE ROLE OF THE CIRCADIAN SYSTEM IN PHOTOPERIODIC TIME MEASUREMENT IN MAMMALS

Michael H. Hastings, Elizabeth S. Maywood, and Francis J.P. Ebling

Department of Anatomy
University of Cambridge
Downing Street
Cambridge CB2 3DY
U.K.

SEASONAL CYCLES OF NEUROENDOCRINE ACTIVITY

Mammals which inhabit seasonal environments have evolved under strong selective pressures to restrict growth and reproduction to particular times of the year. Many neuroendocrine axes are therefore sensitive to seasonal cues and can be activated or suppressed in order to generate adaptive annual cycles. These cycles are of interest for two basic reasons: first they serve as an example of how neuroendocrine function can be controlled and so may provide a model for understanding clinical disorders of those functions. Second, they are an example of biological timing and an important question is the identity of the timing mechanisms which synchronise these cycles.

Three basic strategies to generate seasonal cycles can be identified (Hastings 1989). In some species, especially those which are smaller and short-lived, seasonal cycles of growth and reproduction arise from a direct and immediate response to seasonal changes in the availability of food. These species are opportunistic. In larger, longer-lived species metabolic and reproductive changes occur more slowly and over a longer interval. Consequently, opportunism would not be an adequate strategy and so there is a greater reliance upon timing mechanisms which enable the individual to *anticipate* and so prepare the neuroendocrine axes in advance of favourable environmental conditions. In some species e.g. ground squirrels, deer, this timing relies upon an endogenous circannual clock (Dark et al. 1985, Goss 1984, Gwinner 1986, Joy & Mrosovsky 1982). Consequently, when these animals are held in the laboratory in a constant environment, devoid of seasonal cues, they continue to exhibit spontaneous cycles of growth and reproductive condition with a period length of approximately one year. The phase of these cycles may anticipate the external seasons, if the individual was previously synchronised to field conditions, whereas the cycles may be completely divorced from external time if the individual has no previous experience.

However, circannual rhythmicity is uncommon, and most seasonal species which have been studied fail to exhibit complete and recurrent seasonal rhythms under constant conditions in the laboratory. Seasonal rhythms in these species are only expressed when the animals are exposed to changing environmental cues, the principal cue being daylength or photoperiod. By exposing photoperiodic species to alternating sequences of long daylengths and short daylengths, it is possible to drive complete seasonal cycles of growth and reproduction which mimic those observed in the field. Obviously, the nature of the response to photoperiod varies with species so that in hamsters for example, long daylengths activate the gonadal axis whereas in sheep the same stimulus acutely suppresses reproduction. For other responses however, there is commonality across species, the best example being the control of prolactin secretion. In all species in which it has been studied, regardless of their gonadotrophic response to daylength, the secretion of prolactin is suppressed by short and stimulated by long daylengths. Finally, it should be pointed out that within species, local races will exhibit photoperiodic responses appropriate to their immediate environment. This is most clearly observed in latitudinal clines, where individuals from high latitudes show a greater sensitivity to photoperiodic cues. Given this variability in photoperiodic responsiveness between and within species and individuals, it is more appropriate to talk about photoperiodic traits rather than photoperiodic species (Zucker 1988).

THE PINEAL, MELATONIN AND PHOTOPERIODIC TIME MEASUREMENT

How does the neuroendocrine axis identify changes in daylength? Two observations indicated a role for the pineal in photoperiodic time measurement: first, that the pineal gland of rodents is activated by exposure to darkness, and second, that protracted exposure of Siberian (*Phodopus sungorus*) and Syrian (*Mesocricetus auratus*) hamsters to darkness or short daylengths induced hibernation and gonadal atrophy comparable to that seen in winter. The central importance of the pineal was confirmed by the demonstration that pinealectomy prevented neuroendocrine responses to photoperiod, not only in hamsters, but in all species which exhibit some form of photoperiodic trait (Reiter 1973, Hoffman 1981, Goldman & Darrow 1983, Bittman 1984). The principal product of the pineal is melatonin, and there are two critical aspects to the control of its secretion which are relevant to photoperiodism. First, it is regulated by the circadian clock, such that secretion occurs only at night under a light-dark cycle, and only during subjective night when individuals are in continuous darkness or dim light (Figure 1a, b). Second, because of the nature of entrainment of the circadian clock and the direct masking effect of light on the secretion of melatonin, the duration of the nocturnal peak of melatonin in the circulation is directly proportional to the length of the dark phase (scotoperiod) (Figure 1b). The duration of the endogenous melatonin signal therefore encodes scotoperiod and so is a direct indication of season (Bittman 1984, Hastings 1989, Illnerova 1991, Bartness et al. 1993). Both of these properties: nocturnality and scotoperiodic variation are observed in all species, including Man (Wehr 1991), when sufficiently precise assays and experimental procedures have been applied.

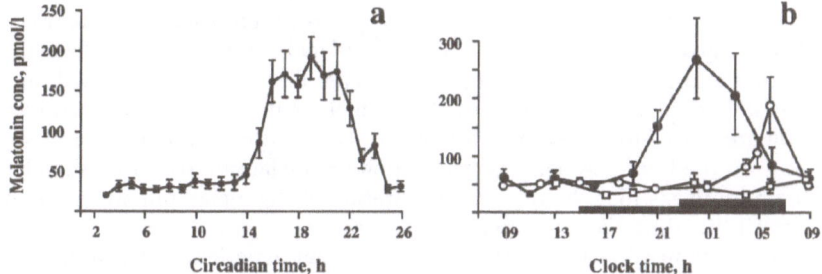

Figure 1 Concentration of circulating melatonin in male Syrian hamsters.
(a) Free-running circadian rhythm of plasma melatonin in animals held in continuous darkness.
Data plotted relative to onset of wheel running, CT 12. Note low levels during subjective day, with
spontaneous rise 4 h after onset of subjective night. (b) Entrained rhythms of serum melatonin in
animals held on long day (open circle) or short day (closed circle). Hours of darkness are indicated
by bars. The open squares indicate basal titres of melatonin in pinealectomised animals. Note that
short days are associated with a prolonged nocturnal melatonin peak. Redrawn from Maywood et
al. (1993).

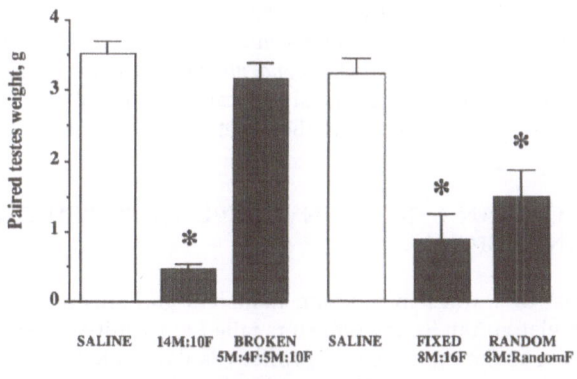

Infusion paradigms

Figure 2. Paired testes weights (mean + SEM) of male Syrian hamsters infused with either saline
(open bars) or melatonin (shaded bars) once daily for 6 weeks. Daily infusions of melatonin lasting
8 or 14 h caused a full short day-like atrophy. Separation of the signal into two 5 h segments in the
"broken" group prevented the gonadal response. Melatonin signals delivered on a random
schedule were as effective as those delivered at the same time of day, every day (Fixed). Redrawn
from Grosse et al. (1993). M= hours of infusion with melatonin, F= hours free of melatonin.

THE IMPORTANCE OF THE DURATION OF THE MELATONIN SIGNAL

The thesis of this paper is that a change in the duration of the melatonin signal is necessary and sufficient for the generation of photoperiodic neuroendocrine responses.

A major goal of research into the functions of melatonin is an understanding of the cellular and molecular bases to the actions of the hormone. This will not be possible unless the formal properties of how melatonin acts upon neuroendocrine function are understood, i.e. what are the critical features of the melatonin signal and how do the timing systems which respond to melatonin read that signal? To test the hypothesis that changes in the duration of the melatonin signal mediate photoperiodic time measurement, rather than present a correlate of these responses, it is necessary to manipulate the signal experienced by animals in a controlled manner independently of photoperiod. This can be achieved by pinealectomising animals to remove the principal endogenous source of hormone, and then infusing them systemically with a solution of melatonin to restore a rhythmic pattern of circulating melatonin at physiological titres. This protocol was first developed by Goldman and colleagues in Siberian hamsters (Carter & Goldman 1983), and Bittman and Karsch in sheep (Bittman 1984), but has now become widely accepted as the most appropriate way of investigating the physiological actions of melatonin in seasonality (Bartness et al. 1993).

When pinealectomised Syrian hamsters held on long photoperiods receive programmed systemic infusions of melatonin once daily for 5 - 6 weeks, it is possible to induce gonadal atrophy typical of exposure to short photoperiods (Grosse et al. 1993, Maywood et al. 1990, 1991, 1992). However, as predicted by the duration hypothesis, the neuroendocrine response is induced only by long-duration infusions (≥ 8 h per 24 h). Above a minimum requirement, the total mass of melatonin infused is not a determining factor. An important feature is the continuity of the signal. If an effective daily infusion of 10 h duration is split into two 5 h infusions, separated by an interval of 4 h during which no hormone is delivered, the gonadal axis fails to respond over the 6 weeks of the study (Figure 2). To be effective, *the long melatonin signal must be continuous*, which demonstrates that the mechanism which times the signal cannot interpolate between exposures to melatonin, and is sensitive both to the onset and offset of the signal.

IS THE CIRCADIAN SYSTEM INVOLVED IN MEASUREMENT OF THE MELATONIN- SIGNAL?

In contrast to studies in Siberian hamsters and sheep, early studies of the secretion of melatonin in Syrian hamsters failed to identify a clear seasonal change in the duration of the endogenous profile (Panke et al. 1980). It was therefore hypothesised that photoperiodic induction in this species depended not upon a change in duration, but on a coincidence between the stable, unchanging melatonin signal and a hypothetical phase of sensitivity to the hormone, the position of which relative to the melatonin signal was determined by photoperiod (Stetson & Watson-Whitmyre 1984). This coincidence model was heavily influenced by studies of the formal properties of photoperiodic induction in non-mammalian species, particularly birds and insects in which a physiological intermediate between the circadian clock and the photoperiodic response, comparable to the mammalian pineal gland, have not been identified. When the scotoperiodic variation of the melatonin signal was finally established in the Syrian hamster (Roberts 1985, Hastings et al. 1987, 1989), the hypothesis was modified to suggest that the increased duration of the signal allowed melatonin to coincide with the phase of sensitivity which occurred close to subjective night. If such a rhythm of sensitivity controlled by the circadian clock or the light:dark cycle is involved in photoperiodic time measurement, then it would be predicted that infusions of melatonin which avoided the phase of sensitivity would be ineffective. The programmed infusion procedure has been used to test this postulate. First, in both hamsters and sheep, pinealectomised animals were given infusions either during the dark or during the light, at non-overlapping phases within the 24 h cycle. Infusions of an appropriate duration were effective, regardless of when relative to the entraining cycle of light and darkness melatonin was delivered (Wayne et al. 1988, Maywood et al. 1990). In a second type of study, melatonin infusions were presented at non-24 h intervals, such that the signal would fall progressively at different portions of the light:dark cycle. Although signals which were delivered at too low a frequency were ineffective (e.g. once every 28 or 32 h in Syrian hamsters), infusions which presented melatonin once every 20 h were even more potent that the control infusions given once every 24 h (Figure 3). The infusions delivered once every 20 h would, over the course of the 6 weeks study, be distributed across all phases of the light:dark cycle, but this did not compromise the neuroendocrine response (Maywood et al. 1992). It is therefore concluded from these studies that *the phase of the light-dark cycle at which appropriate infusions of melatonin are delivered does not influence the neuroendocrine response of the animals.* The timer which reads the melatonin signal operates efficiently at all circadian phases.

IS THE MELATONIN-TIMER SENSITIVE TO THE PHASE OF PREVIOUS MELATONIN SIGNALS?

To reconcile the coincidence hypothesis with the experimental data, it has been suggested that there is a rhythm of sensitivity to melatonin, but that the phase of sensitivity is established by prior melatonin signals, i.e. the animal will be sensitive to melatonin approximately 24 h after experiencing melatonin. This hypothesis has been tested in two ways. First, pinealectomised Syrian hamsters were given daily systemic infusions of melatonin of 10 h duration, but in the experimental group the timing of the infusions alternated between the light and dark phases of successive days, such that no signal overlapped with the phase of its predecessor nor its successor (Maywood et al. 1990). In separate control groups, animals received either morning or evening infusions, coincident with those given to the experimental group, but as a result they received only half of the number of signals, missing the alternate infusions. These control groups did not show gonadal regression in response to the infusions delivered every other day, nor did animals which received saline vehicle. However, the animals which received melatonin signals once a day, alternating between day and night did show gonadal atrophy, equivalent to that observed in a control group which received melatonin during the night, every night. From this result it was concluded that the experimental animals used both components of the alternating pattern of infusion, i.e. day and night signals, to recruit the neuroendocrine response. This indicates that to be read effectively, a melatonin signal does not have to fall in the phase of its predecessor. A similar conclusion was reached following another study in which daily infusions

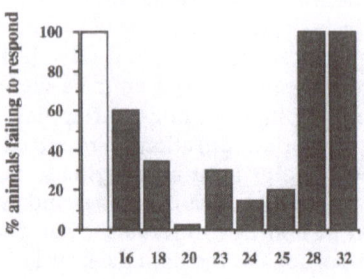

Infusion Period, T hours

Figure 3. The gonadal response of pinealectomised male Syrian hamsters to programmed infusions (10 h duration) of melatonin (shaded bars) or saline depends on the frequency of the infusions. Animals infused once every 28 or 32 h fail to respond, whereas animals receiving an infusion once every 20 to 25 h exhibit gonadal atrophy. Redrawn from Maywood et al. 1992.

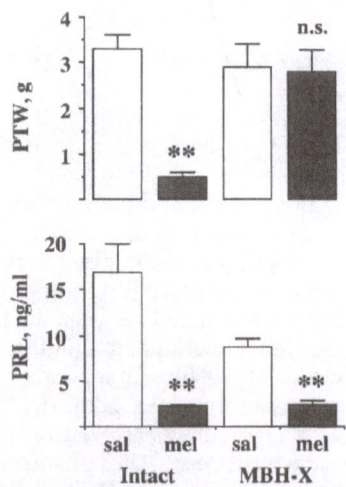

Figure 4. Lesions of the melatonin-binding sites of the mediobasal hypothalamus (MBH-X) block the effect of daily 10 h infusions of melatonin on gonadal size (PTW), but not on the secretion of prolactin. Data are means +SEM of intact or lesioned, pinealectomised animals infused with either saline (open bars) or melatonin (shaded bars) for 6 weeks. Redrawn from Maywood and Hastings, submitted to *Endocrinology*. ** $p<0.01$ vs. saline control, n.s. not significant vs. saline control.

of melatonin were delivered on an irregular schedule (Grosse et al. 1993). In this case, pinealectomised hamsters held on 16L:8D received infusions of melatonin of 8 h duration. Infusions were scheduled to fall within one of three possible 8 h intervals: morning, afternoon and night. The order in which they fell was randomised so that overall the same number of signals (n= 14) fell in each of the three intervals, but there was no consistent pattern from day to day. Nevertheless, these animals showed full gonadal atrophy in response to melatonin (Figure 2 b). Together, the results from these two experiments demonstrate that *the timing mechanism which is sensitive to melatonin is not based upon a rhythm of sensitivity to melatonin established by prior exposure to the hormone.*

ANATOMICAL LOCALISATION OF THE SITES OF ACTION OF MELATONIN

Having established the formal properties of the timer which identified the duration of the melatonin signal, the next important question is its anatomical localisation. A series of studies in rodents, and more recently in sheep, have identified the medial hypothalamus/thalamus as a likely site of action. Local administration of melatonin into various areas of the medial diencephalon can either reproduce or block the effects of photoperiodic cues, depending on the particular protocol involved (Hastings et al. 1988, 1991; Lincoln & Maeda 1992a, b, Malpaux et al. 1993). Conversely, central passive immunisation against melatonin following third ventricular infusions of antiserum against melatonin blocks the inhibitory effect of short photoperiods on serum FSH levels (Bonnefond et al. 1989). The hypothesis that melatonin acts in the diencephalon has been reinforced by autoradiographic studies using [125I]iodomelatonin to reveal sites of melatonin binding in the brain and pituitary (Morgan, this volume). Within the medial hypothalamus/thalamus of the Syrian hamster, iodomelatonin binding sites are present in the paraventricular nucleus of the thalamus (APVT), and the suprachiasmatic nuclei (SCN), preoptic (POA) and medio-basal areas (MBH) of the hypothalamus (Williams et al. 1989). Lesions of the APVT (Ebling et al. 1992) and POA (Maywood, unpublished data) do not block the gonadal response of Syrian hamsters to short photoperiods, demonstrating that they are not necessary for the generation of an endogenous melatonin signal of long duration, nor for the appropriate neuroendocrine interpretation of such a signal. Lesions of the SCN do block the neuroendocrine response to short days, but it is also known that the lesion prevents the generation of an endogenous melatonin signal, and so it is not surprising that photoperiodic time measurement is also compromised (Bittman 1984). The demonstration that melatonin does not act in the SCN comes from studies in which destruction of the SCN of the Syrian hamster does not impair the gonadal response of pinealectomised hamsters to daily schedules of injection or infusion of melatonin (Bittman 1984, Maywood 1990). Furthermore, the SCN of sheep do not contain any detectable binding sites for melatonin (Helliwell & Williams 1992). *The conclusion that melatonin does not act in the SCN is consistent with the conclusion drawn above that the circadian system, which is known to be based in the SCN, is not involved in reading the melatonin signal.*

The remaining site of appreciable binding of melatonin in the hypothalamus of the Syrian hamster lies in the MBH, straddling the ventromedial and dorsomedial nuclei (Williams et al. 1989). Bilateral electrolytic lesions of this area of binding block the effect of short photoperiods on the secretion of gonadotrophin and testicular mass i.e. lesions prevent the typical suppression of gonadotrophic function by short daylengths (Maywood unpublished data). However, the secretion of prolactin in lesioned hamsters is still responsive to short photoperiods and shows the expected decline, confirming that the animals are able to secrete an endogenous melatonin signal with an appropriate long duration, thereby regulating the lactotrophic axis. To test directly whether the lesion impairs the response to melatonin, lesioned animals were pinealectomised and infused daily with melatonin for 10 h. After 6 weeks, they had failed to undergo gonadal regression: testicular mass was high, equivalent to saline-infused controls, as were serum levels of LH (Figure 4). The lesions had therefore impaired the gonadotrophic response to melatonin. Interpretation of these data is subject to the same caution as any other lesion study (Hastings et al. 1991), because it is possible that destruction of the area of the MBH had a non-specific effect, disinhibiting gonadotrophic function and thereby masking any inhibitory response to melatonin. Nevertheless, the presence of iodomelatonin-biding sites in the area does suggest that melatonin acts through the MBH to control reproductive function. Such an action would be selective, because as predicted following the observed decline in prolactin levels on short photoperiod, the infusion of melatonin in lesioned animals was able to hold down prolactin levels: the lesions impaired the response to melatonin of the gonadotrophic but not the lactotrophic neuroendocrine axes. This demonstration of an anatomical dissociation of the sites of action of melatonin is consistent with previous physiological studies which had shown that photoperiod can have differential effects on the secretion of gonadotrophins and prolactin in the Syrian and Siberian hamster (Duncan et al. 1985, Hastings et al. 1989), especially with regard to critical daylengths and the effects of photoperiodic history. Moreover, it is complemented by recent studies in sheep which have revealed that implants of melatonin into the MBH influence both gonadotrophic and lactotrophic function, whereas implants placed into the *pars tuberalis* of the pituitary are more selective, inducing a stronger lactotrophic than gonadotrophic response (Lincoln & Maeda 1992a, b, Lincoln 1994, Malpaux et al. 1993). A picture is emerging in which melatonin probably acts in the MBH to control gonadal function, but acts in the pituitary, explicitly the PT, to control the secretion of prolactin. This hypothesis has received further support from the recent observation that photoperiodic cycles of secretion of prolactin can be induced in rams in which the pituitary has been surgically isolated from the hypothalamus (Lincoln & Clarke 1994). The within- and between-species variability in the effects of daylength upon these two photoperiodic traits may be a direct outcome of the anatomical and physiological separation of the mechanisms of action of melatonin.

CONCLUSION

Melatonin secreted by the pineal gland is the mediator of photoperiodic control over several neuroendocrine axes. The critical effect of photoperiod is to entrain the circadian clock driving the melatonin rhythm such that the duration of the endogenous melatonin signal is a direct reflection of scotoperiod and therefore of season. The effects of season on intact animals can be restored in pinealectomised animals by systemic infusions which restore the circulating pattern of melatonin. The neuroendocrine response to programmed infusions is determined by the duration of the signal. It is therefore established that photoperiodic time measurement depends upon a neural/cellular mechanism which is sensitive to the duration of nocturnal exposure of melatonin. This timing process is sensitive to the uninterrupted duration of exposure to melatonin, not the total time of exposure if the signal is delivered in separate segments. The timing process is equally effective at all phases of the light:dark cycle, refuting the existence of a restricted phase of sensitivity entrained by photoperiod. Furthermore, the timer is equally effective when successive melatonin signals do not follow in a predictable series: there is no evidence for a rhythm of sensitivity, itself entrained by the rhythm of exposure to melatonin. Of the known sites within the medial diencephalon which contain melatonin-binding sites in the Syrian hamster, neither the APVT nor POA are necessary for photoperiodic responses. Integrity of the SCN, which also contain melatonin -binding sites and are the site of the circadian clock, is necessary for responses to photoperiod. This is to be expected because lesions to the SCN will compromise generation of the endogenous melatonin signal and so act as a functional pinealectomy. However, lesions of the SCN do not compromise the ability of the animal to show appropriate gonadal responses to programmed infusions of melatonin, demonstrating that the circadian clock and melatonin-binding sites of the SCN are not essential components of the melatonin-timer. Lesion and implant studies have, however, revealed that the area of MBH containing binding sites for melatonin is an essential component of the photoperiodic time measuring apparatus, and our current working hypothesis is that melatonin acts in the MBH to control gonadotrophic function. Moreover, this is a selective action because lesions to the MBH do not impair photoperiodic regulation of the lactotrophic axis, which is likely to be mediated within the pars tuberalis .

In conclusion, the role of the circadian system in photoperiodic time measurement is to generate an endogenous melatonin signal appropriate to prevailing scotoperiod and thereby encode season. The neuroendocrine mechanism(s) which responds to the melatonin signal is based upon a very precise timer(s), that is independent of the circadian clock. The cells which contain or comprise this timing system have yet to be characterised, but the recent cloning and sequencing of an amphibian melatonin receptor now offers great potential for appropriate anatomical and biochemical studies of the action of this hormone (Ebisawa et al. 1994). Nevertheless, such studies will only be successful if they are conducted in a conceptual framework representative of what is known about the actions of the hormone in the whole animal. The durational model, based on cellular timers independent of the circadian clock, provides such a framework.

ACKNOWLEDGEMENTS

The work of our laboratory is supported by The Wellcome Trust, BBSRC and The Royal Society.

REFERENCES

Bartness, T., Bittman, E.L., Hastings, M.H., Powers, J.B. and Goldman, B.D., 1993, Timed melatonin infusion protocols: what have they told us about seasonal reproduction? *Journal of Pineal Research* 15: 161- 190.
Bittman, E.L. (1984) Melatonin and photoperiodic time measurement: evidence from rodents and ruminants. In: The Pineal Gland, pp. 155- 192, Reiter R.J. (ed.) Raven Press, N.Y.

Bonnefond, C., Walker, A.P., Stutz, J.A., Maywood, E., Juss, T.S., Herbert, J. & Hastings, M.H., 1989. The hypothalamus and photoperiodic control of FSH secretion by melatonin in the male Syrian hamster. *Journal of Endocrinology, Jubilee Edition* 122, 247-254.

Carter, D.S. and Goldman, B.D., 1983, Antigonadal effects of timed melatonin infusion in pinealectomised male Djungarian hamsters (*Phodopus sungorus sungorus*): duration is the critical parameter. *Endocrinology* 113, 1261- 1267.

Dark, J. Pickard, G.E. and Zucker, I., 1985, Persistence of circannual rhythms in ground squirrels with lesions of the suprachiasmatic nuclei. *Brain Research* 332, 201- 207.

Duncan, M.J., Goldman, B.D., Pinto, M.N. and Stetson, M.H., 1985, Testicular function and pelage colour have different critical daylengths in the Djungarian hamster (*Phodopus sungorus sungorus*). *Endocrinology* 116: 424- 430.

Ebisawa, T., Karne, S., Lerner, M.R. and Reppert S.M., 1994, Expression cloning if a high-affinity melatonin receptor from*Xenopus* dermal melanophores. *Proceedings of the National Academy of Sciences*, 91: 6133- 6137.

Ebling, F.J.P., Maywood, E.S., Humby, T. and Hastings, M.H., 1992, Circadian and photoperiodic time measurement in male Syrian hamsters following lesions of the melatonin-binding sites of the paraventricular thalamus. *Journal of Biological Rhythms* 7: 241- 254.

Grosse, J., Maywood, E.S., Ebling, F.J.P. and Hastings, M.H., 1993, Testicular regression in pinealectomised male Syrian hamsters following infusions of melatonin delivered on non-circadian schedules. *Biology of Reproduction* 49: 666- 674.

Goldman, B.D. and Darrow, J.M., 1983, The pineal gland and mammalian photoperiodism. *Neuroendocrinology* 37: 386- 396.

Goss, R.J., 1984, Photoperiodic control of antler cycles in deer VI. Circannual rhythms on altered daylengths. *Journal of experimental Zoology* 230: 265- 271.

Gwinner, E., 1986, "Circannual Rhythms," Springer, Berlin.

Hastings, M.H., 1991, Neuroendocrine Rhythms. *Pharmacology and Therapeutics* 50: 35-71.

Hastings, M.H., Walker, A.P., Powers, J.B., Hutchison, J. Steel, E.A. and Herbert, J., 1989, Differential neuroendocrine effects of photoperiodic history on the responses of gonadotrophins and prolactin to intermediate daylengths in the male Syrian hamster. *Journal of Biological Rhythms.* 4, 335-350.

Hastings, M.H., Walker, A.P. and Herbert, J., 1987, The effect of asymmetric reductions of photoperiod on pineal melatonin, locomotor activity and gonadal condition of male Syrian hamsters. *Journal of Endocrinology.* 114, 221-229.

Hastings, M.H., Walker, A.P., Roberts, A.C. and Herbert, J., 1988, Intra-hypothalamic melatonin blocks photoperiodic responsiveness in the male Syrian hamster. *Neuroscience.* 24, 987-991.

Hastings, M.H., Maywood, E.S., Ebling, F.J.P., Williams, L.M., and Titchener, L., 1991, Sites and mechanism of action of melatonin in the photoperiodic control of reproduction. *Advances in Pineal Research*: 5, 147-157.

Helliwell, R.J.A. and Williams, L.M., 1992, Melatonin binding sites in the ovine brain and pituitary: characterisation during the oestrous cycle. *Journal of Neuroendocrinology* 4: 287- 294.

Hoffman, K., 1981, The role of the pineal gland in the photoperiodic control of seasonal cycles in hamsters. In: "Biological Clocks in Seasonal Reproductive Cycles," Follett B.K. & Follett D.E. eds., Wright, Bristol.

Illnerova, H., 1991, The suprachiasmatic nucleus and rhythmic pineal melatonin production. In: "The Suprachiasmatic Nucleus," D.C. Klein, R.Y. Moore and S.M. Reppert eds., O.U.P., N.Y.

Joy, J.E. and Mrosovsky, N., 1982, Circannual rhythms of molt in ground squirrels. *Canadian Journal of Zoology* 60: 3227- 3231.

Lincoln, G.A., 1994, Effects of placing micro-implants of melatonin in the pars tuberalis, pars distalis and the lateral septum of the forebrain on the secretion of FSH and prolactin, and testicular size in rams. *Journal of Endocrinology* 142: 267- 276.

Lincoln, G.A. and Clarke, I.J., 1994, Photoperiodically-induced cycles in the secretion of prolactin in hypothalamo-pituitary disconnected rams: evidence for translation of the melatonin signal in the pituitary gland. *Journal of Neuroendocrinology*, in press.

Lincoln, G.A. and Maeda, K-I, 1992a, Reproductive effects of placing micro-implants of melatonin in the mediobasal hypothalamus and preoptic area in rams. *Journal of Endocrinology* 132: 210-215.

Lincoln, G.A. and Maeda, K-I, 1992b, Effects of placing micro-implants of melatonin in the mediobasal hypothalamus and preoptic area on the secretion of prolactin and beta-endorphin in rams. *Journal of Endocrinology* 134: 437- 448.

Lincoln, G.A. and R.V. Short, 1980, Seasonal breeding: Nature's contraceptive. *Recent Progress in Hormone Research* 36: 1- 52.

Malpaux, B., Daveau, A. Maurice, F., Gayrard, V. and Thiery, J-C., 1993, Short-day effects of melatonin on luteinizing hormone secretion in the ewe: evidence for central sites of action in the mediobasal hypothalamus. *Biology of Reproduction* 48: 752- 760.

Maywood, E.S., Buttery, R.C., Vance, G.H.S., Herbert, J. and Hastings, M.H., 1990, Gonadal responses of the male Syrian hamster to programmed infusions of melatonin are sensitive to signal duration and frequency but not to signal phase nor to lesions of the suprachiasmatic nuclei. *Biology of Reproduction* 43, 174-182.

Maywood, E.S., Lindsay, J., Karp, J.D., Powers, J.B., Williams, L.M., Titchener, L., Ebling, F.J.P., Herbert, J. & Hastings, M.H., 1991, Occlusion of the melatonin-free interval blocks the short day gonadal response of the male Syrian hamster to programmed melatonin infusions of necessary duration and amplitude. *Journal of Neuroendocrinology* 3: 331-338.

Maywood, E.S., Grosse, J., Lindsay, J.O., Karp, J., Powers, J.B., Ebling, F.J.P., Herbert, J.H. and Hastings, M.H., 1992, The effect of signal frequency on the gonadal response of male Syrian hamsters to programmed melatonin infusions. *Journal of Neuroendocrinology* 4: 37-44.

Maywood, E.S., Hastings, M.H., Max, M., Ampleford, E., Menaker, M. and Loudon, A.S.I., 1993, Circadian and free-running rhythms of melatonin in the blood and pineal gland of free-running and entrained Syrian hamsters. *Journal of Endocrinology* 136: 65-73.

Maywood, E.S. and Hastings, M.H. Lesions of the iodomelatonin-binding sites of the mediobasal hypothalamus spare the lactotrophic but block the gonadotrophic response of male Syrian hamsters to short photoperiod and to melatonin. *Endocrinology*, submitted.

Panke, E.S., Rollag, M.D. and Reiter R.J., 1980, Effects of photoperiod on hamster pineal melatonin concentrations.*Comparative Biochemistry and Physiology* 66A: 691- 693.

Reiter R.J., 1973, Pineal control of a seasonal reproductive rhythm in male golden hamsters exposed to natural light and temperature. *Endocrinology* 92: 423- 430.

Roberts, A.C., Martensz, N.D., Hastings, M.H. and Herbert, J., 1985, Changes in photoperiod alter the daily rhythms of pineal melatonin content and hypothalamic beta-endorphin content and the luteinizing hormone response to naloxone in the male Syrian hamster. *Endocrinology*. 117, 141-148.

Stetson, M.H. and Watson-Whitmyre, M., 1984, Physiology of the pineal and its hormone melatonin in annual reproduction in rodents. In: "The Pineal Gland," Reiter R.J. ed., Raven Press, N.Y.

Wehr, T. 1991 The durations of human melatonin secretion and sleep respond to changes in daylength (photoperiod). *Journal of clinical Endocrinology and Metabolism* 73, 1276- 1280.

Wayne, N. L., Malpaux, B. and Karsch, F.J. 1988, How does melatonin code for daylength in the ewe: duration of nocturnal melatonin release or coincidence of melatonin with a light-entrained sensitive period? *Biology of Reproduction* 39, 66- 75.

Williams, L.M., Morgan, P.J., Hastings, M.H., Lawson, W., Davidson, G. and Howell, H.E., 1989, Melatonin receptor sites in the Syrian hamster brain and pituitary. Localisation and characterisation using 125I iodomelatonin. *Journal of Neuroendocrinology*. 1, 315-320.

Zucker, I., 1988, Seasonal affective disorders: animal models *non fingo. Journal of biological Rhythms* 3: 209- 223.

MELATONIN AND THE VERTEBRATE CIRCADIAN SYSTEM:

PHARMACOLOGICAL EFFECTS

Jennifer R. Redman

Psychology Department
Monash University
Clayton, Victoria, 3168
Australia

The effects of melatonin on the vertebrate circadian system can be studied with two questions in mind: first, to elucidate the physiological role of endogenous melatonin as a chemical messenger relaying information between postulated components of the circadian system (the retina, suprachiasmatic nuclei (SCN) and pineal gland (Gwinner, 1978; Menaker, 1982), and secondly, to investigate the effects on this system of administering melatonin at pharmacological dose levels.

In the first case, one or more components of the circadian system are removed and the effect on circadian rhythmicity assessed. Investigations of the circadian system in many non-mammalian vertebrates have used this approach in order to identify which anatomical components are required for the generation and maintenance of circadian rhythmicity. Therefore, most of the studies in these species have examined the effects of pinealectomy (Px) rather than of melatonin administration (Armstrong and Redman, 1993).

This paper will focus on the second of these two questions, specifically, on the effects of exogenous melatonin given by a variety of administration routes and differences in response to melatonin between various rodent species.

Melatonin has been administered by a number of routes: orally, by injection or infusion, or by slow-release implant. Plasma levels may be elevated chronically or for a limited time only. These different forms of melatonin administration result in very different effects on the circadian system.

Exogenous melatonin has been shown to alter the expression of daily rhythms in the amphibian, tiger salamander, *Ambystoma tigrinum*. Compass orientation using celestial cues enables amphibians such as the tiger salamander to steer in a particular direction. A circadian clock is required to compensate for the earth's 24h rotation. Daily administration of melatonin under constant lighting conditions (both constant light (LL) and constant darkness (DD)) to tiger salamander larvae at 1200h causes the larvae to shift their directional response about 90 degrees clockwise, suggesting that melatonin pulses at this time phase delay the circadian system (Adler and Taylor, 1980).

Amphibians such as the mudpuppy, *Necturus maculosus*, display daily cycles of temperature selection. In this *totally aquatic salamander,* daily injections of melatonin significantly decreased the temperature selected and abolished the normal LD variation in thermoregulatory behaviour (Hutchinson et al., 1979).

The major effects of melatonin administration in some lizard species are summarized in Table 1. Melatonin has been administered both chronically by silastic capsule implants and acutely by injections. There are some differences in the response to exogenous melatonin between the three species of lizard examined so far. However, these variations in response are not as marked as those reported following Px in the same species (see Armstrong and Redman, 1993 for a summary of these findings).

Table 1. Major effects of melatonin administration on the circadian system of some lizard species

Sceloporus olivaceus	Implant - lengthens tau[1], arrhythmia[2]
Sceloporus occidentalis	Implant - tau lengthens[3] Daily injections entrain[4]
Dipsosaurus dorsalis	Implant - tau lengthens[5]

[1]Underwood, 1977   [4]Underwood and Harless, 1985
[2]Underwood, 1979   [5]Janik and Menaker, 1990
[3]Underwood, 1981

Administration via continuous-release silastic capsules lengthens tau in *Sceloporus occidentalis* when held in either LL or DD. Most *Sceloporus olivaceus* respond to melatonin administration in LL similarly although a small number of this species become arrhythmic (Underwood, 1979; Underwood, 1981). Chronic administration of melatonin lengthens tau in intact, blinded, Px and blinded-Px lizards. It appears that neither the retina nor the pineal are the crucial target sites for melatonin's effects on circadian period (Underwood, 1981). Autoradiographic techniques demonstrate that the lizard SCN are the major uptake site of H[3]-melatonin injected intraperitoneally (Joss, 1980) so it is possible that melatonin exerts its circadian effects in lizards at the level of the hypothalamus (Underwood, 1986; Janik, Pickard and Menaker, 1990).

As with the other lizard species examined so far, chronic administration of melatonin via silastic capsules lengthens tau in *Dipsosaurus dorsalis* (Janik and Menaker, 1990). However, the effect was less marked than that reported for *Sceloporus occidentalis*

(Underwood, 1981). It appears that there may be significant differences in the relative importance of the pineal and of circulating melatonin even within lizard genera (Underwood, 1989).

Subcutaneous injections of melatonin given on alternate days to *Sceloporus occidentalis* held in dim LL, entrain the free-running locomotor activity rhythm. In six out of eight lizards, melatonin entrained the rhythm when injection time coincided with or occurred near to the end of the active period. The remaining two lizards entrained when melatonin injections were administered about 2 hours before activity onset (Underwood and Harless, 1985). Before the melatonin injection schedule commenced all lizards free-ran with a period of less than 24h It is not apparent what determines the preferred phase relationship between the time of the melatonin injections and activity onset.

Single injections of melatonin were administered biweekly at varying times across the circadian day to *Sceloporus occidentalis* free-running in dim LL. The phase response curve (PRC) thereby constructed includes phase advances of the activity rhythm between mid-subjective day and early subjective night (CT6-CT15), and phase delays at other times (Underwood, 1986). The shape of the PRC to melatonin in *Sceloporus occidentalis* differs from that described in rats (see below).

In the avian species examined so far, there are significant differences in both the type and the degree of alteration to circadian rhythms observed following melatonin administration. These are summarized in Table 2. It has been suggested that this may reflect differences in the natural frequencies of oscillators within circadian systems and/or variations in coupling strength between these component oscillators (Gwinner,1978; Underwood, 1981). Differences in experimental conditions can make comparison between studies problematic. For example, although Px had similar effects in pigeon and quail, continuous melatonin administration suppressed rhythms and reduced total activity in pigeons in dim LL (Ebihara, Uchiyama and Oshima, 1984), whereas there were no effects in quail in DD (Simpson and Follett, 1981). The differing light intensities, which in themselves alter tau (Aschoff, 1960), could account for this difference in response.

Continuous administration of melatonin via silastic capsules or in the drinking water either shortened tau or induced arrhythmicity in house sparrows, *Passer domesticus*, held in DD (Turek, McMillan and Menaker, 1976; Binkley and Mosher, 1985). Similarly, in the Java sparrow, *Padda oryzivora*, held in LL continuous long-term melatonin administration abolished the activity rhythm and decreased activity levels (Ebihara and Kawamura, 1981).

In LL, continuous melatonin administration via a silastic capsule abolished the endogenous melatonin rhythm in starlings, *Sturnus vulgaris*, and in many cases, the daily rhythm in perch-hopping activity as well (Beldhius, Dittami and Gwinner, 1988; Gwinner et al., 1987). In contrast, the feeding rhythm survived either unchanged or with a decreased tau suggesting that there may be separate control systems for rhythms in perch-hopping and feeding in the starling.

Entrainment by melatonin of free-running rhythms in vertebrates was first demonstrated by Gwinner and Benzinger in 1978. In Px starlings, *Sturnus vulgaris*, free-running in dim LL, injections of melatonin in oil entrained the activity rhythm. In the entrained state, activity time preceded the time of melatonin injections.

Table 2. Major effects of melatonin administration on the circadian system of some avian species.

Quail (*Coturnix* *(Coturnix japonica)* (sighted)	Implant - no effect[1]
Finch (*Carpodacus mexicanus*)	Changed PAD in Px birds entrained to LD cycle[2]
Pigeon (*Columba livia*)	Implant - rhythms suppressed but not abolished[3]
House Sparrow (*Passer domesticus*)	Implant - tau decreases or arrhythmia[4,5]
Java Sparrow (*Padda oryzivora)*	Implant - tau decreases or arrhythmia[6]
European Starling (*Sturnus vulgaris*)	Implant - activity arrhythmic, feeding rhythmic[7,8] Daily injections entrain[9]

[1]Aschoff, 1960
[2]Fuchs, 1983
[3]Ebihara, Uchiyama and Oshima, 1984
[4]Turek, McMillan and Menaker, 1976
[5]Binkley and Mosher, 1985

[6]Ebihara and Kawamura, 1981
[7]Beldhius, Dittami and Gwinner, 1988
[8]Gwinner et al., 1987
[9]Gwinner and Bensinger, 1978

Initially it appeared that the mammalian circadian system was not responsive to exogenous melatonin as attempts to alter circadian activity rhythms in rats with melatonin implants were not successful (Cheung and McCormack, 1982). However, in 1983 it was demonstrated that daily injections of melatonin could entrain free-running running-wheel activity rhythms in a mammal, the laboratory rat (Redman, Armstrong and Ng, 1983). Rats were allowed to free-run in DD until free running rhythms were established. Running-wheel activity and drinking (licks of the water spout) were monitored continuously by computer. Subcutaneous (SC) injections of melatonin (1mg/kg) were then administered at the same time each day until entrainment occurred or until the injection time had coincided with all circadian phases. Injections were then discontinued and the phase of the subsequent free-run confirmed that entrainment to melatonin injections had occurred. Entrainment took place when the time of day of injection coincided with the onset of locomotor activity (CT12), that is, the active phase of the circadian cycle followed the melatonin injection while the quiescent phase preceded it. In addition to running-wheel activity and drinking rhythms, daily injections of melatonin in rats entrain locomotor activity and body temperature rhythms measured concurrently (Beaven, 1991) and pineal N-acetyltransferase (NAT) activity (Illnerová, Trentini and Maslova, 1989).

The rats used in these studies were male, Long-Evans strain which have a tau in DD which is greater than 24h. Entrainment to a daily zeitgeber therefore requires a small phase advance of the circadian system each day. Daily administration of melatonin at the same time each day induces the necessary small phase advances (Armstrong and Chesworth, 1987). If tau is too long, entrainment does not always occur. This suggests that there are limits to the range of entrainment at least in female rats (Thomas and Armstrong, 1988; Armstrong, Thomas and Chesworth, 1989).

A PRC to melatonin has been constructed by administering single injections once every two to three weeks at different circadian times to rats free-running in DD conditions. As would be predicted from the entrainment studies, the melatonin PRC shows small phase advances occurring in a narrow window between CT9 and 11, one to three hours before activity onset. There was only one time at which phase delays resulted (CT18) and this was only one animal (Armstrong and Chesworth, 1987). Although there is now convincing evidence for melatonin induced phase delays of the human circadian system (Lewy et al., 1992), such phase shifts seem more difficult to induce in rats.

More recently a PRC to melatonin in Syrian hamsters was reported (Hastings et al., 1992). As in the rat, the maximum phase advances were late in the subjective day (CT8 and CT10). No phase delays were observed. Injections of ethanolic-saline control vehicle also resulted in phase advances at these times. When melatonin solution or ethanolic-saline was administered by infusion no effects were noted. The authors postulated that the phase shifting effects of melatonin are not specific but are a consequence of the arousing effects of handling and injection procedures. The phase shifting effects of arousing stimuli in hamsters had previously been noted by Mrosovsky (1988) and we have also noted effects of saline injections on activity rhythms in a few rats (Redman, Armstrong and Ng, 1983). In order to determine whether entrainment by melatonin is mediated by alterations in activity levels, we examined the effects of daily injections of melatonin when wheel-running activity was prevented by blocking wheel rotation while still allowing rats access to the running-wheel. The phase of the rhythm when wheel-running resumed indicated that entrainment to melatonin persisted even when running-wheel activity was prevented (Redman and Roberts, 1991).

In an extension of this investigation we have recently been investigating alternative modes of administration of melatonin which avoid any handling (as with injections) or physical attachments (as with infusion pumps). Preliminary results indicate that melatonin (5mg/30ml) given to male rats in the drinking water for 2 hours out of 24h also entrains running-wheel activity rhythms when the 2 hour period coincides with activity onset (Figure 1) while control animals do not entrain. During that 2 hour period rats drink on average 6 ml, that is, they ingest about 1 mg of melatonin. In this protocol, access to either the melatonin solution or to water is controlled electronically so personnel do not need to enter the laboratory at the same time each day and animals are left undisturbed. It seems that, in rats, entrainment by melatonin is not a consequence of activity or of increased arousal.

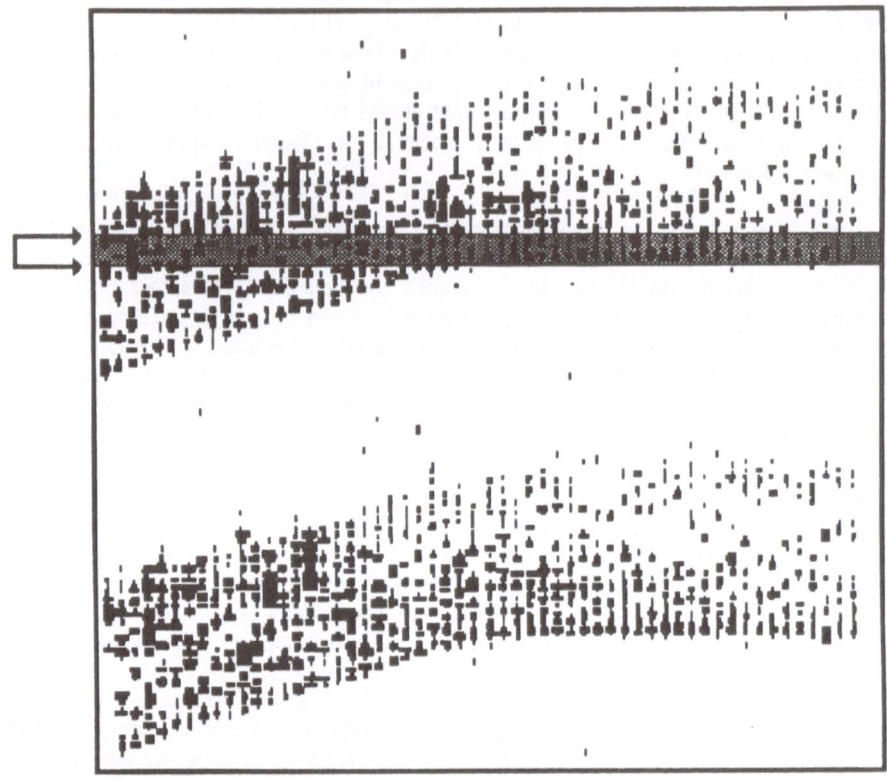

Figure 1. Actogram of wheel-running activity in a male Long-Evans rat. Shaded area with vertical arrows indicates the 2 hour period each day during which the rat had access to the melatonin solution.

The restriction of the entraining effects in rats to one time of day (immediately before activity onset and at no other circadian phase) suggests that the window of time during which the pacemaker is sensitive to melatonin is quite narrow and specific. It should be noted however that the phase at which melatonin injections entrain circadian rhythms in rats is the opposite to that which entrains rhythms in European starlings and iguanid lizards. In *Sturnus vulgaris* and most *Sceloporus occidentalis* entrainment takes place when injections coincide with the beginning of the quiescent period, not the active period (Gwinner and Benzinger, 1978; Underwood and Harless, 1985). This pattern of response to melatonin injections may reflect the fact that starlings and lizards are day-active while laboratory rats are night-active. The daily increase in melatonin levels resulting from the daily injections, although supraphysiological, may mimic the endogenous nighttime peak of pineal melatonin secretion which occurs in all species examined so far. Thus one would expect a quiescent period to follow melatonin injections in diurnal species whereas nocturnal species would become active following an increase in hormone levels. On the other hand there may be species differences in the timing of the melatonin sensitive period that is unrelated to whether the animal is diurnal or nocturnal.

When hamsters were tested with daily injections of exogenous melatonin (1mg/Kg), no entrainment occurred (Armstrong and Menaker, 1984 cited in Armstrong and Redman, 1985). It had been assumed that the sensitive phase to melatonin in the hamster would be similar to that identified in the rat, that is, at activity onset. When no entrainment occurred it was concluded that the hamster circadian system is not sensitive to the entraining effects of daily melatonin injections at this dose. However, in at least one animal (see Armstrong and Redman, 1985 Figure 3(A), p.194) melatonin was not administered at all circadian phases so such conclusions may have been premature.

Several different laboratories have subsequently provided evidence that melatonin may, under certain circumstances, alter and even entrain hamster circadian rhythms. Under certain photoperiods daily injections of melatonin phase advance the locomotor activity rhythm and extend the active period in those Djungarian hamsters (*Phodopus sungorus*) which also demonstrate responses in the reproductive system (Puchalski and Lynch, 1988). Daily 10 hour infusions of melatonin (at about physiological levels) administered in a T-cycle paradigm alter circadian rhythms in young Px Djungarian hamsters (*Phodopus sungorus*) in LL (Darrow and Goldman, 1986). Some saline treated control animals also showed some responses making interpretation difficult. In Px Siberian hamsters, 6 or 8 hour infusions of melatonin entrain wheel-running activity rhythms when the beginning of the infusion coincides with the onset of activity, that is, at a similar phase to that observed in laboratory rats. Px Syrian hamsters, on the other hand, entrain when the start of the infusion occurs in the second half of the activity period (Kirsch, Belgnaoui, Gourmelen and Pévet, 1993).

Given the differing effects on circadian organization observed following melatonin administration in lower vertebrates, including entrainment occurring with atypical phase relationships between activity onset and melatonin administration in some lizards, it might be expected that a similar variation of responses may occur among mammalian species as well.

Currently we are using the experimental protocol of daily injections in constant lighting conditions to examine the effects of melatonin (1mg/Kg) on free-running activity-wheel rhythms in mice. Preliminary results with CBA and NZB strains indicate that the phase at which melatonin injections entrain the rhythms may differ. (See Figure 2).

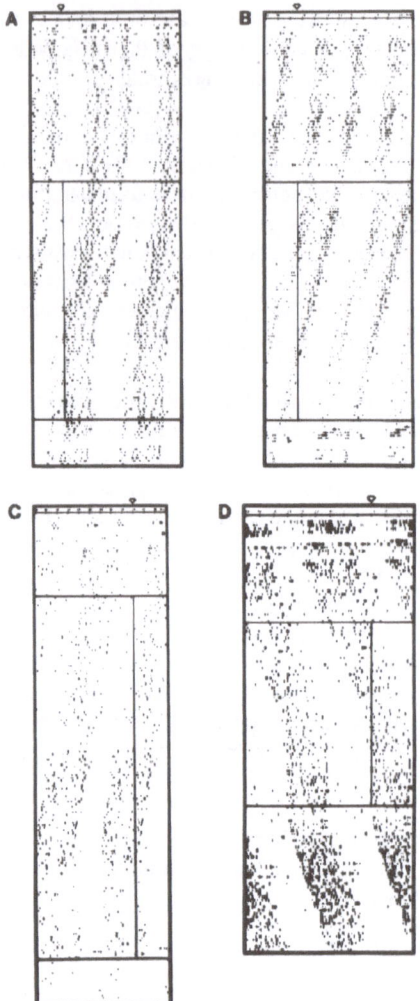

Figure 2. Actograms of wheel-running activity in (A) an NZB mouse given daily injections of either melatonin or (B) ethanolic-saline vehicle, and CBA mice (C) and (D) which entrained to melatonin injections at different circadian times. Arrows indicate time of day of injection. Vertical line indicate number of days of injection.

It is not clear in the melatonin PRC study in lizards whether animals which initially have a tau of less than 24h respond differently from those in which the initial tau is longer than 24h. Similarly, with the limited number of mice we have tested, it is not yet possible to say whether an animals's initial tau can predict the phase of entrainment to daily melatonin injections. One might hypothesise that animals which free-run with a period of less than 24h would show a greater propensity to phase delay to melatonin injections while the opposite should be the case for animals which free-run with periods greater than 24h.

The difference between pharmacological and physiological levels of melatonin is not always obvious. While peripheral administration of melatonin usually results in plasma concentrations which are supraphysiological, it is not known what the corresponding concentrations are in the brain or at the level of the melatonin SCN receptors presumed to mediate the circadian effects of the hormone (Vaněček, Pavlik and Illnerová, 1987). The magnitude of the phase shifts obtained with acute melatonin injections (10 μg per animal) in lizards, a 5 hour delay, was much larger than the maximum shift in rats (50μg/Kg) of a one hour advance, or in humans (0.5mg), also about one hour. If rats and humans are less responsive to the circadian effects of melatonin than lower vertebrates, higher doses may be required to obtain significant phase shifts. We are currently looking at the effects of higher doses of melatonin in rats.

REFERENCES

Adler, K. and Taylor, D.H., 1980, Melatonin and thyroxine: Influence on compass orientation in salamanders, *J. Comp. Physiol.* 136:235.

Armstrong, S.M., and Chesworth, M.J., 1987, Melatonin phase-shifts a mammalian circadian clock, *in*: "Fundamentals and Clinics in Pineal Research", G.P. Trentini, C. De Gaetani and P. Pévet, eds., Raven, New York.

Armstrong, S.M. and Redman, J., 1985, Melatonin administration: Effects on rodent circadian rhythms, *in*: "Photoperiodism, Melatonin and the Pineal", D. Evered and S. Clark, eds., Ciba Foundation Symposium 117, Pitman, London.

Armstrong, S.M. and Redman, J.R., 1993, Melatonin and circadian rhythmicity, *in*: "Melatonin Biosynthesis, Physiological Effects and Clinical Applications", H-S Yu and R.J. Reiter, eds., CRC Press, Boca Raton, Florida.

Armstrong, S.M., Thomas, E.M.V., and Chesworth, M.J., 1989, Melatonin induced phase- shifts of rat circadian rhythms, *in*: "Advances in Pineal Research", R.J. Reiter and S.F. Pang, eds., Libbey and Co, U.K., 265.

Aschoff, J., 1960, Exogenous and endogenous components in circadian rhythms. *Cold Spring Harbor Symp. Quant. Biol.* 25:11.

Beaven, M., 1991, The effect of melatonin on the circadian rhythms of locomotor activity and body temperature in the rat: the search for a non-SCN pacemaker, Unpublished Honours Thesis, Department of Psychology, Monash University.

Beldhius, H.J.A., Dittami, J.P., and Gwinner, E., 1988, Melatonin and the circadian rhythms of feeding and perch-hopping in the European starling, *Sturnus vulgaris*, *J. Comp. Physiol. A*, 164:7 .

Binkley, S. and Mosher, K., 1985, Oral melatonin produces arrhythmia in sparrows, *Experientia* 41:1615.

Cheung, P.W. and McCormack, C.E., 1982, Failure of pinealectomy or melatonin to alter circadian activity rhythm of the rat, *Am. J. Physiol.* 242:R261.

Darrow, J.M. and Goldman, B.D., 1986, Effects of pinealectomy and timed daily melatonin infusions on wheel-running rhythms in Djungarian hamsters, presented at *Society for Neuroscience*, November 9 to 14:843.

Ebihara, S. and Kawamura, H., 1981, The role of the pineal organ and the suprachiasmatic nucleus in the control of circadian locomotor rhythms in the Java sparrow, *Padda aryzivoza*, *J. Comp. Physiol.* 141:207.

Ebihara, S., Uchiyama, K. and Oshima, I., 1984, Circadian organization in the pidgeon, *Columba livia:* the role of the pineal organ and the eye, *J. Comp. Physiol.* 154:59.

Fuchs, J.L., 1983, Effects of pinealectomy and subsequent melatonin implants on activity rhythms in the house-finch *(Carpodacus mexicanus), J.Comp. Physiol.* 153:413.

Gwinner, E., 1978, Effects of pinealectomy on circadian locomotor activity rhythms in European starlings, *Sturnus vulgaris, J.Comp.Physiol.* 126:123

Gwinner, E. and Benzinger, I., 1978, Synchronization of a circadian rhythm in pinealectomized European starlings by daily injections of melatonin, *J. Comp. Physiol.* 127:209.

Gwinner, E., Subbaraj, R., Bluhm, C.K., and Gerkema, M., 1987, Differential effects of pinealectomy on circadian rhythms of feeding and perch hopping in the European starling, *J. Biol. Rhythms* 2(2):109.

Hastings, M.H., Mead, S.M., Vindlacheruvu, R.R., Ebling, F.J.P., Maywood, E.S. and Grosse,J., 1992, Non-photic phase shifting of the circadian activity rhythm of Syrian hampsters: the relative potency of arousal and melatonin, *Brain Res.* 591:20.

Hutchison, V.H., Black, J.J. and Erskine, D., 1979, Melatonin and chlorpromazine: Thermal selection in the mudpuppy, *Necturus maculosus, Life Sci.* 25:527.

Illnerová, H., Trentini, G.P., and Maslova, L., 1989, Melatonin accelerates reentrainment of the circadian rhythm of its own production, *J. Comp. Physiol.* 166:97.

Janik, D.S. and Menaker, M., 1990, Circadian locomotor rhythms in the desert iguana, I. The role of the eyes and the pineal, *J. Comp. Physiol. A* 166:803.

Janik, D.S., Pickard, G.E. and Menaker, M., 1990, Circadian locomotor rhythms in the desert iguana II. Effects of electrolytic lesions to the hypothalamus, *J. Comp. Physiol. A* 166:811.

Joss, J.M.P., 1980, Autoradiographic localisation of sites of uptake of [H^3] melatonin in the brain of a lizard and lamprey, *in:* "Pineal Function", C.D. Matthews and R.F. Seamark, eds., Elservier, Amsterdam.

Kirsch R., Belgnaoui, S. Gourmelen, S. and Pévet, P., 1993, Daily melatonin infusion entrains free-running activity in Syrian and Siberian hamsters, *in:* "Light and Biological Rhythms in Man", L.Wetterberg, ed.,Pergamon Press, Oxford.

Lewy, A.J., Ahmed, S., Latham Jackson J.M., Sack,R.L., 1992, Melatonin shifts human circadian rhythms according to a phase-response curve, *Chronobiol. Int.* 9(5):380.

Menaker, M., 1982 The search for principles of physiological organization in vertebrate circadian systems, *in:* "Vertebrate Circadian Systems Structure and Physiology", J. Aschoff, S. Daan and G. Gross, eds., Springer-Verlag, Berlin.

Mrosovsky, N., 1988, Phase response curves for social entrainment, *J. Comp. Physiol.* 162:35.

Puchalski, W., and Lynch, A., 1988, Daily melatonin injections affect the expression of circadian rhythmicity in Djungarian hamsters kept under a long-day photoperiod, *Neuroendocr.* 48:280.

Redman, J., Armstrong, S.M., and Ng, K.T., 1983, Free-running activity rhythms in the rat: Entrainment by melatonin, *Science* 219:1089.

Redman, J., and Roberts, C.M., 1991, Entrainment of rat activity rhythms by melatonin does not depend on wheel-running activity, *Soc. Neurosci. Abstracts* 17:673.

Simpson, S.M. and Follett, B.K., 1981, Pineal and hypothalamic pacemakers: their role in regulating circadian rhythmicity in Japanese quail, *J. Comp. Physiol.* 144:381

Thomas, E.M.V., and Armstrong, S.M., 1988, Melatonin administration entrains female rat activity rhythms in constant darkness but not in constant light, *Am. J. Physiol.* 255:R237,

Turek, F.W., McMillan, M.P. and Menaker, M., 1976, Melatonin: effects on the circadian locomotor rhythm of sparrows, *Science* 194:1441.

Vaněček, J., Pavlik, A., Illnerová, H., 1987, Hypothalamic melatonin receptor sites revealed by autoradiography, *Brain Res.* 435:359.

Underwood, H., 1977, Circadian organization in lizards: The rate of the pineal organ, *Science* 195:587.

Underwood, H., 1979, Melatonin affects circadian rhythmicity in lizards, *J. Comp. Physiol.* 130:317.

Underwood, H., 1981, Circadian organization in the lizard, *Scelporus occidentalis:* The effect of pinealectomy, blinding, and melatonin, *J. Comp. Physiol.* 141:537.

Underwood, H., 1986, Circadian rhythms in lizards: Phase response curve for melatonin, *J. Pineal Res.* 3:18.

Underwood, H., 1989, The pineal and melatonin: regulators of circadian function in lower vertebrates, *Experientia* 45(10):914.

Underwood, H. and Harless, M., 1985, Entrainment of the circadian rhythm of a lizard to melatonin injections, *Physiol Behav.* 35:267.

CHRONOBIOLOGICAL ACTIVITY OF MELATONIN:

MEDIATION BY GABAERGIC MECHANISMS

D. P. Cardinali, D. A. Golombek, R. E. Rosenstein, B. A. Kanterewicz, M. Fiszman

Department of Physiology, Faculty of Medicine, University of Buenos Aires, CC 243, 1425 Buenos Aires, Argentina

ABSTRACT

In the last years, considerable efforts have been devoted to examine the participation of brain γ-aminobutyric acid (GABA) neurons in circadian phenomena. The importance of gabaergic neurons in circadian organization of brain function is underlined by the fact that almost every single neuron in the suprachiasmatic nucleus contains GABA. In addition, drugs affecting GABA function, like benzodiazepines (BZP) or melatonin, are effective to phase-shift circadian rhythms. Several data point out to a melatonin interaction with GABA-containing neurons in the CNS. Melatonin injection decreases brain GABA concentration, modifies GABA-BZP binding to brain membranes, and increases GABA turnover rate, GABA-induced chloride influx in rat hypothalamus, and the electrophysiological effects of GABA in rabbit cerebral cortex. As other GABA-positive ligands, melatonin inhibited cage-convulsant (TBPS) binding to rat brain membranes. Under long photoperiods, a significant rhythm of GABA turnover was detected in the areas studied (cerebral cortex, preoptic-medial basal hypothalamus, cerebellum and pineal gland) of Syrian hamsters, with maxima at night. Under short photoperiods the synchronization in turnover rate among the remaining regions was lost. This effect was attributed to the different melatonin secretory patterns under both lighting environments. In a number of studies carried out to define the participation of gabaergic mechanisms in behavioral effects of melatonin in rodents, we found that: (1) The administration of the central-type BZP antagonist flumazenil blunted the analgesic response to melatonin in mice, indicating that time-dependent melatonin analgesia was sensitive to central-type BZP antagonism. (2) Flumazenil although unable by itself to modify locomotor activity or induced seizures in rodents, significantly attenuated the inhibitory effects of melatonin. (3) The anxiolytic and pro-exploratory melatonin properties assessed in rats in a plus-maze indicated maximal effects of melatonin at night, with absence of effects at noon and a weak activity at the beginning of the light phase, an effect blunted by administration of flumazenil. (4) In Syrian hamsters, flumazenil inhibited melatonin-induced re-entrainment of locomotor activity and body temperature rhythms of Syrian hamsters after phase-advancing the L:D cycle. Collectively, the results are compatible with the view that melatonin activity on circadian rhythmicity was sensitive to central-type BZP antagonism. Melatonin and BZP seem to have in common an activity on central gabaergic neurons participating in circadian organization.

INTRODUCTION

The pineal gland has been implicated in the control of rhythmic adaptations to daily and seasonal cycles, but the nature of the effect of melatonin on the circadian timing system remains unclear. In the fetus, strong evidence exists for a physiological role of the maternal melatonin signal as a true internal "zeitgeber" (Horton et al., 1992), remnants of which may persist in the adult (Armstrong, 1991). Photoperiodic time measurement in adult as well as in fetal mammals is critically dependent upon the melatonin signal (Goldman and Nelson, 1993).

In the last years, a number of laboratories including ours, have obtained evidence indicating the participation of γ-aminobutyric acid (GABA)-mediated mechanisms in melatonin behavioral and chronobiological effects. The importance of gabaergic neurons in circadian organization of brain function is underlined by the fact that almost every single neuron in the major circadian oscillator in mammals, the suprachiasmatic nucleus (SCN), contains GABA (Moore and Speh, 1993). This paper discusses available the data supporting the hypothesis (Rosenstein and Cardinali, 1990) that GABA neurons may be principal targets of melatonin activity in brain.

BRAIN GABAERGIC MECHANISMS AND CIRCADIAN FUNCTION

GABA is the major inhibitory transmitter in the adult mammalian CNS. It activates $GABA_A$ receptors and inhibits neuronal firing by increasing a Cl^- conductance (Sieghart, 1992). Blockade of $GABA_A$ receptors by bicuculline generates epileptic activity. The receptor-channel complex, which has been sequenced, can be allosterically modulated by drugs such as benzodiazepines (BZP) or barbiturates, known to be therapeutically active in several CNS disorders (e.g., epilepsy, anxiety). In addition to its effect on Cl^- channels, GABA inhibits neuronal activity by activating $GABA_B$ receptors coupled to K^+ channels (Bowery, 1993).

In the CNS, a range of adaptative functions is made possible by the activity of gabaergic inhibitory neurons. Inhibitory projection and local circuit gabaergic neurons play crucial roles in the processing of information within the brain. Roberts (1986) proposed that inhibition enables the CNS to produce variability in behavior and to adjust the extent and rate at which its adaptative options take place.

The $GABA_A$ binding site can be selectively labeled by agonists like ^{3}H-GABA or ^{3}H-muscimol. This site shows both high and low affinity for GABA and its agonists, with K_d values in the nanomolar or low micromolar range, respectively. The low affinity $GABA_A$ recognition site appears to be an antagonist-preferring site, since it can be selectively labeled by specific antagonists, like bicuculline. It is now assumed that GABA exerts its physiological effect by acting at the low affinity binding sites, since it has been demonstrated that micromolar concentrations of GABA are needed to activate the Cl^- channel in electrophysiological experiments and to modulate other binding sites at the

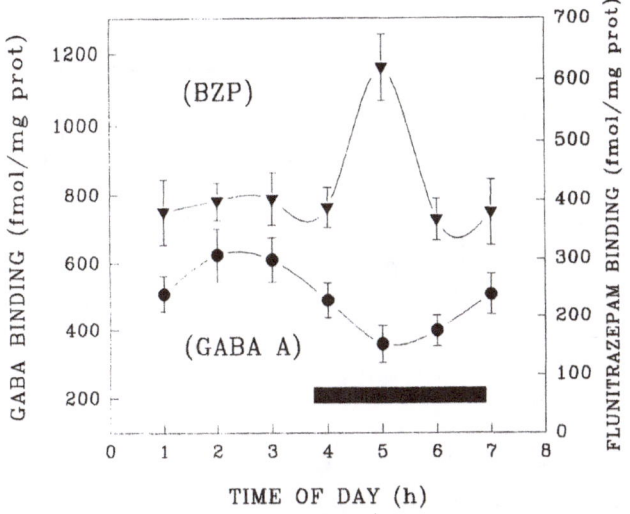

Fig. 1   Diurnal changes in cerebral cortex GABA and BZP binding in rats. Shown
are B_{max} of binding, mean $\pm$ SEM (n= 6-8/point). Redrawn from Acuña-
Castroviejo et al. (1986 a,b).

$GABA_A$ receptor (Sieghart, 1992). The high affinity $GABA_A$ sites presumably represent a
desensitized form of the $GABA_A$ receptor.

Daily fluctuations of GABA levels in the tuberoinfundibular region (Cattabeni et al.,
1978; Casanueva et al., 1984) and SCN (Aguilar-Robledo et al., 1993) have been reported,
as well as in GABA uptake in hypothalamic preoptic area (Barkai et al., 1985). In our
laboratory, Acuña-Castroviejo et al. (1986 a,b) reported a diurnal rhythmicity in the
number of rat brain high affinity $GABA_A$ and BZP receptors (Fig. 1). The decrease in
$GABA_A$ high affinity binding during the night suggested that the GABA synthesis and
release could be stimulated at this time.

Since transmitter content measurements yielded a poor evaluation of transmitter's
dynamics, Kanterewicz et al. (1993) studied the circadian rhythms of GABA turnover rate
in cerebral cortex, preoptic-medial basal hypothalamus (PMBH), cerebellum and pineal
gland of Syrian hamsters kept under either long or short days. In long days (i.e., 14 h of
light per day), GABA turnover in cerebral cortex, PMBH, cerebellum and pineal gland
exhibited a clear phase relationship, showing maximal values towards the first half of the
night. In short days (i.e., 10 h of light per day), rhythmicity of GABA turnover was still
apparent but the phase relationship among the rhythms became lost.

That the changes in GABA turnover described in hamster brain could be associated
with concomitant changes in the principal phenomena linked to $GABA_A$ receptor activation
is shown in Fig. 2. Circadian changes in K_d of low affinity $GABA_A$ binding sites and in the
major physiologically event linked to $GABA_A$ receptor stimulation, that is, the uptake of
$^{36}Cl^-$ by brain synaptoneurosomes, are depicted, together with concomitant changes in
GABA content and GABA turnover rate in hamsters kept under 14L:10L photoperiods.

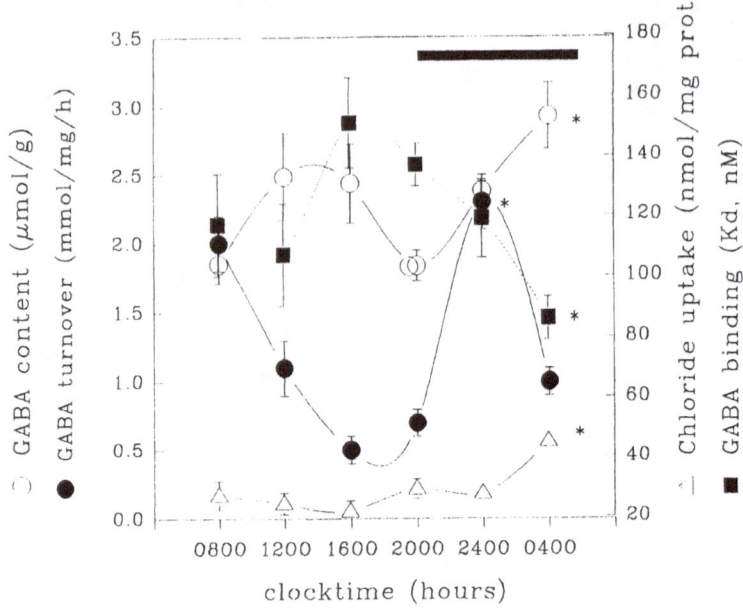

Fig. 2   Diurnal changes in GABA content, GABA turnover rate and ^{36}Cl$^-$ uptake in cerebral cortex of Syrian hamsters kept under 14L:10D. Data on GABA content and turnover redrawn from Kanterewicz et al. (1993).

The term "circadian timing system" is used to describe the brain areas responsible for the generation and regulation of circadian rhythms. In mammals, the function of the circadian timing system is expressed by the complex array of homeostatic adjustments associated with the rest-activity cycle that promotes adaptive behavior. The mammalian circadian timing system consists of visual pathways mediating entrainment, the hypothalamic SCN, and the efferent projections of the SCN to effector systems (Card and Moore, 1991). The principal visual pathway mediating entrainment is the retinohypothalamic tract The pathway is formed from collaterals of optic chiasm axons that also innervate the intergeniculate leaflet (IGL) of the thalamic lateral geniculate complex. The RHT terminates in the ventrolateral portion of the SCN, a zone characterized by the presence of vasoactive intestinal polypeptide-producing neurons, co-extensive with the terminals of neuropeptide Y-producing pathway arising from the IGL and arriving at the SCN through the geniculate-hypothalamic tract (Albers et al., 1991).

Relevant to the present discussion, every single neuron in SCN contains GABA (Moore and Speh, 1993). Moreover, GABA co-exists with neuropeptide Y in IGL neurons, and in the retina, GABA is contained in certain ganglion cells as well as in horizontal cell interneurons (Caruso et al., 1990). Because of this key distribution of GABA, it was logical to postulate the amino acid as the principal neurotransmitter of the circadian system (Moore and Speh, 1993).

The functional participation of gabaergic neurons in the regulation of circadian timing function was indicated by electrophysiological and pharmacological studies. As shown by Liou et al. (1990) and Masson et al. (1991) circadian single unit discharge of SCN neurons is affected by GABA and BZP. Likewise, phase-delays induced by brief pulses of light presented during early subjective night are specifically blocked by bicuculline while phase-advances induced during late subjective light are specifically blocked by diazepam; the administration of baclofen also blocks the phase shifting effect of light, affecting both phase advances and delays (Ralph and Menaker, 1989). These observations together with the neurochemical evidence on daily variations of GABA-mediated phenomena in brain (Fig. 1 and 2), underline the importance of gabaergic neurons in circadian organization.

MELATONIN EFFECT ON BRAIN GABAERGIC FUNCTION

The link between pineal function and brain GABA-mediated mechanisms was suggested by early observations indicating that the pineal gland exerted a depressive influence on CNS excitability (Nir et al., 1968; Romijn, 1978). Pineal activity appeared to be linked to melatonin. since pharmacological doses of the hormone prevented pinealectomy-induced seizures in gerbils, as well as ouabain-induced and kindled convulsions in rats. In mice, melatonin potentiated barbiturate-induced sleep and antagonized pentylenetetrazol-induced, 3-mercaptoproprionic acid-elicited or electroshock-induced convulsions; melatonin also had analgesic properties in this species. Likewise, melatonin ameliorated photoperiodic-induced convulsions in baboons and decreased focal epilepsy in cats (for ref. see Rosenstein and Cardinali, 1990).

The first indication on a possible link between the pineal and brain gabaergic neurons was provided by Anton-Tay (1974), who reported increased GABA levels in rat brain following pinealectomy and depressed levels after melatonin injection. About 10 years later, the subject was re-examined by Lowenstein et al. (1985) and Acuña-Castroviejo et al. (1986 a,b). Pinealectomy increased generally B_{max} GABA high affinity binding and disrupted its normal diurnal rhythmicity. Pineal ablation also blunted the maximum of B_{max} of BZP binding to cerebral cortex membranes observed in control rats at midnight and caused a significant depression of BZP binding at noon. Pinealectomy-induced changes of brain BZP binding were already detectable 3 days after surgery (Lowenstein et al., 1985).

Melatonin treatment counteracted the changes in GABA and BZP binding found after pinealectomy of rats (Acuña-Castroviejo et al., 1986 a,b). As depicted in Fig. 3, the minimal effective dose of melatonin to produce effects was 25 μg/kg. Low concentrations of melatonin did not modify GABA or BZP binding through a direct effect on the binding sites (Acuña-Castroviejo et al., 1986 a,b). However, Marangos et al. (1981) reported direct effects of melatonin on brain BZP binding in vitro at supramicromolar concentrations, much higher than those reaching the brain even after injecting pharmacological doses of melatonin. As other $GABA_A$-positive ligands, melatonin inhibited TBPS binding to rat brain membranes (Niles and Peace, 1990).

The effect of melatonin on brain GABA and BZP receptors was also explored in other laboratories. Kennaway et al. (1988) reported a species differences in melatonin response of brain BZP receptors, low doses of melatonin increasing BZP binding in Wistar but not in Porton rats. In a recent study, Gomar et al. (1993) further refined the mechanisms involved in melatonin action on brain BZP binding. The intracerebroventricular administration of melatonin, β-endorphin and melatonin plus β-endorphin all increased BZP binding to a similar extent and in a dose-related manner. The effect of melatonin was prevented by the simultaneous injection of the specific opioid antagonist naloxone. These results implicated the modulation of melatonin-dependent changes on brain BZP receptors by opioid peptides.

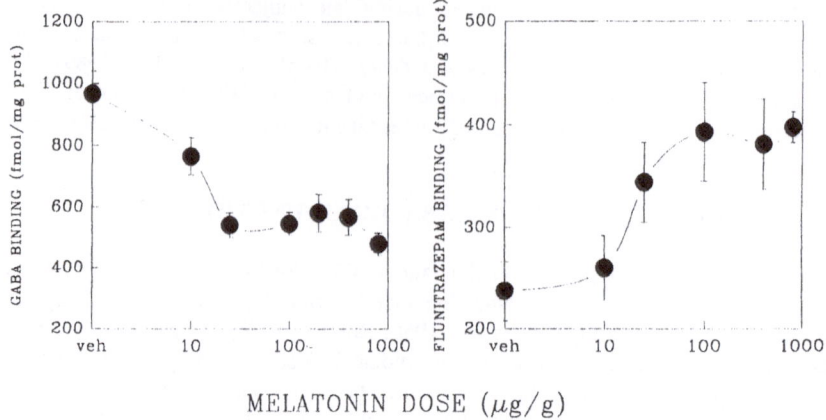

MELATONIN DOSE (μg/g)

Fig. 3   Effect of melatonin on pinealectomy-induced changes in cerebral cortex GABA and BZP binding in rats. Redrawn from Acuña-Castroviejo et al. (1986 a,b).

Direct effects of BZP on melatonin binding were reported. Oxazepam treatment decreased 2-^{125}I-iodomelatonin binding at 2400 h in the hippocampus and medulla-pons of the pinealectomized rats and did not significantly affect the binding in the hypothalamus (Anis et al., 1992).

A single melatonin injection to rats (25 μg/kg) augmented significantly GABA turnover in the hypothalamus, while a dose of 100 μg/kg was needed to affect GABA turnover in cerebral cortex or cerebellum (Rosenstein and Cardinali, 1986). The activity of melatonin on GABA turnover was due, at least in part, to a stimulatory effect on the GABA synthesizing enzyme glutamic acid decarboxylase. In the same study it was found that, in synaptoneurosomes of rat PMBH prepared at late evening, nanomolar concentrations of melatonin augmented ^{36}Cl$^-$ influx by potentiating GABA-induced increase of chloride ion uptake. Evening activity of melatonin was detectable at concentrations 1 order of magnitude greater than those effective at early morning (Rosenstein et al., 1989). These results further supported a link between melatonin and the activity of gabaergic neurons in PMBH.

In experiments on the spontaneous firing activity of single neurons in parietal cortex of anesthetized rabbits Stankov et al. (1992) showed that melatonin and 2-iodomelatonin had GABA-like effects and were able alone, in nanomolar concentrations, to significantly slow the neuronal firing activity. Moreover, both melatonin and 2-iodomelatonin potentiated the effect of GABA on the neuronal activity.

THE ROLE OF GABA-RELATED MECHANISMS IN MELATONIN ACTIVITY

In principle, to demonstrate that a neurotransmitter system is involved in the mediation of a given melatonin effect, two requirements should be fulfilled: (a) the neurotransmitter system should show dynamic changes as a consequence of melatonin injection; (b) functional obliteration of the neurotransmitter system should modify significantly the melatonin effect. A demonstration of this sort does not necessarily imply that the neurotransmitter system is an integral part of the sequence of events triggered by melatonin action; rather, it may only indicate a permissive role in the examined effect.

In addition to the effects of melatonin on GABA neurons above described, brain monoaminergic mechanisms have been explored as possible targets for melatonin action. As demonstrated by Fang and Dubocovich (1990), the activation of melatonin receptors retarded the turnover of norepinephrine (NE) in murine hypothalamus. In the case of the serotoninergic and dopaminergic systems, Alexiuk and Vriend (1991) reported that melatonin injection brought about a substantial inhibition of daytime dopamine synthesis in median eminence, and a less important inhibitions of serotonin turnover rate. It should be noted that monoamine pathways within the brain are not important for melatonin entrainment of circadian rhythmicity in rodents, since the intraventricular injection of 6-hydroxydopamine and 5,7 dihydroxytryptamine, which deplete catecholamines and indoleamines, failed to alter entrainment (Cassone et al., 1986).

In our Laboratory a number of studies were carried out to define the participation of brain gabaergic mechanisms in the behavioral effects of melatonin in rodents. To achieve an effective inhibition of $GABA_A$-mediated mechanisms a rather indirect procedure had to be employed, because the use of $GABA_A$ antagonists, like bicuculline or picrotoxin, was precluded due to the convulsive state produced in the animals. Flumazenil, a BZP receptor antagonist with some inverse receptor agonist activity (File and Pellow, 1986), was employed. By interacting with the BZP binding site inverse receptor agonists reduce the GABA-induced Cl⁻ ion flux.

In a study aiming to determine whether melatonin-induced analgesia could be inhibited by flumazenil (Golombek et al., 1991a), a significant diurnal variation in the pain threshold, with an increase in latency during the dark phase of the daily photoperiod, was observed. Melatonin exhibited maximal analgesic effects at late evening. The administration of flumazenil at 2000 h, although unable by itself to modify pain threshold, blunted the analgesic response to melatonin, indicating that time-dependent melatonin analgesia was sensitive to impairment of $GABA_A$-mediated mechanisms.

In subsequent studies, Golombek et al. (1991b; 1992) analyzed the inhibitory effects of flumazenil on melatonin-induced depression of locomotor behavior and 3-mercaptopropionic acid seizures. The administration of flumazenil, although unable by itself to modify locomotor activity or seizures, significantly attenuated the inhibitory effects of melatonin. Golombek et al. (1993) also examined the anxiolytic and pro-exploratory melatonin properties assessed in rats using a plus-maze procedure. Melatonin displayed maximal effects at night, with absence of effects at noon and a weak activity at the beginning of the light phase, an effect also blunted by administration of flumazenil.

Several interpretations could be entertained to explain the efficacy of flumazenil to blunt melatonin activity in the several behavioral paradigms tested. Firstly, melatonin could be bound to BZP sites on the GABA A receptor complex, an effect counteracted by flumazenil (this possibility is not supported by binding data of labeled BZP in vitro, Acuña-Castroviejo et al., 1986a). Secondly, flumazenil could displace melatonin from melatonin receptor sites in brain; recent data on direct effects of BZP on melatonin binding endorse this view (Anis et al., 1992). Thirdly, melatonin could need an intact endogenous GABAergic "tone" to affect behavior, a situation somewhat perturbed by the administration of flumazenil. A last possibility is that the action of melatonin could need an intact endogenous BZP "tone", which could be effectively blocked by flumazenil. Further experiments are needed to disclose among these various possibilities. Figure 4 summarizes

Table 1.   Inhibition by flumazenil of melatonin-induced resynchronization of wheel-running activity and body temperature in Syrian hamsters after a phase-advance of 6 h in L:D cycle.

treatment	wheel running activity	body temperature
none	6.5 ± 0.3	6.7 ± 0.4
vehicle	5.8 ± 0.4	6.4 ± 0.5
melatonin (1 mg/kg)	4.3 ± 0.4 [*]	4.1 ± 0.5 [*]
melatonin (1 mg/kg) + flumazenil (5 mg/kg)	6.6 ± 0.3	6.9 ± 0.6
flumazenil) 5 mg/kg)	6.7 ± 0.4	6.6 ± 0.6

Two months-old male Syrian hamsters, weighing 120-150 g, were raised in our colony under a 14L:10D photoperiod. For telemetric monitoring of temperature and activity a Dataquest III system (Mini Mitter DQ-III, Mini Mitter Co., Inc. Sunriver, OR) was employed. After a baseline recording of the rhythm, a 6-hour advance of the L:D cycle was made (day D). Injections of melatonin, flumazenil or vehicle (100 μl DMSO) were made on days D -2, D -1 and D, at the expected time of lights off after the phase shift. Shown are the means ± SEM. Asterisks designate significant differences, ANOVA, Dunnett's t test.

the data obtained in the several behavioral experiments employing flumazenil above discussed.

It is well documented that melatonin influences circadian rhythmicity in several species including man (Arendt et al., 1987; Lewy et al., 1991), probably by acting on the SCN. SCN single-unit activity in vitro was inhibited by melatonin superfusion during the late subjective day in a dose-dependent manner but not at other times of day, supporting the view that melatonin acts directly on SCN neurons (Shibata et al., 1989). Indeed, melatonin directly resets the rat SCN circadian clock in vitro (McArthur et al., 1991).

In view of the results obtained by employing flumazenil in melatonin-injected animals, we considered it worthwhile to examine the extent to which flumazenil could affect the melatonin-induced re-entrainment of locomotor activity and body temperature rhythms of Syrian hamsters after phase-advancing the light-dark cycle by 6 h (Table 1). In vehicle-injected and untreated controls, about 1 day per hour of phase advance was needed to resynchronize the rhythms. The administration of melatonin brought about a significant decrease of resynchronization time, an effect prevented by the previous administration of flumazenil. The results were compatible with the view that melatonin activity on circadian rhythmicity was sensitive to central-type BZP antagonism (Golombek and Cardinali, 1993). Melatonin and BZP may have in common an activity on central gabaergic neurons participating in circadian organization.

The link of melatonin and GABA-mediated mechanisms in brain was examined in other Laboratories. Guardiola-Lemaitre et al. (1992) studied the combined effects of diazepam and melatonin in four plates test and the tail suspension test for anxiolytic activity in mice. In the former test anxiolytics increase the number of punished crossings and in the latter increase the duration of immobility of mice suspended by the tail. In both tests, the combined treatment with melatonin and diazepam caused a significant increase in the

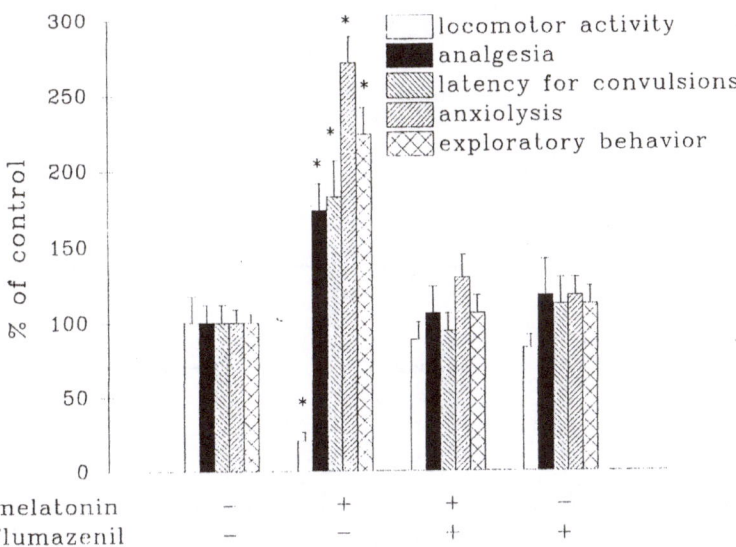

melatonin	−	+	+	−
flumazenil	−	−	+	+

Fig. 4   Antagonism by flumazenil of melatonin behavioral activity. Flumazenil (0.5 mg/kg) was administered 15 min before melatonin. Results were expressed as percent changes as compared to controls. Data from Golombek et al., 1991 a,b; 1992b; 1993).

number of punished crossings, whereas each treatment alone was without effect. The results indicated that melatonin enhanced the anxiolytic actions of diazepam.

Dubocovich et al. (1990) examined the anti-immobility effect of the selective melatonin receptor antagonist, luzindole, in the behavioral despair test in mice. Luzindole reduced the time of immobility in a dose-dependent manner, the effect being more pronounced at midnight than at noon. The anti-immobility effect of luzindole was prevented by the administration of melatonin, a finding which suggested that endogenous melatonin plays a role during swimming in the C^3H/HeN mouse behavioral despair test.

CONCLUDING REMARKS

Collectively, the results above discussed are compatible with the view that melatonin activity on circadian rhythmicity is sensitive to impairment of $GABA_A$-receptor function. Melatonin and BZP seem to have in common an activity on central gabaergic neurons participating in circadian organization.

Circadian rhythms are already present in the fetus. At a certain stage of pre-natal hypothalamic development (around 30 weeks of gestation in humans) the fetus becomes responsive to maternal circadian signals. Photoperiodic information is transferred from mother to their fetuses during gestation. Maternal melatonin is known to be essential for the transfer of this prenatal photoperiodic information. The duration of the daily melatonin signal, expressed as an elevation of serum melatonin levels in the maternal circulation, may convey day length information to the fetus (Horton et al., 1992).

It is interesting that, differing from the adult, GABA plays in the fetus and early postnatal life the role of an excitatory transmitter (Cherubini et al., 1991). At this age,

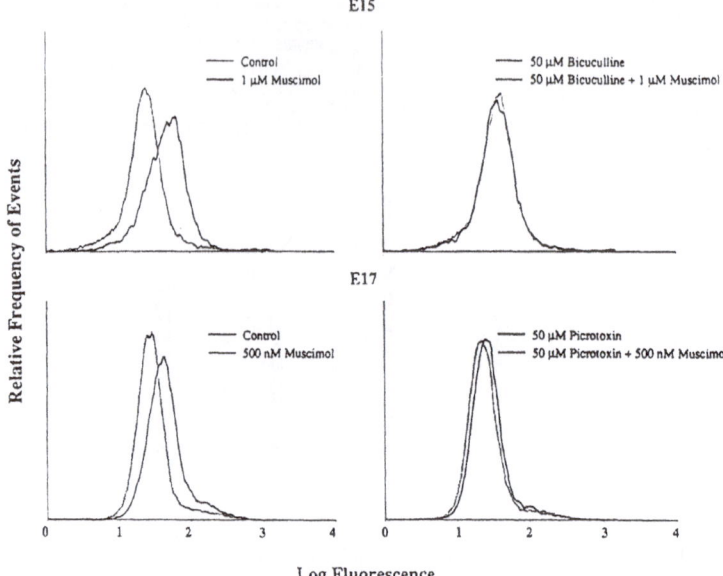

E15

E17

Log Fluorescence

Fig. 5   GABA$_A$ receptor-mediated depolarization in rat striatal cell suspensions at embryonic age E15 and E17. Experiments were carried out by employing a fluorescence-activated cell sorter and the negatively charged fluorescent indicator oxonol. A shift to the right indicated depolarization. Reprinted with permission from Fiszman et al. (1993).

synaptically released or exogenously applied GABA depolarizes and excites the neuronal membranes. GABA acting on GABA$_A$ receptors during early development provides most of the excitatory drive, whereas excitatory glutamatergic synapses are quiescent. One of us by using imaging techniques found that in fetal and neonatal hippocampal and striatal neurons that activation of GABA$_A$ receptors produced depolarization (Fig. 5). Since blockade of GABA$_A$ receptors by bicuculline inhibited developmental growth of neurons in culture, GABA$_A$ receptors may provide the tonic excitatory drive needed for growth and differentiation of neurons at a fetal age. To what extent maternal melatonin affects these phenomena in the fetal circadian timing system is presently being examined.

ACKNOWLEDGEMENTS

These studies were supported by grants from the Community of the European Countries, Bruxelles (CCE contract n° C11* 0636), Consejo Nacional de Investigaciones Científicas y Técnicas and Sandoz Foundation for Gerontological Research, Sao Paulo (grant 095/93).

REFERENCES

Acuña-Castroviejo, D., Lowenstein, P.R., Rosenstein, R., and Cardinali, D.P., 1986a, Diurnal variations of benzodiazepine binding in rat cerebral cortex: disruption by pinealectomy, J Pineal Res, 3:101-109.

Acuña-Castroviejo, D., Rosenstein, R.E., Romeo, H.E., and Cardinali, D.P., 1986b, Changes in gamma-aminobutyric acid high affinity binding to cerebral cortex membranes after pinealectomy or melatonin administration to rats, Neuroendocrinology, 43:24-31.

Aguilar-Robledo, R., Verduzco-Carbajal, L., Rodriguez, C., Mendes-Franco, J., Moran, J., and Peres de la Mora, M., 1993, Circadian rhythmicity in the GABAergic system in the suprachiasmatic nuclei of the rat, Neurosci Lett, 157:199-202.

Alexiuk, N.A. and Vriend, J., 1991, Effects of daily afternoon melatonin administration on monoamine accumulation in median eminence and striatum of ovariectomized hamsters receiving pargyline, Neuroendocrinology, 54:55-61.

Anis, Y., Nir, I., Schmidt, U., and Zisapel, N, 1992, Modification by oxazepam of the diurnal variations in brain ^{125}I-melatonin binding sites in sham-operated and pinealectomized rats, J Neural Transm Gen Sect, 89:155-166.

Anton-Tay, F., 1974, Melatonin: effects on brain function, Adv Biochem Psychopharmacol, 11:315-324.

Arendt, J., Aldhous, M., English, J., Marks, V., and Arendt, J.H., 1987, Some effects of jet-lag and their alleviation by melatonin, Ergonomics, 30:1379-1393.

Armstrong, S.M., 1991, Entrainment of vertebrate circadian rhythms by melatonin, Adv Pineal Res, 5:259-266.

Barkai, A.I., Potegal, M., and Kowalik, S., 1985, Decline of GABA uptake in the hamster preoptic area following light onset, J Neurochem, 44:987-989.

Bowery, N.G., 1993, GABA$_B$ receptor pharmacology, Annu Rev Pharmacol Toxicol, 33:109-147.

Card, J.P. and Moore, R.Y., 1991, The organization of visual circuits influencing the circadian activity of the suprachiasmatic nucleus, in: "Suprachiasmatic Nucleus. The Mind's Clock", Klein, D.C., Moore, R.Y., and Reppert, S.M., eds., Oxford University Press, New York, pp.51-76.

Caruso, D.M., Owczarzak, M.T., and Poucho, M.G., 1990, Colocalization of substance P and GABA in retinal ganglion cells: A computer-assisted visualization, Vis Neurosci, 5:389-394.

Casanueva, F., Apud, J.A., Masotto, C., Cocchi, D., Locatelli, V., Racagni, G., and Muller, E.E., 1984, Daily fluctuations in the activity of the tuberoinfundibular GABAergic system and prolactin levels, Neuroendocrinology, 39:367-370.

Cassone, V.M., Chesworth, M.J., and Armstrong, S.M., 1986, Entrainment of rat circadian rhythms by daily injection of melatonin depends upon the hypothalamic suprachiasmatic nuclei., Physiol Behav, 36:1111-1121.

Cattabeni, F., Maggi, A., Monduzzi, M., De Angellis, L., and Racagni, G., 1978, GABA: circadian fluctuations in rat hypothalamus, J Neurochem, 31:565-567.

Cherubini, E., Gaiarsa, J.L., and Ben-Ari, Y., 1991, GABA: An excitatory transmitter in early postnatal life, Trends Neurosci, 14:515-519.

Dubocovich, M.L., Mogilnicka, E., and Areso, P. M, 1990, Antidepressant-like activity of the melatonin receptor antagonist, luzindole (N-0774), in the mouse behavioral despair test, Eur J Pharmacol, 182:313-325.

Fang, J.M. and Dubocovich, M.L., 1990, Activation of melatonin receptor sites retarded the depletion of norepinephrine following inhibition of synthesis in the C3H/HeN mouse hypothalamus, J Neurochem, 55:76-82.

File, S.E. and Pellow, S., 1986, Intrinsic actions of the benzodiazepine receptor antagonist Ro 15-1788, Psychopharmacol, 88:1-11.

Fiszman, M.L., Behar, T., Lange, G.D., Smith, S. V., Novotny, E.A., and Barker, J.L., 1993, Embryonic and early postnatal hippocampal cells respond to nanomolar concentrations of muscimol, Dev Brain Res, 73:243-251.

Goldman, B.D. and Nelson, R.J., 1993, Melatonin and seasonality in mammals, in: "Melatonin.Biosynthesis, Physiological Effects, and Clinical Applications", Yu, H. and Reiter, R.J., eds., CRC Press, Boca Raton, pp.225-252.

Golombek, D.A. and Cardinali, D.P., 1993, Melatonin accelerates re-entrainment after phase advance of the light-dark cycle in Syrian hamsters. Antagonism by flumazenil, Chronobiol Int 10:435-441.

Golombek, D.A., Escolar, E., Burin, L., Brito Sanchez, M.G., and Cardinali, D.P., 1991a, Time-dependent melatonin analgesia in mice: inhibition by opiate or benzodiazepine antagonism, Eur J Pharmacol, 194:25-30.

Golombek, D.A., Escolar, E., Burin, L., Brito Sanchez, M.G., and Cardinali, D.P., 1992a, Chronopharmacology of melatonin: inhibition by benzodiazepine antagonism, Chronobiol Int, 9:124-131.

Golombek, D.A., Escolar, E., and Cardinali, D.P., 1991b, Melatonin-induced depression of locomotor activity in hamsters: time dependency and inhibition by the central type benzodiazepine antagonist Ro 15-1788, Physiol Behav, 49:1091-1098.

Golombek, D.A., Fernandez Duque, D., Burin, L., Brito, Sanchez, M.G., and Cardinali, D.P., 1992b, Time-dependent anticonvulsant activity of melatonin in hamsters, Eur J Pharmacol, 210:253-258.

Gomar, M.D., Castillo, J.L., del, Aguila, C.M., Fernandez, B., and Acuna-Castroviejo, D, 1993, Intracerebroventricular injection of naloxone blocks melatonin-dependent brain [^{3}H]-flunitrazepam binding, Neuroreport, 4:987-993.

Guardiola-Lemaitre, B., Lenegre, A., and Porsolt, R. D, 1992, Combined effects of diazepam and melatonin in two tests for anxiolytic activity in the mouse, Pharmacol Biochem Behav, 41:405-408.

Horton, T.H., Ray, S.L., Rollag, M.D., Yellon, S.M., and Stetson, M.H., 1992, Maternal transfer of photoperiodic information in Siberian hamsters. V. Effects of melatonin implants are dependent on photoperiod, Biol Reprod, 47:291-296.

Kanterewicz, B.I., Golombek, D.A., Rosenstein, R.E., and Cardinali, D.P., 1993, Diurnal changes of GABA turnover rate in brain and pineal gland of Syrian hamsters, Brain Res Bull, 31:661-666.

Kennaway, D.J., Royles, P., Webb, H., and Carbone, F., 1988b, Effects of protein restriction, melatonin administration, and short daylength on brain benzodiazepine receptors in prepubertal male rats, J Pineal Res, 5:455-467.

Lewy, A., Sack, R.L., and Latham, J., 1991, Melatonin and the acute suppressant effect of light may help regulate circadian rhythms in humans, Adv Pineal Res, 5:285-293.

Liou, S.Y., Shibata, S., Albers, H.E., and Ukei, S., 1990, Effects of GABA and anxiolytics on the single unit discharge of suprachiasmatic neurons in rat hypothalamic slices, Brain Res Bull, 25:103-107.

Lowenstein, P.R., Rosenstein, R.E., and Cardinali, D.P., 1985, Melatonin reverses pinealectomy-induced decrease of benzodiazepine binding in rat cerebral cortex, Neurochem Int, 7:675-681.

Marangos, P.J., Patel, J., Hirata, F., Sonhein, D., Paul, S.M., Skolnick, P., and Goodwin, F., 1981, Inhibition of diazepam binding by tryptophan derivatives including melatonin and its brain metabolite N-acetyl-5-methoxykynurenamine, Life Sci, 29:259-267.

Masson, R., Biello, S.M., and Harrington, M.E., 1991, The effects of GABA and benzodiazepines on neurons in the suprachiasmatic nucleus (SCN) of Syrian hamsters, Brain Res, 552:53-57.

McArthur, A.J., Gillette, M.U., and Prosser, R. A, 1991, Melatonin directly resets the rat suprachiasmatic circadian clock in vitro, Brain Res, 565:158-161.

Moore, R.Y. and Speh, J.C., 1993, GABA is the principal neurotransmitter of the circadian system, Neurosci Lett, 150:112-116.

Niles, L.P. and Peace, C.H., 1990, Allosteric modulation of t-[^{35}S] butylbicyclo-phosphothionate binding in rat brain by melatonin, Brain Res Bull, 24:635-638.

Nir, I., Behroozi, K., Assael, M., Ivriani, I., and Sulman, F.G., 1968, Changes in the electrical activity of the brain following pinealectomy, Neuroendocrinology, 4:122-127.

Ralph, M.R. and Menaker, M., 1989, GABA regulation of circadian responses to light. I. Involvement of GABA$_A$-benzodiazepine and GABA$_B$ receptors, J Neurosci, 9:2858-2865.

Reiter, R.J., 1991, Pineal melatonin: Cell biology of its synthesis and of its physiological interactions, Endocrine Rev, 12:151-180.

Roberts, E., 1986, What do GABA neurons really do? They make possible variability generation in relation to demand, Exp Neurol, 93:279-290.

Romijn, H., 1978, The pineal, a tranquillizing organ?, Life Sci, 23:2257-2274.

Rosenstein, R.E. and Cardinali, D.P., 1986, Melatonin increases in vivo GABA accumulation in rat hypothalamus, cerebellum, cerebral cortex and pineal gland, Brain Res, 398:403-406.

Rosenstein, R.E. and Cardinali, D.P., 1990, Central gabaergic mechanisms as target for melatonin activity, Neurochem Int, 17:373-379.

Rosenstein, R.E., Estevez, A., and Cardinali, D.P., 1989, Time-dependent effect of melatonin on glutamic acid decarboxylase activity and ^{36}Cl$^-$ influx in rat hypothalamus, J Neuroendocrinol, 1:443-447.

Shibata, S., Cassone, V.M., and Moore, R.Y., 1989, Effects of melatonin on neuronal activity in the rat suprachiasmatic nucleus in vitro, Neurosci Lett, 97:140-144.

Sieghart, W., 1992, GABA$_A$ receptors: Ligand-gated Cl$^-$ ion channels modulated by multiple drug-binding sites, Trends Pharmacol Sci, 13:446-450.

Stankov, B., Biella, B., Panara, C., Lucini, V., Capsoni, S., Fauteck, J., Cozzi, B., and Fraschini, F., 1992, Melatonin signal transduction and mechanism of action in the central nervous system: Using the rabbit cortex as a model, Endocrinology, 130:2152-2159.

NEW MOLECULES ACTIVE AT THE MELATONIN RECEPTOR LEVEL

Franco Fraschini and Bojidar Stankov

Chair of Chemotherapy, University of Milan
Via Vanvitelli 32, 20129 Milano, ITALY

INTRODUCTION

Melatonin (5-methoxy-*N*-acetyltryptamine) is a hormone synthesized and secreted by the pineal gland on a circadian basis, with elevated peripheral blood levels at night. The interest in its sites and mechanism of action has been growing since its isolation and synthesis (Lerner et al., 1958). Melatonin is considered an important internal *Zeitgeber*, translating the photoperiodic information from the environment (Reiter, 1991) and the circadian melatonin signal is decoded by high-affinity binding sites in the Central Nervous System (Stankov et al., 1993). Melatonin plays a role in regulating the seasonal reproductive competence in most of the mammals living in the temperate zones and influences the circadian rhythmicity in reptiles, birds and mammals (Armstrong and Redman, 1993). The exact role of melatonin in human is less explicit, but recent research demonstrated presence of melatonin receptors in the human circadian clock (Reppert et al., 1988), as well as the ability of humans to respond to the melatonin signal with a well-defined circadian phase-response curve (Lewy et al., 1992). Changes in the melatonin rhythm and/or peripheral blood levels are now thought to be implicated in a number of patho/physiological conditions, such as the seasonal affective disorder (Wetterberg wt al., 1990), regulation of the sleep-wake cycles (Wehr, 1991), puberty (Waldhauser and Steger, 1989), jet-lag (Petrie et al., 1989), and reproduction (Voordouw et al., 1992).

From the structure-activity studies with melatonin analogues it appeared that the *N*-acetyl and 5-methoxy substituents are necessary for biological activity and binding affinity at the receptor level (Heward and Hadley, 1975; Krause and Dubocovich, 1991; Dubocovich, 1988); moreover, the indole nucleus apparently could be substituted with a naphthalenic (Yous et al., 1992) or amidotetraline ring (Copinga et al., 1993). We have previously reported that substitution with bromine at C-$_2$ position of the indole nucleus yielded an agonist showing extremely high affinity for the melatonin receptor (Duranti et

The Pineal Gland and Its Hormones
Edited by F. Fraschini *et al.*, Plenum Press, New York, 1995

al., 1992). This compound was more potent *in vivo* and *in situ* than either melatonin or 2-iodomelatonin. Recent experiments suggested that *N*-cyclopropanoyl substitution can lead to an increased affinity for the melatonin receptor expressed in the cells of pars tuberalis, isolated from ovine pituitaries (Yous et al., 1992).

Therefore, we investigated the consequences of chemical modifications at C_{-2}, C_{-5} and *N*-acyl positions, alone or in combination, on the binding affinity and the activity of the generated compounds, in order to acquire a better understanding of the structural requirements for binding to the melatonin receptor, and prerequisites in terms of biological activity.

MATERIALS AND METHODS

Chemistry

The synthesis of the novel melatonin analogues has been described in detail elsewhere (Spadoni et al., 1993). The modified Madelung synthesis (Houlihan et al., 1981) was adopted for the preparation of 2-substituted indoles, starting from *N*-(2-methyl-5-methoxyphenyl) alkanamides or benzamides. The 2-substituted indoles were coupled with 1-dimethylamino-2-nitroethylene to give the required nitrovinylindoles in the conditions previously described for related compounds (Buchi and Mak, 1977). The syntheses were completed by reducing the nitrovinylindoles with lithium aluminum hydride ($LiAlH_4$) and acylating the resulting crude tryptamines with acetic anhydride or cyclopropanecarbonyl chloride in the presence of triethylamine (TEA).

Compound 4j was obtained by acylating the 5-methoxytryptamine with cyclopropanecarbonyl chloride. Compound 4h was prepared by direct bromination with *N*-bromosuccinimide of 4j according to a reported method (Duranti et al., 1992).

^{1}H NMR, IR and elemental analysis data of the new compounds were found in accord with the assigned structures.

Binding assays

The source of the animals, the characterization of the melatonin receptor in the quail brain, the isolation of the crude membrane preparation, enriched with melatonin receptors, used in the present study has been described in details elsewhere (Cozzi et al., 1993; Stankov et al., 1991).

Determination of the biological activity *in vivo*

For the *in vivo* studies, the Syrian hamster gonadal regression model (Duranti et al., 1992; Reiter, 1983) was employed. Briefly, male sexually mature Syrian hamsters (five per group), held under constant photoperiod conditions of 14:10 LD, lights off 20:00 h were treated with the tested compounds alone, in a dose of 200 μg/animal/day at 17:00 h, or 200 μg at 16:30 h, followed by 20 μg melatonin at 17:00 h, for six weeks. The negative controls were given equal volumes of saline; the positive controls received 200 μg or 20 μg melatonin at 16:30 and 17:00 h, respectively. At the end of the experimental period, the combined testes and seminal vesicle weights were recorded. Analogs that induced partial gonadal regression (testes and seminal vesicle weights equal to 50% or less, compared to the respective positive control) were considered weak agonists. Compounds that had no influence *per se*, but were able to block the effect of subsequent melatonin treatment were evaluated as antagonists.

Determination of the biological activity *in situ*

The *in situ* electrophysiological studies employed the rabbit parietal cortex as a model system and the basic approach have been described in details elsewhere (Stankov et al., 1992). Additionally, in separate series of experiments, saturation concentrations of GABA antagonists phaclofen and bicuculline ($1E^{-4}$ M) were applied, starting before, and continuing during the drugs application, in order to block the possible effects of the compounds on the GABA-receptor complex and allow for a delineation of their principal activities at the level of the melatonin receptor.

Determination of the biological activity *in vitro*

Determination of the effects of all compounds on the forskolin-stimulted cAMP accumulation in neonatal rat adenopituitaries (Stankov et al., manuscript in preparation) was used as a means to determine the activity of the compounds at the receptor level.

RESULTS AND DISCUSSION

The affinities of the synthesized molecules and their apparent biological activities, determined independently by using the experimental model systems described above, in comparison to previously developed compounds, and the native indole are reported in Table.

The greater part of the new analogues expressed an increase in their apparent binding affinity. This increment ranged from average (4a, 4e, 4g, 4h) to very dramatic (4d). A number of compounds, however, expressed apparent loss of affinity, the worst decrement being registered in the case of 4f. Among the substituted tryptamines examined, the affinity for the melatonin receptor decreased with the bulk and the lypophilicity of the alkyl substituents, as shown by the trend: methyl (4a) > isopropyl (4b) > cyclohexyl (4c). From a careful analysis of the data, it became clear that only few substitutions at C_{-2} of the indole ring have resulted in increased affinity: halogenation, methylation and introduction of aromatic ring (4a, 4d, 4k, 4r). Moreover, only these substitutions (4e, 4g, 4h, 4m) were able to counteract to a great extent the decrease in the affinity resulting from introduction of cyclopropyl at R_1. In other words, introduction of cyclopropyl at the *N*-acyl of the lateral chain in all cases led to an average ($2.2 E^{-9}$ for 4j *vs* $1.1 E^{-9}$ for melatonin) or significant ($2.1 E^{-10}$ for 4h *vs* $5.8 E^{-11}$ for 2-bromomelatonin) decrease in the apparent binding affinity of the new compound. This is in clear contrast to the reported affinity value of this molecule (4j: R=H; R_1=cyclopropyl) obtained in a different system, *i.e.* crude membrane preparation enriched in melatonin receptors, isolated from ovine pars tuberalis (Yous et al., 1992). This obvious discrepancy could possibly be due to a so far unsuspected diversity in the structural organization of the high-affinity melatonin receptor in the brain and the pars tuberalis, or to methodological differences in the evaluation of the affinity constants. The kinetic, pharmacological and biochemical characteristics of the high-affinity melatonin receptor in the quail brain and the rabbit cortex are very similar (Stankov et al., 1992); a comparison between these receptor parameters in the sheep pars tuberalis and the avian brain has given essentially the same result (Sugden and Chong, 1991).

The substitution of the 5-methoxy group with a hydrogen led to a dramatic loss of affinity (4n), that could be counteracted by introduction of a halogen (4o) or phenyl (4p) substituent at C_{-2} of the indole ring.

It is reasonable to suppose that the phenyl group can be more easily accommodated in a lipophilic pocket of the receptor site, covering the C_{-2} position, than the cyclohexyl group, which has larger dimensions. The increase in the affinity is probably due to a local negative, rather polarizable, charge around the C_{-2} position of the indole nucleus. A similar effect probably occurs with the C_{-2}-halogen derivatives, where the affinity, increasing along the series Cl, Br, I, could be related to a polarizability effect: the more polarizable an atom, the more effectively it binds to the receptor. Of course, the size of the receptor pocket near the C_{-2} plays a role in the complex formation, but the binding affinity to the receptor site can also be influenced by the electronic charge density around the C_{-2}, which may establish a receptor point-interaction, by means of a secondary electrostatic interaction. These ideas are consistent with a recent QSAR study of a series of previously synthesized melatonin analogues (Lewis et al., 1990) good correlation between the molecular polarizability and the affinity.

The extended sequences of experiments regarding the biological activity of the newly-synthesized compounds included three approaches, *in situ in vitro* and *in vivo*, commonly accepted as model systems. All experimental series confirmed that the greater part of the compounds expressed agonist activity. Of all molecules, 4d and 4g only expressed apparent antagonist activity in two of the models. The summary of the results is reported in the Table. Figure 1 shows examples of the data obtained in the Syrian hamster gonadal regression model. Melatonin in both doses of 20 and 200 μg/animal/day, induced complete gonadal regression within six weeks (results of 20 μg dose shown only). The agonist-activity compounds generated the same response, and an excess dose of 200 μg, administered 30 minutes before melatonin (20 μg), was not able to prevent the melatonin-induced gonadal regression (data not shown). On the contrary, two of the molecules tested (4d and 4g) had no effect alone in a dose of 200 μg, but efficiently prevented the aftermath of 20 μg melatonin, injected 30 minutes later. Therefore, their effect was initially evaluated as antagonistic.

However, those compound induced complete gonadal regression in a dose of 20 μg. It should be pointed out here that high doses of melatonin (1000 - 2500 μg/animal/day) do not induce gonadal regression (Chen, 1981), supposedly because of long-term down-regulation of the melatonin receptor. The higher (5-20 times) affinity of 4g and 4d for the melatonin receptor presumably has required lower doses ($\approx$ 10 times) to produce similar "antagonistic" effect. Obviously these compounds are melatonin agonists at the receptor level (see cAMP data), but their high affinity allows to use them eventually as potential melatonin antagonists *in vivo*. These data once more underline the necessity of effectuating both *in vivo* and *in vitro* tests, when evaluating the biological activity of new compounds.

Compounds 4b, 4e and 4j given alone, induced partial gonadal regression, but were unable to prevent the impact of melatonin, and their behavior was evaluated as weak agonistic. The only molecule that expressed no effect alone and did not influence the results of the subsequent melatonin administration was 4f.

Table 1. Relative binding affinities and biological activity of 2-, 2,5- and 2,6-substituted-5-Methoxy-*N*-acyltryptamines for the melatonin receptor

Compound	R	R_1	R_2	R_3	Ki (1/Ka)@	SHGRM	RPCM	cAMP
4a	CH_3	CH_3	H	OCH_3	4.3×10^{-10}	Agonist	Agonist	Agonist
4b	isopropyl	CH_3	H	OCH_3	4.3×10^{-9}	Weak Agonist	Weak Agonist	Agonist
4c	cyclohexyl	CH_3	H	OCH_3	5.3×10^{-9}	Agonist	Agonist	Agonist
4d	phenyl	CH_3	H	OCH_3	5.7×10^{-11}	Antagonist	Mixed Activity*	Agonist
4e	CH_3	cyclopropyl	H	OCH_3	6.3×10^{-10}	Weak Agonist	Weak Agonist	Agonist
4f	isopropyl	cyclopropyl	H	OCH_3	1.8×10^{-8}	No Effect	No Effect	Weak Agonist
4g	phenyl	cyclopropyl	H	OCH_3	2.4×10^{-10}	Antagonist	Mixed Activity**	Agonist
4h	Br	cyclopropyl	H	OCH_3	2.1×10^{-10}	Agonist	Agonist	Agonist
4j	H	cyclopropyl	H	OCH_3	2.2×10^{-9}	Agonist	Weak Agonist	Agonist
4k	Br	CH_3	H	OCH_3	5.8×10^{-11}	Agonist	Agonist	Agonist
4l	Br	CH_3	Br	OCH_3	6.7×10^{-11}	Agonist	Agonist	Agonist
4m	I	cyclopropyl	H	OCH_3	1.0×10^{-10}	Agonist	Agonist	Agonist
4n	H	CH_3	H	H	1.0×10^{-7}	NT	Agonist	NT
4o	Br	CH_3	H	H	4.0×10^{-9}	Agonist	Agonist	Agonist
4p	phenyl	CH_3	H	H	2.1×10^{-9}	Agonist	Agonist	Agonist
4q	H	CH_3	Cl	OCH_3	2.2×10^{-9}	Agonist	Agonist	Agonist
4r	I	CH_3	H	OCH_3	2.1×10^{-11}	Agonist	Agonist	Agonist
melatonin	H	CH_3	H	OCH_3	1.1×10^{-9}	Agonist	Agonist	Agonist

@ K_i's are in moles/L and are the means of five to nine indipendent determinations, derived from calculations using non-linear fitting strategies (LIGAND). 2-[^{125}I]-iodomelatonin was used as a labelled ligand in the series of saturation and competition experiments. The Kd was calculated simultaneously from the series of saturation experiments, and the mean value was 20 pM. SHGRM, Syrian hamster gonadal regression model; RPCM, rabbit parietal cortex model; cAMP, Inhibnition of forskolin-stimulated cAMP accumulation in neonatal rat adenopituitaries. *Behaves as antagonist in 80%, and **60% of the tested neurons, respectively; NT, not tested.

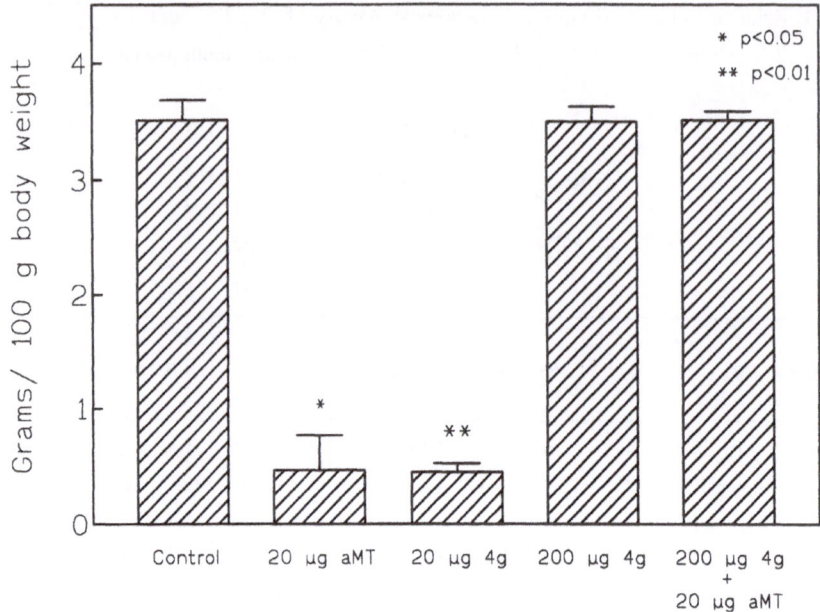

Figure 1. Examples of the effects of late-afternoon treatment with melatonin (aMT) and a putative antagonist (4g) in two different doses (20 and 200 μg/animal/day) on the paired testes weights of Syrian hamsters kept in 14:10 LD photoperiod. Note that 4g expresses agonistic activity at the low dose, while clearly antagonistic properties emerge with the higher dose.

     The cAMP data confirmed that all compounds are melatonin receptor agonists at the receptor level (see Table). The compound 4f that had no activity in the SHGR model behaved as a weak melatonin agonist *in vitro* in the cAMP assay. Thus, the number of melatonin receptor agonist is steadily growing. No data on a melatonin receptor antagonist that behaves as such in all systems available has been published to date.

     In conclusion, three important points emerge from the present study. First, the *N*-acetyl group is important for both, affinity for the receptor and the biological activity. Clearly, cyclopropyl substitution has always led to a decreased affinity and somewhat weaker activity, suggesting that there is probably a small hydrophobic pocket in the receptor site, limiting the size of the *N*-acyl group. Second, methylation, halogenation and substitution with an aromatic ring at C_{-2} of the indole nucleus has led to a dramatic increase in the binding affinity; the biological activity, however, varied greatly, according to the model used, and while all compound are agonist at the receptor level, *in vivo* antagonistic activity was obtained in the cases of introduction of a full aromatic ring. Isopropyl substitution at C_{-2} position has brought a decrease in both, affinity and agonist biological activity. Third, the 5-methoxyl group of melatonin is a prominent factor for the binding affinity, but is not an essential requirement for agonist biological activity.

REFERENCES

Armstrong, S.M. and Redman, J.R., 1993, Melatonin and circadian rhythmicity, in: "Melatonin: Biosynthesis, Physiological Effects and Clinical Application", H-S. Yu and R.J. Reiter, Eds., CRC Press, Boca Raton, chapter 8, 187.

Buchi, G. and Mak, C.P., 1977, Nitro olefination of indoles and some substituted benzenes with 1-Dimethylamino-2-nitroethylene, *J. Org. Chem.* 42: 1784.

Chen, H.I., 1981, Melatonin: failure of pharmacological doses to induce testicular atrophy in the male golden hamster, *Life Sci.* 28: 767.

Copinga, S., Tepper, P.G., Grol, C.J., Horn, A.S. and Dubocovich, M.L., 1993, 2-Amido-8-methoxytetralins: a series of nonindolic melatonin-like agents, *J. Med. Chem.* 36: 2891.

Cozzi, B., Stankov, B., Viglietti-Panzica, C., Capsoni, S., Aste, N., Lucini, V., Fraschini, F. and Panzica, G.C., 1993, Distribution and characterization of melatonin receptor in the brain of the Japanese quail, *Coturnix japonica*, *Neurosci. Lett.* 150: 149.

Dubocovich, M.L., 1988, Luzindole (N-0774): a novel melatonin receptor antagonist, *J. Pharmacol. Exp. Ther.* 246: 902.

Duranti, E., Stankov, B., Spadoni, G., Duranti, A., Lucini, V., Capsoni, S., Biella, G. and Fraschini, F., 1992, 2-Bromomelatonin: synthesis and characterization of a potent melatonin agonist, *Life Sci.* 51: 479.

Heward, C.B. and Hadley, M.E., 1975, Structure activity relationships of melatonin and related indoleamines, *Life Sci.* 17: 1167.

Houlihan, W.J., Parrino, V.A. and Uike, Y., 1981, Lithiation of N-(2-Alkylphenyl)alkanamides and related compounds. A modified Madelung indole synthesis, *J. Org. Chem.* 46: 4511.

Krause, D.N. and Dubocovich, M.L., 1991, Melatonin receptors, *Annu. Rev. Pharmacol. Toxicol.* 31: 549.

Lerner, A.B., Case, J.D., Takahashi, Y., Lee, T.H. and Mori, W., 1958, Isolation of melatonin, the pineal gland factor that lightens melanocytes. *J. Am. Chem. Soc.* 80: 2587

Lewis, D.F.V., Arendt, J. and English, J., Quantitative structure-activity relationships within a series of melatonin analogs and related indolealkylamines, *J. Pharmacol. Exp. Ther.* 252: 370.

Lewy, A.J., Ahmed, S., Latham, J.M. and Sack, R.L., 1992, Melatonin shifts human circadian rhythms according to a phase-response curve, *Chronobiol. Int.* 9: 380.

Petrie, K., Conaglen, J.V., Thompson, L. and Chamberlain, K., 1989, Effect of melatonin on jet lag after long haul flights, *Br. Med. J.* 298: 705.

Reiter, R.J., 1983, Melatonin as the hormone that mediates the effects of the pineal gland on neuroendocrine-reproductive axis of the Syrian hamster, in: "The Pineal Gland and Its Endocrine Role; J. Axelrod, F. Fraschini and G.P. Velo, Eds, Plenum Press, New York-London, 317-330.

Reiter, R.J., 1991, Pineal melatonin: cell biology of its synthesis and of its physiological interactions, *Endocr. Rev.* 12: 151.

Reppert, S.M., Weaver, D.R., Rivkees, S.A. and Stopa, E.G., 1988, Putative melatonin receptors in human biological clock. *Science* 242: 78.

Spadoni, G., Stankov, B., Duranti, A., Biella, G., Lucini, V., Salvatori, AA. and Fraschini, F., 1993, 2-Substituted 5-methoxy-*N*-acyltryptamines: Synthesis, binding affinity for the melatonin receptor and evaaluation of the biological activity, *J. Med. Chem.* 36: 4069.

Stankov, B., Biella, G., Panara, C., Lucini, V., Capsoni, S., Fauteck, J., Cozzi, B. and Fraschini, F., 1992, Melatonin signal transduction and mechanism of action in the central nervous system: using the rabbit cortex as a model, *Endocrinology* 130 (4): 2152.

Stankov, B., Cozzi, B., Lucini, V., Fumagalli, P., Scaglione, F. and Fraschini, F., 1991, Characterization and mapping of melatonin receptors in the brain of three mammalian species: rabbit, horse and sheep, *Neuroendocrinology* 53: 214.

Stankov, B., Fraschini, F. and Reiter, R.J., 1993, The melatonin receptor: distribution, biochemistry, and pharmacology, in: " Melatonin: Biosynthesis, Physiological Effects and Clinical Application, H-S. Yu and R.J. Reiter, Eds., CRC Press, Boca Raton, chapter 7, 155-186.

Sudgen, D. and Chong, N.W.S., 1991, Pharmacological identity of 2-[125I]iodomelatonin binding sites in chicken brain and sheep pars tuberalis, *Brain Res.* 539: 151.

Voordouw, B.C.G., Euser, R., Verdonk, R.E.R., Alberda, B.TH., De Jong, F.H., Drogendijk, A.C., Fauser, B.C.J.M. and Cohen, M., 1992, Melatonin and melatonin-progestin combinations alter pituitary-ovarian function in women and can inhibit ovulation, *J. Clin. Endocrinol. Metab.* 74: 108.

Waldhauser, F. and Steger, H., 1986, Changes in melatonin secretion with age and pubescence, *J. Neural Transm. Suppl.* 21: 183.

Wehr, T.A., 1991, The durations of human melatonin secretion and sleep respond to changes in daylenght (photoperiod), *J. Clin. Endocrinol. Metab.* 73: 1276.

Wetterberg, L., Beck-Friis, J. and Kiellman, B.F., 1990, Melatonin as a marker of a subgroup of depression in adults, in: " Biological rhythms, mood disorders, light therapy and the pineal gland; M. Shafii and S.L. Shafii, Eds, American Psychiatric Press, Washington, p. 69.

Yous, S., Andrieux, J., Howell, H.E., Morgan, P.J., Renard, P., Pfeiffer, B., Lesieur, D. and Guardiola-Lemaitre, B., 1992, Novel naphthalenic ligands with high affinity for the melatonin receptor, *J. Med. Chem.* 35: 1484.

DEVELOPMENT OF MELATONIN ANALOGS

Philippe Delagrange, Pierre Renard, Daniel Henri Caignard,
and Béatrice Guardiola-Lemaître

I.R.I.S.
6, Place des Pléiades
92415 Courbevoie Cedex
France

INTRODUCTION

The nocturnal secretion of melatonin from the pineal gland is regulated by circadian and seasonal variations in daylength mediated via visual projections to the suprachiasmatic nucleus (SCN) which functions as a circadian clock in mammals. Melatonin in turn may regulate daily and / or seasonal physiological processes, in part at least, through feedback to the SCN or to other neuronal inputs. The modification (daylength variations, ageing) or the lack (blindness) of one parameter could alter the behaviour and result in the desynchronization, temporary (shiftworkers, jet lag, Seasonal Affective Disorders) or definitive (elderly, blind people) of different circadian rhythms, and in particular the wake / sleep cycle.

Experiments in animals and in humans have shown that melatonin is a potent "chronobiotic" which can resynchronize the dissociated wake / sleep cycle and is consequently suggested for use in the treatment of sleep disorders (Armstrong et al., 1991).

CHEMISTRY

The purpose is to synthesize melatonin analogs with a longer half-life than melatonin and a better bioavailability when orally administered. The synthesis of compounds has been carried out where an indole ring, a target of metabolic degradation is replaced by bioisosteric systems such as naphtalene, benzofuran, benzothiophene or benzimidazole.

The Pineal Gland and Its Hormones
Edited by F. Fraschini *et al.*, Plenum Press, New York, 1995

The elements of pharmacomodulation mainly concern the acylamino group of the side chain and the ether function located in the 5 position of the indole ring (Yous et al., 1992; 1994).

This programme has already led to the synthesis of more than one hundred and fifty compounds. These compounds are all evaluated on the melatonin binding sites of the ovine pars tuberalis (PT) membranes using $[^{125}I]$-2-iodo melatonin (65 pM) as a ligand according to the method described by Morgan et al., 1989. The drugs are tested in triplicate at varying concentrations (10^{-14}-10^{-4} M final concentration). Initially, binding parameters for a 1-site mass-action model were determined, after a while, we observed after testing about fifty compounds for which the single-site binding model was inadequate for describing the results obtained on some of them. Subsequently, data from individual experiments were normalized, pooled and analysed by a 4-parameter logistic regression to obtain an IC_{50} or by a 5-parameter regression model which describes biphasic inhibition and gives two inhibition constants K_H and K_L (H. Howell et al., in press). These binding experiments allow us to point out some structural activity relationships (Yous et al., in press). The replacement of the indole moiety of melatonin ($IC_{50} = 1.6.10^{-10}$ M) by a napthalene ($IC_{50} = 7.6.10^{-11}$ M) gives a similar binding whereas only a slight decrease of potency is observed for benzothiophene ($IC_{50} = 1.2.10^{-9}$ M) or benzofurane ($IC_{50} = 1.7.10^{-9}$ M) isosters. On the other hand, the benzimidazole analog has a poor affinity ($IC_{50} = 5.3.10^{-6}$ M).

Melatonin
$IC_{50} = 1.6.10^{-10}$ M

S 20098
$IC_{50} = 7.6.10^{-11}$ M

X = S $IC_{50} = 1.2.10^{-9}$ M
X = O $IC_{50} = 1.7.10^{-9}$ M

$IC_{50} = 5.3.10^{-6}$ M

The modifications of the side chain induce in some cases a biphasic binding profile. Nevertheless, the number of compounds exhibiting this kind of binding is not yet sufficient to establish a structure relationship on this characteristic. In order to compare these compounds to the others, we assumed that the K_H for the high affinity binding site or state could be assimilated to the IC_{50} for the other compounds.

Following this assumption, we observe the role played by the N-acyl side chain size for the binding affinity. For the linear acyl substituent, the optimal length corresponds to the ethyl, propyl and butyl whereas the affinity is lower for the methyl and pentyl and weak for the hexyl. Concerning the cycloalkyl substituents, the best affinity is observed for the cyclopropyl and the cyclobutyl while the bulkier groups cyclopentyl and cyclohexyl have weak affinities.

	Alkyl substituents	
R	IC_{50} (M)	K_H (M)
CH_3	$7.6.10^{-11}$	
$-C_2H_5$		$2.8.10^{-12}$
nC_3H_7	$3.4.10^{-12}$	
nC_4H_9		6.10^{-12}
nC_5H_{11}		$8.1.10^{-10}$
nC_6H_{13}	$> 10^{-5}$ M	

	Cycloalkyl substituents	
R	IC_{50} (M)	K_H (M)
C_3H_5		$6.8.10^{-14}$
C_4H_7		$4.2.10^{-13}$
C_5H_9	$3.1.10^{-7}$	
C_6H_{11}	$4.4.10^{-7}$	

	Aromatic substituents
R	IC_{50} (M)
C_6H_5	$2.5.10^{-6}$
CH_2-(C_6H_5)	1.10^{-5}
$(CH_2)_2$-C_6H_5	$2.6.10^{-6}$
$(CH_2)_3$-C_6H_5	$2.4.10^{-6}$

These results predict that the receptor site would have a hydrophobic pocket of a relatively small size which is important to ligand binding.

It is known that the melatonin activity is partially dependent on the 5-methoxy group (Heward and Hadley, 1975). In order to confirm this observation on the naphtalenic analogs, the 7-methoxy group is replaced by an hydrogen atom. This modification decreases the binding affinity, confirming the results obtained on the indole nucleus.

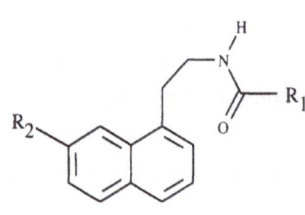

R_1	R_2	IC_{50}	K_H
C_3H_5	OCH_3		$6.8.10^{-14}$ M
C_3H_5	H	$8.4.10^{-8}$ M	
C_4H_7	OCH_3		$4.2.10^{-13}$ M
C_4H_7	H	$9.8.10^{-7}$ M	

Among all the compounds synthetized, some of them exhibit a clear biphasic binding curve (figure 1). This is the case for example for the napthalenic analogs of melatonin substituted by an unsaturated acyl substituent (-CH = CH-CH$_3$, $K_H = 3.2.10^{-15}$ M, $K_L = 7.1.10^{-8}$ M) (Howell et al, in press).

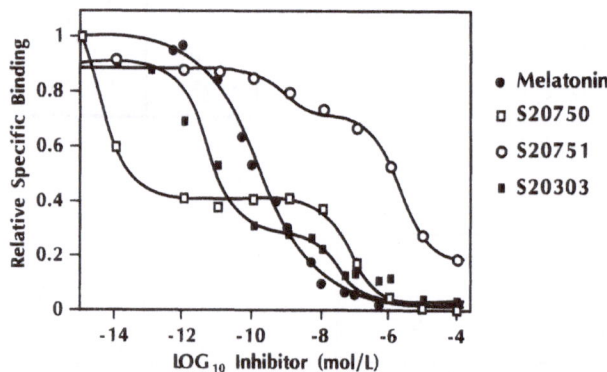

Figure 1 : Biphasic inhibition of [^{125}I] Iodomelatonin binding to ovine PT membranes by melatonin and various melatonin analogs. The chemical modifications of the acylamino group (R) of the side chain for these naphtalene compounds are the following :

S 20750 : R : -CH = CH-CH$_3$; S 20751 : R : -(CH$_2$)$_4$-CH$_3$; S 20303 : R : -(CH$_2$)$_3$-CH$_3$

About twenty of the analogs exhibit a similar profile. These data suggest either the presence of two distinct receptor subtypes in ovine pars tuberalis cells, or a detection of the receptor in two different affinity states. Further experiments using guanine nucleotides with a view to modulating the proportion of receptors in the high and low affinity state should clarify this point.

This first stage of screening led to the identification of thirty-two compounds more potent than melatonin on the ovine pars tuberalis binding sites and twenty-one compounds exhibiting biphasic binding curves.

The binding studies are completed by in vitro experiments on ovine pars tuberalis culture to characterize the activity of all these compounds on the cAMP production induced by forskolin. The drugs are tested for their agonist / antagonist activity following the method described by P. Morgan et al., (1989). The drugs (1.10^{-5} M) are tested alone and are compared to melatonin (1.10^{-9} M) for their inhibition potency of forskolin (1.10^{-6} M) stimulated cAMP production or in combination with melatonin to detect an antagonist activity. These experiments are carried out in triplicate. Two different cAMP indexes are then determined, one for the drug itself cAMP[D] and the other for the drug in combination with melatonin cAMP[D/M]. They are calculated from the following formulae :

cAMP[D] = ([F] - [F/D]) / ([F] - [F/M])

cAMP[D/M] = ([F] - [F/D/M]) / ([F] - [F/M])

where

[F] : cAMP levels in cells stimulated with 1 μM forskolin

[F/D] : cAMP levels in cells stimulated with 1 μM forskolin in the presence of 1 or 10 μM drug

[F/M] : cAMP levels in cells stimulated with 1 μM forskolin in the presence of 1 or 10 nM melatonin

[F/M/D] : cAMP levels in cells stimulated with 1 μM forskolin in the presence of 1 or 10 nM melatonin and 1 or 10 μM drug.

These experiments are carried out only at one concentration giving consequently only semi-quantitative information.

A compound which fully mimicks melatonin has cAMP[D] and cAMP [D/M] equal to 1 : the table hereunder summarizes the criteria fixed to define the compounds.

cAMP[D] index	cAMP[D/M] index	activity*
> 0.8	> 0.8	Potent agonist
0.4 < < 0.8	> 0.8	Weak agonist
0.4 < < 0.8	< 0.8	Partial agonist
< 0.4	0.2 < < 0.8	Weak antagonist
< 0.4	< 0.2	Potent antagonist
< 0.4	> 0.8	No activity

* activity on the melatonin receptor coupled to cAMP transduction

It appears that the replacement of the indole group by naphtalene, benzofurane, benzothiophene or benzimidazole isosters does not modify the agonist character of these compounds. The modifications of the side chain as previously described do not bring about major modifications of the character of the activity. On the contrary, the replacement of the 7-methoxy group by hydrogen not only reduces the activity but also in some cases gives an antagonistic activity as revealed by the quite low cAMP[D] and cAMP[D/M] indexes.

R_1	R_2	cAMP [D] index	cAMP [D/M] index]
C_3H_5	OCH_3	1.02	1.03
	H	0.32	0.28
C_4H_7	OCH_3	0.46	0.71
	H	- 0.1	0.01

In the cAMP studies on ovine PT cell culture, seven compounds have been characterized as antagonists.

The agonist or antagonist properties have been confirmed on the *Xenopus laevis* melanophores. In this model, melatonin induces an aggregation of pigment granules in the cultured melanophores (Sugden, 1991) with a maximal effect at 10 nM. The drugs are tested for their activity at concentrations ranging from 10^{-13} M to 10^{-4} M alone for the agonistic property or after aggregation of the pigments by melatonin 10^{-8} M (maximal effect).

For all the compounds tested on this model, the agonist or antagonist activities determined on the OPT models are confirmed. For example, we observed the same consequences of the replacement of the methoxy group in position 7 by a hydrogen atom : the compounds become antagonist.

R_1	R_2	Xenopus laevis
C_3H_5	OCH_3	Agonist
	H	Antagonist
C_4H_7	OCH_3	Agonist
	H	Antagonist

CHRONOBIOTIC PROPERTIES

According to Armstrong (1991), "a chronobiotic is a chemical substance capable of therapeutically re-entraining short-term dissociated or long-term desynchronized circadian rhythms or prophylactically preventing their disruption following environmental insult". Such a compound must have some direct effects on the organism circadian clock, the SCN, and be active in animal models of circadian rhythm disturbances.

S 20098 (N[2-(7-methoxynapht-1-yl) ethyl]) acetamide), the strict naphtalenic analog of melatonin has been tested for these different criteria, in vitro, on SCN neurons in binding and in electrophysiology experiments, in vivo in different models of circadian rhythm disturbances.

This melatoninergic compound has two times a higher affinity than melatonin for the ovine PT binding site ($IC_{50} = 7.6 \pm 3.1.10^{-11}$ M) (Yous et al., 1992) and 17 times for high affinity states of the chick retina binding site ($K_{iH} = 2.1 \pm 1.0.10^{-12}$ M) (Dubocovich et al., 1993). It behaves like an agonist on three different models : the ovine PT cell culture (Yous et al., 1992), the *Xenopus* melanophore pigment aggregation and the rabbit retina (Dubocovich et al., 1993).

SCN Binding

S 20098 has been tested for its ability to displace ^{125}I-2-Iodo melatonin from its binding sites in the SCN (Bonnefond et al., 1993).

These experiments were done on SCN sections of two month-old male Long Evans rats killed by decapitation one hour before lights were turned off. Slices were incubated in Tris-HCl buffer containing 25 pM [^{125}I]-2-iodo melatonin (Amersham, specific activity : 2000 Ci/mmol) and increasing concentrations from 10^{-12} to 10^{-6} M of either non labelled S 20098 or melatonin for two hours at room temperature. Autoradiograms were obtained after apposing the films for four days.

Increasing concentrations of S 20098 or melatonin led to a total displacement of the binding when concentrations reached 10^{-9} M. S 20098 and melatonin have the same potency to displace 2-iodo melatonin from its binding sites. The same properties are observed on the ovine pars tuberalis.

SCN Electrophysiology

The SCN neuronal activity in brain slices is acutely depressed by melatonin application only during late subjective day (Mason and Brooks, 1988, Shibata et al., 1989). Moreover, the SCN neurons recorded in vitro exhibit a circadian rhythm in discharge activity which peaks during the subjective light phase of the circadian cycle and can be maintained for up to three successive days (McArthur et al., 1991). This circadian peak can be displaced by a pharmacological challenge. The effects of S 20098 are compared to melatonin both on the spontaneous discharge and on the circadian rhythm of the neuron activity (Delagrange et al., 1993; Mason et al.,1993).

Methods. Male Lister rats were housed in a 12hr light-12hr dark cycle (LD 12:12), with computer monitoring of locomotor activity in a dedicated holding room for assessment of animal entrainment to the LD cycle. At random times during different phases of the circadian cycle, animals were anesthetized with halothane and decapitated. The brain was trimmed to a block containing the hypothalamus and coronal hypothalamic slices (450-500 μm) containing the SCN cut with a mechanical chopper. Slices are transferred to a brain slice chamber and maintained at $35\pm0.5°$ C at the gas / fluid interface between warm humidified O_2 - CO_2 (95 % : 5 %) and artificial cerebrospinal fluid using a peristaltic pump. Extracellular recordings from single neurons were obtained with single-barrel micropipettes containing 2 M NaCl. Micropipettes are guided to the SCN under visual control with the aid of a stereo-microscope. The discriminative action potential spikes were integrated over successive five-second episodes. For studies on the effect of melatonin or S 20098 on spontaneous discharge activity the neurons were recorded for as long as possible. Melatonin, S 20098, 2-Iodo melatonin and 5HT were successively applied in the artificial cerebrospinal superfusate. Between each application the medium in the brain slice chamber was completely replaced.

On the other hand, for studies on the circadian rhythm in spontaneous discharge activity, neurons were recorded only during 5-10 min before hunting another one. For each slice, we determine on the first day the time peak of activity. Then the drug is applied either at different concentrations to establish a dose response curve or at different circadian time (CT) to determine the phase response curve. The second day under each condition, the phase shift of the peak of discharge activity was determined in direction (advance or delay) and in time. Each hypothalamic slice is used for one condition.

Spontaneous Discharge Activity. Application of the melatonin analog S 20098 suppressed the discharge activity of 37 % spontaneously SCN active neurons, increased the discharge of 5 % and had no activity on 58 % of neurons. The table below summarizes the results obtained with melatonin, 2-iodo melatonin, serotonin and S 20098.

	Suppression	Activation	Non responsive
S 20098	16 cells	2 cells	43 cells
Serotonin (5HT)	23 cells	4 cells	12 cells
Melatonin	32 cells	3 cells	57 cells
2-iodomelatonin	9 cells	0 cell	17 cells

The inhibitory potency of S 20098 and melatonin are quite similar and they both showed a window of sensitivity between CT8 and CT12 corresponding to the subjective light / dark transition. The application of S 20098 (10^{-6} M) at CT07 had no effect on the spontaneous discharge activity. The excitatory effect of both S 20098 or melatonin observed in a small proportion of neurons could be due to a direct or indirect effect on interneurons.

Circadian Discharge Rhythm

The spontaneous discharge activity is at a maximum during the subjective light phase and peaks at CT7 over the three consecutive days. When S 20098 or melatonin are delivered over one hour between CT 08.00-12.00 on the second day of recording, they both advance the peak of discharge activity. This advance is also maintained the day after. A phase response curve (PRC) indicating the shift in time of the peak in SCN neuronal discharge rhythm (figure 2) is obtained by delivering for one hour the drug at different CT. This advance effect is observed only between CT8 and CT12 with a maximum (3.5-4h advance) at CT10 and is dose dependent beginning at 10^{-10} M (2h advance) and reaches the maximum effect at 10^{-9} M. These experiments prove that S 20098 like melatonin is able to reset in vitro the SCN activity.

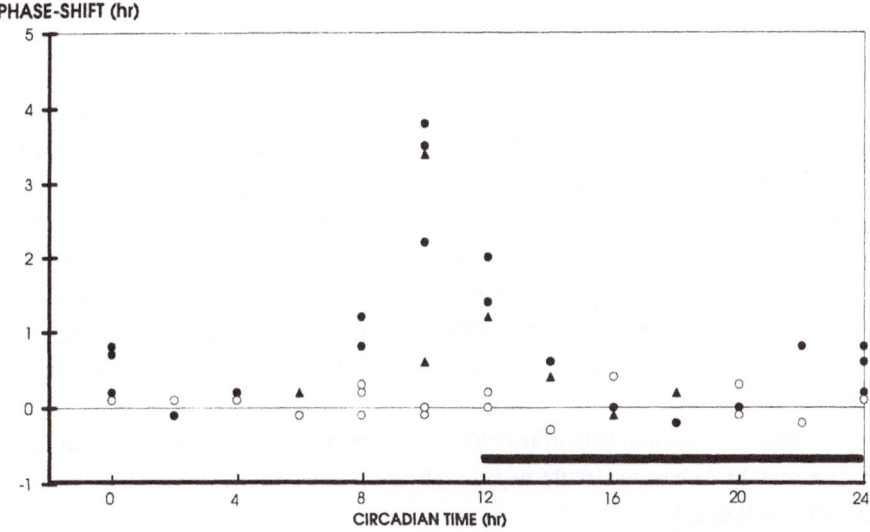

Figure 2 : Phase response curve indicating the shift in time of peak in SCN neuronal discharge rhythm on rat hypothalamic slice induced by 1 hr delivery of vehicle O, melatonin (10^{-8} M) ● and S 20098 (10^{-8} M) ▲ The temporal window of sensitivity for S 20098 or melatonin occurs between CT08-12.

In Vivo Effect On Three Models Of Circadian Rhythm Disturbances

To confirm the entrainment effect in vivo, we have tested the activity of S 20098 in different models of circadian rhythm disturbances in rats. These models are considered to reproduce at least partly the disrupted circadian wake / sleep activity observed in humans (Armstrong, 1991) :

- free-running rats maintained in constant darkness (as a therapeutic model of non-24 hour wake / sleep disorders);
- eight-hour phase advance shift (as a therapeutic model of shift work wake / sleep disorders);
- a negative phase angle (as a model of delayed sleep phase syndromes).

Free-running. In the free-running conditions, rats are first habituated to a 12:12 LD cycle until stable entrained rhythms are established. The locomotor activity and the body temperature are recorded by transmitters implanted under anesthesia.

Rats are then placed in constant very dim, red light conditions (L 0.1 lux) to induce free-running rhythms. When this rhythm is well established, rats are treated every day at the same time with S 20098 (100 µg.kg^{-1}, IP) or the vehicle (DMSO). S 20098 re-entrains both the locomotor activity and body temperature in all the rats treated. This effect appears when the time of injections coincides with the onset of locomotor activity (Delagrange et al., 1993). The same effect is observed in the case of wheel running activity records (Bonnefond et al., 1993). In both experiments, the vehicle has no effect.

The free-running model reveals the ability of S 20098 to re-entrain a circadian rhythm but not the possibility of occuring on the direction of re-entrainment.

8-hr Phase Advance. This activity is tested on the eight hours phase advance model. Rats habituated to a 12:12 LD cycle are submitted to an eight hours advance of the LD cycle. On the day of the phase shift, rats are injected just before dark onset and for a number of days following the shift with S 20098 or melatonin. The control or vehicle treated rats re-entrain by delaying their activity rhythm. On the contrary, S 20098 (1 mg.kg^{-1} IP) or melatonin (1 mg.kg^{-1}IP) re-entrain by advancing the rhythm. The effect of S 20098 is dose dependent with 100 % of re-entrainment at 100 µg.kg^{-1} and requires seven days of injections for all rats to respond (Redman and Guardiola-Lemaître, 1993).

S 20098 is therefore able to re-entrain a rhythm and to modify the direction of re-entrainment. In the next model, the negative phase angle, the ability of S 20098 to induce a new rhythm is tested.

Negative Phase Angle. Rats are maintained in constant darkness (DD) and free-running conditions for several months. When returned to a 12:12 LD cycle, some animals show a lag in the onset of activity behind the onset of darkness by three-four hours, showing the negative phase angle (PAD) (Armstrong, 1991). The animals which exhibited a large, negative PAD were selected to take part in the experiments. The stability of the PAD is followed for at least 23 days.

S 20098 or melatonin both at 1 mg.kg^{-1} S.C. given 0.5 h before the dark onset advance over approximatively nine days the onset of activity towards the onset of darkness. At cessation, the activity onset delays over approximatively eleven days but does not reach the initial PAD (Armstrong et al., 1993).

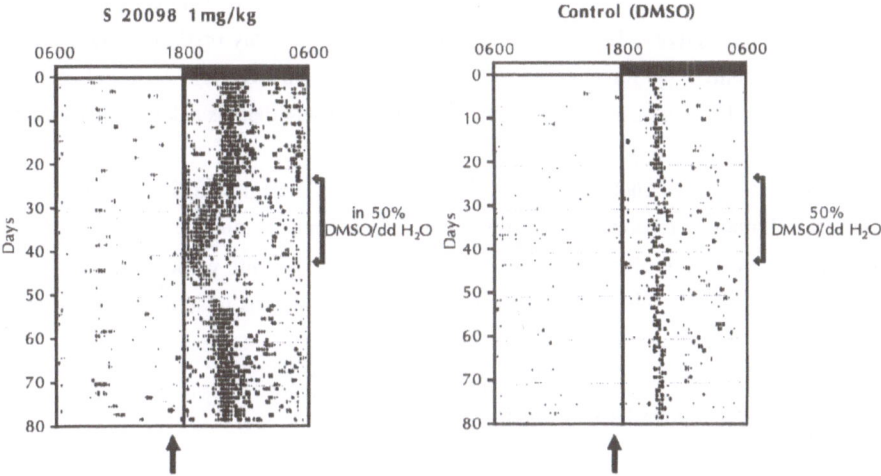

Figure 3 : Actograms of two rats treated with S 20098 (1 mg.kg^{-1}) or the vehicle (DMSO 50 %). Rats were injected in the last 30 minutes of the light phase (indicated by vertical arrow actogram) for twenty-two days (indicated by horizontal arrows on right-hand side of actogram). Light on 06.00 - light off 18.00.

The lack of sedative effects of the compounds at pharmacological active doses observed in behavioural studies has been confirmed in EEG studies (Tobler et al., 1994).

Adult, male Sprague-Dawley rats are housed individually and maintained under LD 12:12 for twenty days. Electrodes for EEG, EMG and a termistor for cortical temperature (TCRT) are implanted under anesthesia seven to nine days before the experiments. The rats are recorded for the first six hours of the twelve hours light period on four separate days. The first day served as a baseline followed by the IP injection of either S 20098

(3 mg.kg^{-1}) or vehicle (50 % DMSO) on day two and day six. On day eight, all animals received melatonin (3 mg.kg^{-1} IP). All data were analysed for eight-second episodes and for each episode EEG spectra in range of 0.25-25 Hz and the integrated full wave rectified EMGs were determined. The vigilance states waking, REM sleep (rapid eye movement sleep) and non-REM sleep were determined on the basis of the EEG and EMG records.

The effects of the factors "time" and "conditions" on the vigilance states and EEG power spectrum were assessed by an analysis of variance (ANOVA) and a two-tailed paired t-test. Compared to the vehicle injection, S 20098 or melatonin reduced the power density in non-REM sleep in the low frequency range (1-8 Hz) but did not affect the vigilance states and the brain temperature. These results demonstrate that both S 20098 and melatonin, at doses where they have a resynchronizing effect, have no hypnotic properties.

To summarize, S 20098 a potent melatoninergic agonist, 1) in vitro displaces ^{125}-I-2-iodo melatonin from its binding sites in the SCN, 2) in vitro induces a shift in time of peak in SCN neuronal activity, 3) in vivo entrains the locomotor activity rhythm in free-running conditions, modifies the direction of re-entrainment after the LD cycle shift and corrects the anomaly of locomotor activity rhythm in conditions where there is a delay in relation to the dark onset.

All these results obtained in different animal models are relevant for the therapeutic use of a "chronobiotic" compound (Armstrong, 1991). Their therapeutic interest could be used in the treatment of circadian sleep disorders such as shift work sleep disorders, delayed sleep phase syndromes or irregular sleep / wake patterns relating to blindness or ageing.

Circadian Disturbances Due To Ageing

Circadian rhythms deteriorate with ageing (Van Gool and Mirmiran, 1986). The amplitude of the rhythms reduces, leading first to a decrease of the stability and finally to a total disruption whereby the rhythms become aperiodic (Satinoff et al., 1993).

Many reasons have been suggested for these disturbances, such as degeneration, reduced activity of the SCN neurons, alteration of light imput to SCN and decrease of melatonin synthesis. Considering all these reasons, a chronobiotic should improve some aspects of the circadian rhythm disorders. For that reason, S 20242 [N-[2-(7-methoxy napht-1-yl) ethyl]] propionamide a potent melatonin agonist on the ovine PT model (K_H = $2.8.10^{-12}$ M) has been tested on body temperature and locomotor activity rhythms in aged rats (Koster van Hoffen et al., 1993).

Young (6-8 months), middle-aged (17-20 months) and old (27-30 months) male Brown Norway rats were implanted with EMG electrodes and a subcutaneously placed thermistor to record the circadian rhythms of wake / sleep activity and body temperature. Rats are housed in a 12:12 LD cycle and are chronically treated for fourteen days at the dark onset with S 20242 (1 mg.kg^{-1}, S.C.) The amount of activity is measured by integration of the EMG activity. The 24 hour values of the Chi-square periodogram, the light / dark differences and the hourly values are compared by means of ANOVA.

As expected, there is a significant reduction in the amplitude and stability of body temperature and activity in both middle-aged and old rats. S 20242 treatment restores the stability of the body temperature in both middle-aged and old rats but does not improve the locomotor activity rhythm. S 20242 increases the amplitude of body temperature rhythm in middle-aged rats to the level of young rats but does not change the amplitude in the young and aged groups.

The fact that S 20242 increased both amplitude and stability of the circadian rhythm of only body temperature in middle-aged rats and only the stability in old rats could be the consequence of a loss of melatonin receptors and / or preferential influence analog for some subtypes of receptors and could explain the lack of beneficial effect on activity.

Another explanation could be that this analog does not interfere directly on the clock, but acts at a projecting area level, for example on the preoptic area of the hypothalamus.

In any case, such properties should be very useful in treating circadian rhythms disturbances due to ageing, particularly in Alzheimer patients where waking episodes during night and daytime sleepiness are probably caused partly by a decreased of circadian rhythmicity (Ancoli-Israel et al., 1989; Witting et al., 1993).

CONCLUSION

To conclude, this programme has led to the synthesis of potent and specific melatoninergic compounds characterized either as agonists or antagonists. Two agonists S 20098 and S 20242, have shown chronobiotic properties in different in vitro and in vivo models. S 20098 has been selected for therapeutic development. The antagonists are new tools which will be helpful in clarifying the role played by the melatoninergic system in different physiopathological processes. Lastly, the binding data obtained for these compounds has put into evidence the heterogeneity of melatonin receptors, suggesting either the possibility of independent receptor subtypes or different affinity states at the same receptor or a combination of both.

Acknowledgements

The author gratefully acknowledges the secretarial assistance of Miss Orla De Bhal and Miss Michelle Félicier in preparing this chapter.

REFERENCES

Ancoli-Israel, S., Parker, L., Sinae, R., Fell, R.L. and Kripke, D.F., 1989, Sleep fragmentation in patients from a nursing home, *J. Gerontol.* 44:M18-M21.

Armstrong, S.M., 1991, Treatment of sleep disorders by melatonin administration, *Advances in Pineal Research* 6, 263-274, eds. Foldes and R. Reiter. London : John Libbey.

Armstrong, S.M., McNulty, O.M., Guardiola-Lemaitre, B. and Redman, J.R., 1993, Successful use of S 20098 and melatonin in an animal model of delayed sleep phase syndrome [DSPS], *Pharmacol. Biochem. Behav.* 46, 45-49.

Bonnefond, C., Martinet, L., Lesieur, D., Adam, G. and Guardiola-Lemaitre, B., 1993, Characterization of S 20098 : a new melatonin analog., *Melatonin And The Pineal Gland - From Basic Science To Clinical Application*, eds. Y. Touitou, J. Arendt and P. Pévet, 123-126. Amsterdam : Elsevier Science Publishers.

Delagrange, P., Guardiola, B. and Mason, R.,1993, The effects of the new melatonin receptor agonist S 20098 on rat suprachiasmatic nucleus circadian clock neurones in vitro, *Sleep Res.* 22, 614.

Delagrange, P., Mason, R., Redman, J., Defrance, R., Mocaër, E. and Guardiola-Lemaître B.,1993, Effect of S 20098, a potent melatonin agonist on rat temperature and activity circadian rhythm, *Sleep Res.* 22, 616.

Depreux, P., Lesieur, D., Ait Mansour, H., Morgan, P., Howell, H.E., Renard, P., Caignard, D.H., Pfeiffer, B., Delagrange, P., Guardiola, B., Yous, S., Demarque, A., Adam G and Andrieux, J., 1994, Synthesis and structure-activity relationships of novel naphtalenic an bioisosteric related amidic derivatives as melatonin receptor ligands, *J. Med. Chem.* 37, 3231-3239.

Dubocovich, M., 1985, Characterization of a Retinal Melatonin Receptor, *J. Pharamocol. and Exp. Ther.* G., 395-401.

Dubocovich, M.L., North, P.C., Oakley, N.R. and Hagan, R.H., 1993, A potent naphthalenic melatonin receptor agonist, *Melatonin and the pineal gland - from basic science to clinical application*, eds. Y. Touitou, J. Arendt and P. Pévet, 123-126. Amsterdam : Elsevier Science Publishers.

Guardiola-Lemaitre, B., 1991, Development of melatonin analogs, *Advances in Pineal Research* 5, 351-353, eds. J. Arendt and P. Pévet, London : John Libbey

Heward, C.B., Hadley, M.E., 1975, Structure activity relation-ships of melatonin and related indoleamines, *Life Sci.*, 1167-1178.

Howell, H.E., Guardiola, B., Renard, P. and Morgan, P.J., 1994, Naphtalenic ligands reveal melatonin binding site heterogeneity Endocrine 2. *In press.*

Koster van Hoffen, G.C., Mirmiran, M., Bos, N.P.A., Witting, W., Delagrange, P. and Guardiola-Lemaitre, B., 1993, Effects of a novel melatonin analog on circadian rhythms of body temperature and activity in young, middle-aged and old rats, *Neurobiol. Aging*, 14, 565-569.

McArthur, A.J., Gillette, M.U. and Prosser, R.A., 1991, Melatonin directly resets the rat suprachiasmatic circadian clock in vitro. *Brain Res.* 565, 158-161.

Mason, R. and Brooks, A., 1988, The electrophysiological effects of melatonin and a putative melatonin antagonist (N-acetyl-tryptamine) on rat suprachiasmatic neurones in vitro, *Neurosci. Lett.*, 95, 296-301.

Mason, R., Delagrange, P. and Guardiola. B., 1993, Melatonin and S 20098 mediated resetting of rat suprachiasmatic circadian clock neurones in vitro, *Brit. J. Pharmacol.*, 108, 32.

Morgan, P.J., Lawson, W., Davidson, G. and Howell, H.E., 1989, Melatonin inhibits cyclic AMP production in cultured ovine pars tuberalis cells, *J. Molecular Endocrinol.* 3, R5-R8.

Morgan, P.J., Williams, L.M., Davidson, G. and Howell, H.E., 1989, Melatonin receptors on ovine pars tuberalis : characterization and auoradiographical localization., *J. Neuroendocrinol.* 1, 1-40.

Redman, J.R. and Guardiola-Lemaitre, B., 1993, The melatonin agonist : S 20098 : Effects on the rat circadian system, *Melatonin And The Pineal Gland* - From Basic Science To Clinical Application, eds. Y. Touitou, J. Arendt and P. Pévet, 127-130, Amsterdam : Elsevier Science Publishers.

Satinoff, E., Li, H., Tcheng, T.K., Mc Arthur A.J., Medanic M. and Gillette M.U., 1993, Do the suprachismatic nuclei oscillate in old rats as they do in young ones ?, *Am. J. Physiol.*, 265, R1216-R1222.

Shibata, S., Cassone, V.M. and Moore, R.Y., 1989, Effects of melatonin on neuronal activity in the rat suprachiasmatic nucleus in vitro, *Neurosci. Lett.*, 97, 140-144.

Sugden, D., 1983, Psychopharmacological effects of melatonin in mouse and rat, *J. Pharmacol. Exp. Ther.* 227, 587-591.

Tobler, I., Jaggi, K. and Borbely, A.A., 1994, Effects of melatonin and the melatonin receptor agonist S 20098 on the vigilance states. EEG spectra and cortical temperature in the rat., *J. Pineal Res.*, 16, 26-32.

Van Gool, W.A. and Mirmiran, M., 1986, Aging and circadian rhythms, *Progress In Brain Research* 70 : Aging of the Brain and Alzheimer's Disease., 255-279. Amsterdam : Elsevier Science Publishers.

Witting, W., Mirmiran, M., Bos, N.P.A. and Swaab, D.F., 1993, Effect of light intensity on diurnal sleep-wake distribution in young and old rats, *Brain Res. Bull.* 30; 157-162.

Yous, Y., Andrieux, J., Howell, M.E., Morgan, P.J., Renard, P., Pfeiffer, B., Lesieur, D. and Guardiola-Lemaitre, B., 1992, Novel naphtalenic ligands with high affinity for the melatonin receptor. *J. Med. Chem.* 35, 1484-1486.

PRECLINICAL AND CLINICAL PHARMACOLOGY OF MELATONIN AND ITS ANALOGS

Francesco Scaglione[1], Daniele Esposti[2], Marco Mariani[3], Germana Demartini[1], Valeria Lucini[1], Franco Fraschini[1]

[1] Dept. Pharmacology, University of Milan, 20129 Milano, Italy
[2] Inst. Human Physiology II, University of Milan, 20133 Milano, Italy
[3] Geriatric Inst. Redaelli, Vimodrone, Italy

INTRODUCTION

Melatonin is considered the main hormone of the vertebrate pineal gland. Melatonin seems to acts as an endogenous Zeitgeber, the chemical expression of darkness (Reiter, 1991), since its rhythmic release in the peripheral blood and the cerebrospinal fluid at night is probably involved in the resetting of the endogenous biological clock (Arendt, 1988).

Besides the well established function of melatonin for the timing of reproduction in seasonal breeding mammals (Arendt, 1988), there is increasing evidence for a more general physiological role of this substance. Its oncostatic properties (Bartsch and Bartsch, 1981; Blask, 1993) and the recently described scavenger activity (Tan et al., 1993) lead to hypothesize that melatonin may be an antagonist of cell damage induced by oxigen, being particularly important in aging processes.

The exact role of melatonin in human is not completely known, but recent research showed the presence of melatonin receptors in the human circadian clock (Reppert et al., 1988), as well as the ability of humans to respond to the melatonin signal with a circadian phase-response curve (Lewy et al., 1992). Alterations in the melatonin rhythm and/or circulating levels are now thought to be involved in several patho-physiological conditions, such as the seasonal affective disorders (Wetterberg et al., 1990), regulation of the sleep-wake cycles (Wehr,1991), puberty (Waldhauser and Steger, 1986), jet-lag (Petrie et al., 1989), and reproduction (Voordouw et al, 1992). Recently, the synthesis of some melatonin analogs with different receptor affinity, according to the chemical modification of the primary indole structure, has been reported (Dubocovich, 1988; Stankov et al., 1993). The results of the limited number of structure-activity studies showed that the methyl group at C-5 and the acetamidoethyl group at C-3 are probably necessary for the expression of high affinity and biological activity. Additional halogenation at C-2 led to an increased affinity of the analog for the receptor (Dubocovich, 1988; Duranti et al., 1992). With the development in the field of chronobiology and the recently developed studies aimed to investigate the clinical perspectives of melatonin, a better understanding of the fate and distribution of the exogenous hormone and its analogs may accelerate the growth of knowledge concerning its possible therapeutic use.

The Pineal Gland and Its Hormones
Edited by F. Fraschini *et al.*, Plenum Press, New York, 1995

PHARMACOKINETIC PROFILE OF MELATONIN

Melatonin, after oral administration in humans was proven to reach peak serum levels at about 30-60 minutes after the intake depending of the oral preparation and regardless of nutritional status (Wetterberg et al., 1978; Vakkuri et al.,1985; Aldhous et al., 1985). Valdhauser et al., (1984) found that orally administered melatonin causes a very rapid elevation in serum melatonin levels with a plateau persisting for 1.5 hours and absorbtion of ingested melatonin can be extremely different between subjects. Moreover, pharmacokinetics of melatonin resulted to be influenced by a marked first pass metabolism (Lane and Moss, 1985).The elimination of serum melatonin in humans appeared to follow a first order kinetics (Vakkuri et al., 1985) as observed also in rats (Reppert and Klein, 1978; Gibbs and Vriend, 1981) and in the rhesus monkey (Reppert et al., 1979) after administration of ^{3}H melatonin.

\After oral administration of pharmacological doses (80-100 mg) in man, melatonin reached peak serum levels 300-3700 times and six hours after the intake still 10-30 times higher than physiological serum concentrations at night (Wetterberg et al., 1978; Waldhauser et al., 1984; Vakkuri et al., 1985). Therefore it can be concluded that the oral doses employed to evoke physiological responses have been far too large if nocturnal endogenous melatonin concentrations are used as reference. After an oral dose of 2 mg given as different preparations, plasma melatonin concentrations remained at or above endogenous night time levels for 3-4 hours (Aldhous et al., 1985)

The human serum half- life was showed to range from 2 to 44 minutes after oral administration of 80-100 mg (Wetterberg et al., 1985; Waldhauser et al., 1985; Vakkuri et al., 1985) and about 30-40 minutes after oral administration of 2 mg (Aldhous et al 1985). After injection of 5 or 10 ug of melatonin in healthy subjects, melatonin clearance from the blood showed a biphasic pattern, corresponding to a bi-exponential decay with distribution phase mean half-life (T½α) 1.35 minutes and metabolic phase mean half-life (T½β) 28.4 minutes.

The mean apparent volume of distribution at steady- state was 36.7 and the systemic clearance was 966 ml/min (Claustrat et al., 1989; Mallo et al., 1990). These data compared to the recalculated results of Iguchi et al., (1982) showed shorter half-life and higher systemic clearance. During melatonin infusion (20 mg for 5 hours) in healthy volounteers the plasma melatonin concentrations reached the steady-state between 60 and 120 minutes after begining the treatment. The mean plasma melatonin concentration at the plateau was comparable to physiological nocturnal levels (50-120 pg/ml) wich gave a mean systemic clearance of 932-970 ml/min and a mean systemic apparent volume of distribution of 52-63 l (Claustrat et al., 1989; Mallo et al., 1990). At the end of the infusion, plasma melatonin resulted cleared in a bi-exponential manner, as after the bolus injection.

The difference in the results concerning the half-life of melatonin may be , at least in part, due to different route of administration, different method for the determination of melatonin, but mainly to the calculation of this parameter (distribution phase or elimination phase).

The results of some studies performed in laboratory mammals showed that melatonin, after systemic injection, rapidly penetrates into brain and cerebrospinal fluid (CSF) (Vitte et al., 1988; 1990). The heterogeneous distribution and the partial retention of melatonin in the brain evidenced by these results are compatible with the hypothesis of a central action of the hormone mediated via binding sites. Recent results obtained in one healthy volounteer (Le Bars et al., 1991) after bolus i.v. administration of ^{14}C melatonin, showed that this substance readily crosses the blood brain barrier, interacts with brain structures and quickly disappears from the brain, suggesting rapid diffusion and turnover.

Melatonin is metabolized through different ways in the various species. In humans and rodents is metabolized in the liver by 6-hydroxymethylation, and excreted as coniugates of sulfuric (70-80%) and glucuronic acids (5%) (Kveder and McIsaac, 1961). Sulfatoximelatonin is the main plasma and urinary metabolite (Kveder and McIsaac, 1961; Kopin et al.,1961; Arendt et al.,1985; Bojkowski et al., 1987). Morover, in humans and rodents it has been shown that exogenous melatonin demethylation led to the formation of N-acetyl serotonin-glucuronate and sulfate (Leone and Silman, 1984). Exogenous melatonin after oral administration of pharmacological doses (80-100 mg) appeared to be excreted as unchanged in a minimal extent (Wetterberg et al., 1978; Vakkuri et al., 1985).

Recently, some experiments were performed in our laboratories in order to evaluate the effect of age on melatonin pharmacokinetics in humans. Moreover, the pharmacokinetic profile of melatonin after oral ingestion of different doses were studied.

The experiments were conducted in 6 male voluunteers aging from 75 to 92 years, orally treated in the morning with 20 mg of melatonin in single dose and in 10 male volunteers aging from 23 to 31 years, orally treated in the morning with 20, 40, 80 mg of melatonin in three separate experiments, performed after a wash out period of 4 weeks. Every subject gave his informed consent. Blood samples were collected at different times (10, 20 30 40 60 minutes, 1.5, 3, 4.5 and 6 hours) after melatonin intake. Melatonin concentration in serum was determined by radioimmunoassay according to Fraser et al., (1983) with slight modifications, as described by Stankov et al.,(1990). Pharmacokinetic parameters were calculated using a computed two compartment pharmacokinetic model (Gomeni, 1984). The obtained results (Table 1) showed that in elderly melatonin maximal serum concentrations are slightly lower than those found in adults treated with the same oral dose of melatonin but Tmax values resulted significantly higher (1.59 ± 0.66 vs 0.83 ± 0.41 hours). Moreover, the absorbtion half- life ($T_{1/2}$ abs), the metabolic phase half-life ($T_{1/2}\beta$), the area under the concentration-time curve (AUC) and the volume of distribution ($Vd\alpha$) resulted markedly affected by aging, being significantly higher in elderly than in adult subjects (Table 1).

Table 1 - Pharmacokinetic parameters of melatonin in adults and aged healthy volunteers, after oral administration.

DOSE (mg)	ADULTS 20	ELDERLY 20
T ½ abs (h)	0.27 ± 0.09	0.34 ± 0.16*
-- α (h)	0.45 ± 0.11	0.55 ± 0.2
-- β (h)	1.01 ± 0.13	1.59 ± 0.66**
AUC µg/l h	57.62 ± 14.45	79.13 ± 33.6*
T-max (h)	0.83 ± 0.41	1.13 ± 0.76*
C-max (µg/l)	35.52 ± 6.5	28.39 ± 10.18
Vd α (l)	58 ± 22	75.31 ± 18*
Vd β (l)	537 ± 170	584 ± 154

* = p < 0.05   ** = p > 0.01

In adult volounteers, melatonin, after oral treatment at different single doses, reached dose-dependent serum peak concentrations (Cmax) ranging from 35.52 ± 6.5 to 115.9 ± 35.5 μg/l for doses of 20, 40 and 80 mg (Table 2).Likewise, T½β, AUC, Vdα and Vdβ values significantly increased with increasing of the dose.

Table 2 - Pharmacokinetic parameters of melatonin orally given at different doses in healthy volunteers

DOSE (mg)	20	40	80
T ½ abs (h)	0.27 ± 0.09	0.19 ± 0.06	0.24 ± 0.05
-- α (h)	0.45 ± 0.11	0.46 ± 0.2	0.48 ± 0.09
-- β (h)	1.01 ± 0.13	1.34 ± 0.66	$1.54 \pm 0.1*$
AUC μg/l h	57.62 ± 14.45	84.39 ± 31.6	$180.53 \pm 65.3**$
T-max (h)	0.83 ± 0.41	0.67 ± 0.26	0.80 ± 0.26
C-max (μg/l)	35.52 ± 6.5	57.33 ± 12	$115.9 \pm 35.5**$
Vd α (l)	58 ± 22	61.31 ± 18	70.84 ± 23
Vd β (l)	537 ± 170	684 ± 180	786 ± 210

$* = p < 0.05$ $** = p > 0.01$

PHARMACOKINETICS OF MELATONIN ANALOGS

From the limited number of structure-activity studies with melatonin analogs performed in the last years, it has been shown that both the N-acetyl and 5 methoxy substituents are necessary for biological activity and binding affinity at the receptor level (Flaugh et al., 1979; Krause and Dubocovich, 1991).Moreover the indole nucleus could be substituted with a naphtalenic ring (Yous et al., 1992). Recently it has been shown that methylation, halogenation, and substitution with an aromatic ring at C2 of the indole nucleus has led to a remarkable increase in the binding activity. The biological activity varied greatly and clear agonist properties were obtained in the case of methylation or halogenation. Isopropyl substitution at C2 resulted in a decrease in both affinity and agonist biological activity (Spadoni et al., 1993). Among the various melatonin analogs, the halogenated compounds 2-iodomelatonin and 2-bromomelatonin was proven to be highly active agonists (Duranti et al., 1992; Stankov et al., 1993). The high biological activity of 2-iodomelatonin and 2-bromomelatonin, could be related not only to an increased binding affinity, and possibly longer half-lives, but also to the better distribution due to the higher lipophilicity of these compounds in comparison to melatonin.

Metabolism and pharmacokinetics of 2-iodomelatonin were studied in rats (Stankov et al., 1993) and in humans. The primary metabolism studies were performed in male rats intraperitoneally treated with 2-^{125}I iodomelatonin (2.5 uCi/animal). The TLC analysis of serum samples collected 30 minutes after treatment and extracted with chloroform, revealed eight radioactive spots. The 46% of the radioactivity was recovered with Rf values equal to that of 2-^{125}I iodomelatonin and 9% with Rf values equal to that of melatonin. The TLC data were influenced by the apparently high background, inherent to the method when relatively low quantities of the labeled compound are used. Therefore high sensitivity HPLC analysis was performed in order to determine the pharmacokinetics and metabolism. The results obtained with HPLC analysis showed that in serum collected from rats intraperitoneally treated with 2-iodomelatonin 30 mg and sacrificed at different time after treatment, the major part of the compounds present in the samples is represented by 2-iodomelatonin. The major metabolite appeared to have the retention time equal to that of melatonin. The HPLC approach precluded collection of the effluent for further analysis, therefore it was impossible to judge whether the major metabolite was structurally identical

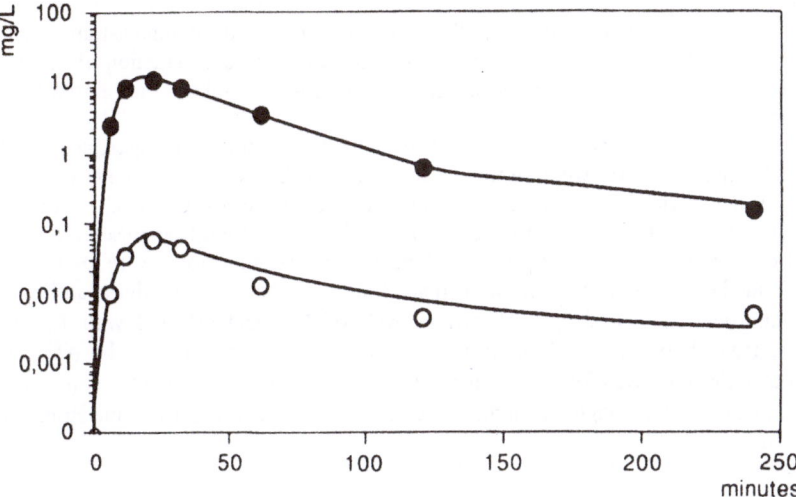

Figure 1 - Pharmacokinetic profiles of 2-iodomelatonin (closed symbols) and its major metabolite (open symbols, RT = 4.89 min) in the peripheral plasma of rats, following single intraperitoneal administration of 30 mg 2-iodomelatonin (Stankov et al., 1993, with permission).

to the native indole. On the basis of HPLC data, pharmacokinetics analysis were effectuated, evaluating the data for 2-iodomelatonin and its major metabolite. A computerized approach applying three exponential model fitted to data, using a peeling algorithm was employed for 2-iodomelatonin. The data for the main metabolite were analyzed using Powell minimisation algorithm, data being fitted with a two exponential model. The pharmacokinetic profile of 2-iodomelatonin and its main metabolite is reported in Figure 1 and Table 3 (Stankov et al., 1993).

Table 3 - Pharmacokinetic parameters of 2-iodomelatonin and its major metabolite (RT = 4.89) in the peripheral plasma of rats treated with 2-iodomelatonin (30 mg/animal, i.p.) (Stankov et al., 1993, with permission).

DRUG	T ½ elimination (min)	T ½ absorption (min)	Cmax (mg/l)	Tmax (min)
2IM	61.75	5.65	12.37	20
MM*	17.06	13.27°	72.00^	30

* The major metabolite, having RT = 4.89 min.

° Denotes the apparent formation time; Cmax, peak concentration, Tmax, time to peak concentration.

^ in μg/L

  2-iodomelatonin resulted rapidly absorbed from the site of administration, with an apparent high distribution volume, which could be related to accumulation of the drug in particular compartments. This is a feasible explanation, taken the extremely high lipophilicity of the compound.

  Preliminary experiments aimed to investigate the pharmacokinetics of 2-iodomelatonin in humans were performed in 2 adult healthy volunteers orally treated with 10 mg in the morning. As control, 4 healthy volunteers were orally treated with the same dose of melatonin. Every subject gave the informed consent. Blood samples were collected at different times after dosing and 2-iodomelatonin determinations were performed by HPLC method. The results showed that 2-iodomelatonin is readily absorbed from gastrointestinal tract reaching peak serum levels of 3.58 and 4.1 μg/l with T_{max} values ranging from 2 to 3 hours. Moreover, 2-iodomelatonin is characterized by AUC values lower than those showed for melatonin orally given at the same dose. On the contrary, $T\frac{1}{2}\alpha$ $T\frac{1}{2}\beta$ and $Vd\alpha$ resulted higher than those found with the same oral dose of melatonin (Table 4).

Table 4 - Pharmacokinetic parameters of 2-iodomelatonin and melatonin orally administered in adult healthy volunteers.

	2-IODOMELATONIN		MELATONIN
DOSE (mg)	1st subject 10	2nd subject 10	4 subjects 10
T ½ abs (h)	0.64	1.3	0.27 ± 0.09
-- α (h)	1.75	0.87	0.45 ± 0.11
-- β (h)	2.07	1.59	1.01 ± 0.13
AUC μg/l h	7.93	11	30.30 ± 14.45
T-max (h)	2	3	0.83 ± 0.41
C-max (μg/l)	2.55	3.16	18.52 ± 6.5
Vd α (l)	86	72	58 ± 22
Vd β (l)	836	796	537 ± 170

The pharmacokinetics of 2-bromomelatonin were preliminarly studied in rats and in humans. Adult rats were intraperitoneally treated with 20 mg/kg in the morning and sacrificed at different times after dosing. The determination of 2-bromomelatonin was performed in serum samples by HPLC and pharmacokinetics analysis were effectuated as previously described. The results appeared similar to those obtained using 2-iodomelatonin (data not shown).

The preliminary results obtained in 2 healthy volunteers orally treated in the morning with 10 mg of 2-bromomelatonin, compared with those obtained in 4 subjects treated with melatonin following the same experimental protocol are shown in Table 5. 2-bromomelatonin appeared to reach maximal serum levels of 4.1 and 3.58 ug/l with Tmax values of 1.5 hours for both voolunteers. AUC values appeared lower than those found with the same oral dose of melatonin while T½α, T½β and Vdα values appeared higher (Table 5).

Table 5 - Pharmacokinetic parameters of 2-bromomelatonin and melatonin after oral administration in adult healthy volunteers.

| DOSE (mg) | 2-BROMOMELATONIN | | MELATONIN |
| | 1st subject | 2nd subject | 4 subjects |
	10	10	10
T ½ abs (h)	0.9	0.79	0.27 ± 0.09
-- α (h)	0.64	0.97	0.45 ± 0.11
-- β (h)	2.8	2	1.01 ± 0.13
AUC µg/l h	12.7	9.5	30.30 ± 14.45
T-max (h)	1.5	1.5	0.83 ± 0.41
C-max (µg/l))	4.1	3.58	18.52 ± 6.5
Vd α (l)	80	75	58 ± 22
Vd β (l)	804	754	537 ± 170

CONCLUSIONS

Melatonin is well absorbed from gastrointestinal tract in laboratory mammals as well as in humans.

The elimination phase of melatonin, after oral administration, is biphasic with an α half-life of about 0.45 hours, not influenced by the dose, and a β half-life of about 1 hour, influenced by the dose, like AUC and Tmax (Table 2).

This behaviour is probably related to the metabolic saturation of the catabolic pathways.

Likewise the differences observed between adult and aging volunteers in terms of elimination half-life are likely due to the metabolism.

Our results are in agreement with those obtained by some authors (Waldhauser et al., 1984; Vakkuri et al., 1985).

The difference existing in pharmacokinetic parameters reported in various papers (Wetterberg et al., 1978; Iguchi et al., 1982; Waldhauser et al., 1984; Aldhours et al., 1985;

Vakkuri et al., 1985; Mallo et al., 1990) could be due to several factors such as doses, routes of administration, pharmaceutical preparations.

The relative water-insolubility of melatonin can influence its dissolution into the gastrointestinal tract and consecutively its absorption. The halogenated melatonin analogs show a pharmacokinetic behaviour similar to that of melatonin with a larger elimination half-life.

The preliminary results here reported of course, must be confirmed in further appropriate studies. In conclusion, melatonin and its analogs may be considered as quickly absorbed compounds. Their distribution is rapid and extensive, taking into account their high liposolubility, low protein binding (Cardinali et al., 1992) and the existence of specific tissular binding sites (Cardinali et al., 1979; Niles et al., 1979; Laudon and Zizapel, 1986; Stankov et al., 1991; Spadoni et al., 1993).

REFERENCES

Aldhous, M., Franey, C., Wright, J. and Arendt, J., 1985, Plasma Concentrations of Melatonin in Man Following Oral Absorption of Different Preparations, *Br. J. Clin. Pharmacol.* 19: 517.

Arendt, J., Bojkowski, C., Franey, C., Wright, J. and Marks, V., 1985, Immunoassay of 6-hydroxymelatonin Sulfate in Human Plasma and Urine. Abolition of the Urinary 24 Hour rhithm with Atenolol, *J. Endocrinol. Metab.* 60: 1166.

Arendt, J., 1988, Melatonin, *Clin. Endocrinol.*, 29:173.

Bartsch, H. and Bartsch, C., 1981, Effect of Melatonin on Experimental Tumors under Different Photoperiods and Time of Administration, *J. Neural Transm.*, 52: 269.

Blask, D.E., 1993, Melatonin in Oncology, in "Melatonin: Biosynthesis, Biological Effects, and Clinical Applications," H.S. Yu, R.J. Reiter, eds, *CRC Boca Raton*, p 447.

Bojkowski, C.J., Arendt, J.,Shih, M.C. and Markey, S.P., 1987, Melatonin Secretion in Humans Assessed by Measuring its Metabolite 6-Sulphatoxymelatonin, *Clin. Chem.* 33:1343.

Cardinali, D.P., Lynch, H.S. and Wurtman, R.J., 1972, Binding of Melatonin to Human and Rat Plasma Proteins, *Endocrinology*, 91: 1213.

Cardinali, D.P., Vacas, M.I. and Estevez Boyer, E., 1979, Specific Binding of Melatonin in bovine brain, *Endocrinology*, 105: 437.

Claustrat, B., Le Bars, D., Brun, J., Thivolle, P., Mallo, C., Arendt, J. and Chazot, G., 1989, Plasma and Brain Pharmacokinetic Studirs in Humans after Intravenous Administration of Cold or ^{11}C labelled Melatonin, in "Advances in Pineal Research" 3, R.J. Reiter, S.F. Pang, eds, *John Libbey & Co Ltd*, London, p 305.

Dubocovich, M.L., 1988, Pharmacology and Function of Melatonin Receptors, *FASEB J.*, 2: 2765.

Duranti, E. Stankov, B., Spadoni, G., Duranti , A., Lucini, V., Capsoni, S., Biella, G. and Fraschini, F., 1992, 2-Bromomelatonin: Synthesis and Characterization of a Potent Melatonin Agonist, *Life Sc.*, 51: 479.

Flaugh, M.E., Crowell, T.A., Clemens, J.A. and Sawyer, B.D., 1979, Synthesis and Evaluation of the Antiovulatory Activity of a Variety of Melatonin Analogues, *J. Med. Chem.* 22: 63.

Fraser, S., Cowen, M., Franklin, M. Franley, C. and Arendt,J., 1983, Direct Radioimmunoassay for Melatonin in Plasma, *Clin. Chim. Acta* 29: 386.

Gibbs, F.D. and Vriend, J., 1981, The Half-Life of Melatonin Elimination From Rat Plasma, *Endocrinology* 109: 1796.

Gomeni, R., 1984, Pharm: an Interactive Graphic Program for Individual and Population Pharmacokinetic Parameter Estimation, *Comput. Biol. Med.* 14: 25

Iguchi, H., Kato, K.I. and Ibayashi, H., 1982, Melatonin Serum Levels and Metabolic Clearance rate in Patients with Liver Cirrhosis, *J. Clin. Endocrinol. Metab.* 54: 1025.

Kopin, I.J., Pare, C.M.B., Axelrod, J. and Weissbach, H., 1961, The Fate of Melatonin in Animals, *J. Biol. Chem.* 236: 3072.

Krause, D.N. and Dubocovich, M.L., 1991, Melatonin Receptors, *Annu. Rev. Pharmacol. Toxicol.* 31: 549

Kveder, S. and McIsaac, W.M., 1961, The Metabolism of Melatonin (N-Acetyl-5-Methoxytryptamine) and 5-Methoxytryptamine, *J. Biol. Chem.* 236: 3214.

Lane, H.A. and Moss, H.B., 1985, Pharmacokinetics of Melatonin in Man: First Pass Hepatic Metabolism, *J. Clin. Endocrinol. Metab.* 61: 1214.

Laudon, M. and Zizapel, N., 1986, Characterization of Central Melatonin Receptors Using ^{125}I-melatonin, *FEBS Lett.*, 197: 9.

Le Bars, D., Thivolle, P., Vitte, P.A., Bojkowski, C., Chazot, G. and Arendt, J., 1991, PET and Plasma Pharmacokinetic Studies after Bolus Intravenous Administration of [^{11}C] Melatonin in Humans, *Int. J. Rad. Appl. Instrum.* B. 18: 357.

Leone, R.M. and Silman, R.E., 1984, Melatonin Can Be Differentially Metabolized in the Rat to Produce N-Acetylserotonin in Addition to 6-hydroxy-melatonin, 114: 1825.

Lewy, A.J., Ahmed, S., Latham, J.M. and Sack, R.L., 1992, Melatonin Shifts Human Circadian Rhythms According to a Phase-Response Curve, *Chronobiol. Int.*, 9: 382.

Mallo, C., Zaidan, R., Galy, G., Vermeulen, E., Brun, J., Chazot, G. and Claustrat, B., 1990, Pharmacokinetics of Melatonin in Man after Intravenous Infusion and Bolus Injection, *Eur. J. Clin Pharmacol.* 38: 297.

Niles, L.P., Wong, Y.W., Mishra, R.K. and Brown, G.M., 1979, Melatonin Receptors in Brain, *Eur. J. Pharmacol.*, 55: 219.

Petrie, K., Conaglen, J.V., Thompson, L. and Chamberlain, K., 1989, Effect of Melatonin on Jet Lag after Long Haul Flights, *Br. Med. J.*, 298: 705.

Reiter, R.J., 1991, Pineal Melatonin: Cell Biology of its Synthesis and of its Physiological Interactions, *Endocr. Rev.* 12: 151.

Reppert, S.M. and Klein, D.C., 1978, Trandport of Maternal [^{3}H]Melatonin to Sukling rats and the Fate of [^{3}H]Melatonin in the Neonatal Rat, *Endocrinology* 102: 582.

Reppert, S.M., Perlow, M.J., Tamarkin, L., and Klein, D.C., 1979, A Diurnal Melatonin Rhythm in Primate Cerebrospinal Fluid, *Endocrinology* 104: 295.

Reppert, S.M., Weaver, D.R., Rivkees,S.A. and Stopa, E.G., 1988, Putative Melatonin Receptors in Human Biological Clock, *Science*, 242: 78.

Spadoni, G., Stankov, B., Duranti, A., Biella, G. Lucini, V. Salvadori, A. and Fraschini, F., 1993, 2-Substituted 5-Methoxy-N-Acyltryptamines: Synthesis, Binding Affinity for the Melatonin Receptor, and Evaluation of the Biological Activity, *J. Med. Chem.*, 36: 4079.

Stankov, B., Lucini, V., Mariani, M., Scaglione, F., Demartini,G. and Fraschini, F., 1990, Alpha-1 Adrenoceptor Involvement in the Control of Melatonin secretion in the Golden Hamster, *J. Pineal Res.* 9:21.

Stankov, B., Fraschini, F., and Reiter, R.J., 1991, Melatonin Binding Sites in the Central Nervous System, *Brain Res. Rev.*, 16: 245.

Stankov, B., Gervasoni, M. Scaglione, F., Perego, R., Cova, D., Marabini, L. and Fraschini, F., 1993, Primary Pharmaco-toxicological Evaluation of 2-Iodomelatonin, a Potent Melatonin Agonist, *Life Sc.*, 53: 1357.

Tan, D.X., Chen, D., Poeggeler, B., Manchester, L.C. and Reiter, R.J., 1993, Melatonin: a Potent, Endogenous Hydroxyl Radical Scavenger, *Endocrine J.* 1: 57.

Vakkuri, O., Leppäluoto, J. and Kauppila A., 1985, Oral Administration and Distribution of Melatonin in Human Serum, Saliva and Urine, *Life Sci.* 37: 489.

Vitte, P.A., Harthe, C., Lestage, P., Claustrat, B. and Bobillier, P., 1988, Plasma, Cerebrospinal Fluid, and Brain Distribution of ^{14}C Melatonin in rat: a Biochemical and Autoradiographic Study, *J. Pineal Res.* 5: 437.

Vitte, P.A., Harthe, C., Pevet, P. and Claustrat, B., 1990, Brain Autoradiographic Study in the Golden Hamster after Intracarotid injection of [^{14}C] Melatonin, *Neurosci. Lett.* 110: 1.

Waldhauser, F., Waldhauser, M., Lieberman, H.R., Deng, M.H., Lynch, H.J. and Wurtman, R.J., 1984, Bioavailability of Oral Melatonin in Humans, *Neuroendocrinology* 39: 307.

Waldhauser, F. and Steger, H.,1986, Changes in Melatonin Secretion with Age and Pubescence, *J. Neural Transm.* Suppl. 21: 183.

Wehr, T.A., 1991, The Durations of Human Melatonin Secretion and Sleep Respond to Changes in Daylenght (Photoperiod), *J. Clin. Endocrinol. Metab.,* 73:1276.

Wetterberg, L., Eriksson, O., Friberg, Y. and Vangbo, B., 1978, A Simplified Radioimmunoassay for Melatonin and its Application to Biological Fluids, *Clin. Chim. Acta* 86: 169.

Wetterberg, L., Beck-Friis, J. and Kiellman, B.F., 1990, Melatonin as a Marker of a Subgroup of Depression in Adult, in "Biological Rhythms, Mood Disorders, Light Therapy and the Pineal Gland," M. Shafii, S.L. Shafii, eds, *American Psychiatric Press*, Washington, p 69.

Voordouw, B.C.G., Euser, R., Verdonk, R.E.R., Alberda, B.T.H., De Jong, F.H., Drogendijk, A.C., Fauser, B.C.J.M. and Cohen, M., Melatonin and Melatonin-Progestin Combinations Alter Pituitary-Ovarian Function in Women and Can Inhibit Ovulation., *J. Clin. Endocrinol. Metab.,* 74: 108.

Yous, S., Andrieux, J., Howell, H.E., Morgan, P.J., Renard, P., Pfeiffer, B., Lesieur, D. and Guardiola-Lemaitre, B., 1992, Novel Naphtalenic Ligands with High Affinity for the Melatonin Receptor, *J. Med. Chem.* 35: 1484.

SIGNIFICANCE OF MELATONIN IN HUMANS

Josephine Arendt

School of Biological Sciences
University of Surrey
Guildford
Surrey, GU2 5XH
UK

INTRODUCTION

During the last decade, interest in the possible clinical therapeutic actions of melatonin has increased dramatically. Experiments designed to assess the effects of melatonin in humans have had, as one objective, the clarification of the physiological role of melatonin and the pineal gland in man. However whilst we now know a great deal about the influence of exogenous melatonin on human physiology, its endogenous function remains speculative and based essentially on correlative observations.

Classical endocrine approaches to determination of pineal function, i.e. extirpation of the gland and observation of consequent effects, followed by replacement of the presumed active principle, are evidently not possible in healthy humans. No specific inhibitor of melatonin synthesis is available and studies in humans have relied heavily on peripheral measurement of pineal metabolites to assess function. Pineal tumours though rare provide the only situation where extirpation of the gland is performed, although for specific types of tumour radiation therapy is the method of choice (1). Surgery of the pineal is a major undertaking, given its position deep within the brain, and assessment post surgery always has to consider damage to surrounding brain tissue.

Pinealectomy in humans removes virtually all plasma melatonin (2). Melatonin may nevertheless be synthesised outside the pineal for example in the retina and the gut, be capable of acting locally, but be metabolised before reaching the general circulation. Other consequences of the operation consist of diffuse neurological problems which do not add up to a consistent functional effect as yet and may be more related to non-specific effects of surgery. As in other species it is almost certainly necessary to challenge the organism by, for example, abrupt shifting of the light dark cycle or major changes in applied artificial daylength, in order to show pineal mediated effects. So far there are no such reports although the author is aware of attempts to maintain pinealectomised humans in environmental isolation in order to study circadian function. One potentially useful challenge is suggested by the sleep and temperature response to very short and very long days in healthy human volunteers (see below) which may be pineal mediated(3).

In animals (with species variability) the functions of the pineal via melatonin secretion are considered to be the transmission of information concerning daylength or photoperiod for the organisation of seasonal responses and the transmission of information concerning phase and strength of the daily light dark cycle for appropriate synchronisation of the circadian system (4). Pineal effects are then manifested in the responses of light-dark dependent physiological and behavioural rhythms such as sleep, temperature, reproduction etc. Clearly the importance of the pineal in human physiology depends on the importance of light in human physiology. This brief review will consider the relationships of human seasonality and circadian rhythms to the light-dark cycle and melatonin.

DOES HUMAN PHYSIOLOGY RESPOND TO PHOTOPERIOD CHANGE?

In all species studied to date, in normal entrained conditions, melatonin is produced during the dark phase. In most species its secretion is related to the length of the night: the longer the night the longer the duration of secretion (4). This has been particularly well demonstrated in sheep where melatonin rises within a few minutes of lights off and in photoperiods of more than around 14 hours of light does not decline until lights on. Light of suitable intensity and duration serves to entrain the rhythm (generated in the suprachiasmatic nucleus) and, in sufficiently short photoperiods, to suppress secretion at the beginning and/or the end of the dark phase. This duration change is the critical determinant of photoperiodic responses in many species (5).

In humans, for many years it has proved difficult to demonstrate a change in duration of melatonin secretion with daylength even in polar regions, although very small changes have been reported (6). In structured temperate or polar environments there is a small phase delay of the melatonin rhythm in winter compared to summer of approximately 1-2 hours (7, 8), sometimes accompanied by a slight increase in duration of secretion (6, 9, 10), but with no change in amplitude. Basal (daytime) concentrations of both melatonin (in plasma) and aMT6s (in urine) are reported to have a bimodal pattern with peaks in summer and winter in a temperate zone (52°N), (10).

If however humans are kept strictly in darkness for 14 hours per day for a period of two months, the melatonin secretion pattern expands to cover much of the dark period and concomitantly, in extended photoperiods of 14 hours of light, the rhythm contracts to less than 10 hours (3). Thus in appropriate experimental conditions humans show a duration change like animals and the essential prerequisite for a pineal dependent photoperiodic response is present. As in diurnal animals duration of sleep increased in long nights and the profile of the body temperature rhythm changed as a function of imposed artificial daylengths.Whether or not human seasonality is influenced by the natural photoperiod remains an open question. A major recent survey suggested that annual rhythms in conception, demonstrable over large numbers of births, are latitude dependent, and that both photoperiod and environmental temperature play a role in the timing of human reproduction.(11). However there is good evidence that at the same latitudes, conception timing can be very different in the USA and in the UK, hence factors other than photoperiod (such as social cues) must be very important (12).

There are also seasonal influences on the onset of menarche with a maximum number of first menstruations in summer (12). Conception of dizygotic twin pregnancies is more common in the summer months during continuous light in Finland and increased endometrial hyperplasia and anovulatory cycles are reported in winter, associated with an increase in daytime melatonin secretion (13). Many factors other than reproduction show seasonal variations in humans. The evidence of residual human photoperiodism is accumulating but whether or not it is of real significance is debateable..

Seasonal affective disorder (SAD, winter depression) is another manifestation of human seasonality which was originally thought to be photoperiod (and hence duration of melatonin) dependent (14). Patients suffering from SAD are apparently more seasonal than the general population in terms of reproduction (conception time), body weight etc. There is however very little evidence for changes in melatonin duration in SAD patients, although the phase delay seen in normal subjects in winter may be present and exaggerated (15). Bright light treatment of sufficient duration and intensity is consistently successful in alleviating depression in SAD, however neither in a normal environment nor in constant routine conditions is a change in melatonin secretion necessary for its therapeutic effect. (16, 17). Conceivably the immediate activating effects of bright light in terms of, for example, increased body temperature, are more relevant than longer term photoperiodic effects.

Our behaviour, particularly that of urban indoor workers, is such that in 'normal' circumstances it is likely that we experience and perceive only very small changes in natural daylength during the year. For example in the UK it is fairly typical to rise around 0700h (well after dawn in the summer) and to be inside a dwelling from around 1900h (well before dusk in the summer) and thus not be exposed to natural sunlight for the entire photoperiod in summer.

DOES HUMAN PHYSIOLOGY RESPOND TO APPLIED CHANGES IN DURATION OF MELATONIN?

Few experiments have been performed in such a way as to answer this question. The administration of 2mg melatonin as a fast release preparation at 1700h gives plasma melatonin levels that remain elevated at or above night-time concentrations for at least 14 hours (18, 19) by artificially extending the endogenous night-time rise. Repeated daily dosing at 1700h is in consequence similar (but not identical in view of the supraphysiological exogenous peak concentrations) to the long duration melatonin profiles during imposition of short photoperiod. The effects of prolonged exposure (one month) to such long duration melatonin in an otherwise normal environment are manifested primarily on the sleep-wake cycle (18) and on endogenous melatonin and prolactin timing (19). There were phase-advances in both endogenous prolactin and melatonin rhythms (see below) and a small increase in duration of endogenous melatonin, but no significant changes in anterior pituitary hormones other than prolactin (18, 19, 20). There were no changes in self-rated mood (20), thus reinforcing the evidence for a lack of involvement of melatonin in SAD. Subjects reported increased 'fatigue', or actual sleep, earlier in the evening (3 hours after the dose) than when taking an identical placebo, together with a non-significant trend to early waking. This effect was significant in the group (N=12) as a whole only after 4 days, although some individuals are sensitive to an acute low dose of melatonin even in these conditions. It would be reasonable to assume that the behavioural effects of a normal environment, i.e. evening domestic lighting and social contacts, act to counter the melatonin induced advance in sleepiness or sleep.

Subsequent work using acute low (5mg) dose melatonin administration and more sophisticated multiple sleep latency tests indicated a substantial, time-dependent increase in sleep propensity 2-3 hours after the dose (21). The effect was more rapid in the early evening compared to the morning and the time course of effect after administration at 1700h was very similar to that previously reported for self rated fatigue (18). Although endogenous melatonin levels were not measured in this study it is possible that only administration times in the mid- to late-light phase increased duration of melatonin, rather than inducing two peaks- the endogenous night time peak and a secondary day-time peak.

These experiments underline the problem of differentiating photoperiodic and circadian effects of melatonin if indeed such a distinction is meaningful. Is the extended time of sleepiness or sleep, during long duration melatonin treatment a circadian or a photoperiodic effect? Using 2-5mg fast-release preparations, the initial plasma peak greatly surpasses physiological night-time concentrations and it is conceivable that the observed effects are purely pharmacological and not related to endogenous function. However similar changes in the timing of endogenous melatonin are seen after 4 days of timed daily treatment with physiological amounts of melatonin (22). Recently several groups have used either infusion or very low dose fast-release preparations of melatonin which induce physiogical night-time levels to investigate their acute effects (23, 24, 25). A single-dose administration protocol cannot be considered to be a 'photoperiodic' manipulation however, given that the onset of detectable photoperiodic responses in animals (for example the decrease of prolactin secretion in short days) takes at least 3 days following an abrupt change of photoperiod or of initiating daily melatonin administration (26).

The conclusions to be drawn at present are that increasing duration of melatonin secretion changes the timing and the duration of sleep, endogenous melatonin, prolactin and the circadian profile of body temperature in humans.

EFFECTS OF BRIGHT LIGHT IN HUMANS-ARE THEY RELATED TO MELATONIN?

Full suppression of melatonin production at night in humans requires 2-3000 lux white or full-spectrum light (27), although partial suppression can be acheived with 2-300lux, with considerable individual variations in sensitivity to low light levels (28). This observation led to a reevaluation of the function of light in human circadian physiology, to new treatments for depression (SAD) and to an increasing perception of the more general importance of bright as opposed to dim light (eg domestic intensity-3-500lux) in humans.

It is clear that the human circadian system can be manipulated (resynchronised, phase-shifted, entrained to very long or very short light-dark cycles) by sufficient intensity and duration of bright (2-10000 lux) light (29, 30, 31). The lower limits of intensity/duration have yet to be defined however. Whether or not the suppression of melatonin is essential to the phase-shifting mechanism is a question of some importance. There are no studies relating, for example, the amount of phase shift induced by a given light manipulation to the degree of suppression of plasma melatonin on an individual basis. Likewise there are no studies in humans where melatonin has been administered at the same time, i.e.in the presence of bright-light to evaluate the effects on the circadian response (the acute effects of bright light on body temperature are however counteracted by simultaneous melatonin treatment, see below). In sheep there is evidence that the behavioural effects of light (on rest-activity) override the effects of concomitant melatonin administration (32). In humans it appears that melatonin administration in the circadian evening, timed to phase advance, can at least partially counteract bright-light in the circadian morning, timed to phase delay (33, 34).

Recently very interesting work has sought to link bright light and melatonin by their effects on body temperature. The endogenous peak of melatonin coincides within approximately one hour with the nadir of the body temperature rhythm. Bright light administered at night significantly and acutely raises body temperature (35). Simultaneous melatonin administration counteracts this effect. Melatonin administered in the morning in divided doses acutely lowers body temperature (36). The authors of this latter study calculated that 40% of the amplitude of the body temperature rhythm could be due to the night time rise in melatonin. 5mg melatonin administered at 1700h also acutely lowers body temperature. This accompanied by a decrease in self-rated alertness and a phase advance of the endogenous melatonin rhythm onset the following day (25). The degree of temperature suppression correlates to the degree of phase shift obtained. It would be of considerable interest to perform the same manipulation using bright light to induce acute phase shifts and assess increases in body temperature, in order to evaluate any relationships. Presumably this data already exists in the files of those groups using body temperature as a phase marker of the circadian system and bright light to induce phase shifts, but to the authors knowledge it has not been published. If these threads of evidence are drawn together they suggest that changes in core body temperature may be a common pathway in the production of a phase shift, and that the effects of light and melatonin are linked by their opposing effects on body temperature.

Bright light however is by no means essential to maintain synchronisation of human circadian rhythms. For example on the British Antarctic Survey Base of Halley, light intensity does not exceed 500 lux (average approimately 300 lux) during the winter period of sundown and whilst subjects' light sensitivity increases during winter it does not appear that melatonin can be completely suppressed by 300 lux (37). The Base routine is highly structured, with defined wake-up (except during night-shifts), work and meal times etc. Halley personnel with very few exceptions remain synchronised to 24h throughout this period (38). In contrast, as reported by Kennaway and coworkers (38), on a Greenpeace base with no social impositions, but with a maximum light intensity higher than that of Halley, all 4 subjects overwintering free-ran during sundown. Thus social, non-photic zeitgebers must be powerful enough to synchronise human rhythms in the absence of light sufficient to completely suppress melatonin secretion. As yet there is no evidence that non-photic zeitgebers act through melatonin. On the other hand it seems quite likely that body temperature, via changes ('masking') due to activity during the day and imposed sleep time at night, may be of importance in maintaining synchronisation.

ARE THE CIRCADIAN PHASE SHIFTING EFFECTS OF MELATONIN 'PHYSIOLOGICAL'?

We have seen that acute administration of pharmacological (2-5mg) amounts of fast release melatonin lowers body temperature, phase shifts endogenous melatonin and in a non stimulatory environment consistently lowers alertness and increases sleep propensity. The use of physiological doses of melatonin either by infusion or by very low dose fast-release also induces phase shifts according to a phase response curve (both advances and delays) and increases feelings of tiredness in a suitably quiet environment.(22, 23, 24). The fact that these doses give plasma levels compatible with night-time melatonin secretion suggests that the observed effects are indeed physiological. However there remains at least one problem of interpretation. The maximum phase shifts induced by melatonin are found at a time in the subjective afternoon, when normally the organism does not produce high levels of endogenous melatonin. (This conceptual problem incidentally is also true of the light phase response curve, where maximum shifts are found in the middle of the night), (30).

Melatonin onsets in humans in a normal urban environment with domestic lighting in the evening and in the early morning are usually between 2100 and 2300h and offsets between 0600 and 0800h. It seems likely that, in the majority of humans with a free-running circadian period greater than 24h, bright natural light in the morning may serve to phase advance the circadian system each day to maintain synchrony with 24h, and that the concomitant inferred phase advance of melatonin is an integral part of the mechanism (see Lewy, this volume). A small phase advance of evening melatonin onset of less than one hour (without morning bright light) according to published phase-reponse curves, would not theoretically effect a sufficient phase advance of the circadian system to maintain synchrony in subjects whose tendency to delay is more than a few minutes per day. Patients with delayed sleep phase insomnia treated with exogenous melatonin (5 mg) at 2200h show significant phase advances of the sleep-wake cycle, but the magnitude of the shift is not quite sufficient to maintain a normal sleep onset time (39).

Be that as it may, melatonin in doses of 2-5mg, suitably timed to phase advance or to phase delay, undoubtedly is able to hasten adaptation to phase shift as seen in jetlag and shift work (40, 41, 42). Its principle and most rapid effect is to improve sleep, daytime alertness and, in simulated studies, performance (33, 34). These changes are accompanied by more rapid resynchronisation of the circadian system as assessed by marker rhythms (endogenous melatonin in particular). Complete resynchronisation is however not necessary for a complete restoration of sleep quality, alertness and performance (33, 34). In one reported case study (43) and in 4 further blind subjects (44), daily melatonin treatment was able to significantly improve synchronisation of the sleep-wake cycle to the 24h day without synchronising the endogenous melatonin rhythm.

Thus it appears that melatonin has (at least) two effects, 1) rapid effects on the timing, quality and latency of sleep, daytime alertness and performance and 2) a hastening of the resynchronisation of strongly endogenous circadian rhythms. It appears to be most effective when all zeitgebers are acting in the same direction.

The question as to whether these are pharmacological or physiological effects is of no therapeutic importance.

ENDPIECE

Whether or not these actions of melatonin relate to its reported ability to act as an anti-proliferative agent in certain types of cancer, as an antioxidant, as an anti-aging hormone (see Blask and Reiter, this volume) and indeed in other circumstances where rhythmicity does not appear to be the primary source of pathology, is an unanswered question. It is conceivable that 'optimum' relationships between the multitude of endogenous seasonal and circadian rhythms are essential to health and longevity. Melatonin may, as a coordinator of biological rhythms, act to maintain structured rhythmicity and thereby influence many aspects of health.

REFERENCES

1. M.B. Horowitz, Central nervous system germinomas, a review. *Arch.Neurol.* 48:652 (1991).
2. E.A. Neuwelt and A.J. Lewy, Disappearance of plasma melatonin after removal of a neoplastic pineal gland, *N. Engl. J. Med.* 19:1132 (1983).
3. T.A. Wehr, The durations of human melatonin secretion and sleep respond to changes in daylength (photoperiod),*J.Clin.Endocrinol.Metab.* 73:1276 (1991).
4. J. Arendt, The pineal gland: basic physiology and clinical implications, in: De Groot: Endocrinology, 3rd Edition, W.B.Saunders and Co. Philadelphia, (1994), in press.
5. L. Tamarkin, C.J. Baird and O.F.X. Almeida, Melatonin: a coordinating signal for mammalian reproduction, *Science* 227:714 (1985).
6. J. Beck-Friis, D. von Rosen, B.F. Kjellman, J.G. Ljungen and L. Wetterberg, Melatonin in relation to body measures, sex, age, season and the use of drugs in patients with major affective disorders and healthy subjects, *Psychoneuroendocrinol.* 9:261 (1984).
7. H. Illnerova, P. Zvolsky and J. Vanecek, The circadian rhythm in plasma melatonin concentration of the urbanised man, the effect of summer and winter time, *Brain Res.* 328:186-189 (1985).
8. J. Broadway, J. Arendt and S. Folkard, Bright light phase shifts the human melatonin rhythm during the Antarctic winter, *Neurosci. Lett.* 79:185 (1987).
9. I. Makkison and J. Arendt, Melatonin secretion on two different Antarctic Bases (68°S and 75°S), Proceedings of the 7th meeting of the European Society for Chronobiology,Marburg,Germany,1991, *J. Interdisciplinary Cycle Res.* 22:149 (1991).
10. C. Bojkowski and J. Arendt, Annual changes in 6-sulphatoxymelatonin excretion in man. *Acta Endocrinol.* 117:470 (1988).
11. T. Roenneberg and J. Aschoff, Annual rhythms in human reproduction: I. Biology,sociology or both? *J. Biol. Rhythms* 5:195 (1990).
12. A.S. Parkes. Patterns of Sexuality and Reproduction. Oxford University Press, Oxford, (1976).
13. A. Kauppila, A. Kivela, A. Pakarinen and O. Vakkuri, Inverse seasonal relationship between melatonin and ovarian activity in humans in a region with a strong seasonal contrast in luminosity, *J. Clin. Endocrinol. Metab.* 65:823 (1987).
14. N.E. Rosenthal, D.A. Sack, J.C. Gillin, A.J. Lewy, F.K. Goodwin, Y. Davenport, P.S. Mueller, D.A. Newsome and T.A. Wehr, Seasonal affective disorder. A description of the syndrome and preliminary findings with light therapy, *Arch. Gen. Psychiatr.* 41:72 (1984).
15. A.J. Lewy, R.L. Sack, L.S. Miller, and T.M. Hoban, Anti-depressant and circadian phase-shifting effects of light. *Science* 235:352 (1987).
16. S. Checkley, The relationship between biological rhythms and the affective disorders, *in*: J. Arendt, D.S. Minors, J. Waterhouse, (eds) Biological rhythms in clinical practice. Butterworth, London (1989).
17. A. Wirz-Justice, K. Krauchi, P. Graw, J. Arendt, J. English, H-J. Haug, G. Leonhardt, and D.P. Brunner, Are circadian rhythms involved in the pathophysiology of SAD and its treatment by light? Society for Research on Biological Rhythms, Florida, 4-8th May 1994, Abstr. 149.
18. J. Arendt, A.A. Borbely, C. Franey, and J. Wright, The effects of chronic, small doses of melatonin given in the late afternoon on fatigue in man: a preliminary study, *Neurosci. Letts.* 45:317 (1984).
19. J. Wright, M. Aldhous, C. Franey,J. English, and J. Arendt, The effects of exogenous melatonin on endocrine function in man, *Clin. Endocr.* 24:375 (1986).
20. J. Arendt, C. Bojkowski, S. Folkard, C. Franey, D.S. Minors, J.M. Waterhouse, R.A. Wever, C. Wildgruber and J. Wright, Some effects of melatonin and the control of its secretion in man, *in*: Evered D, Clark S (eds): Ciba Foundation Symposium 117. Photoperiodism, melatonin and the pineal, pp. 266-283. London: Pitman, (1985).
21. O. Tzischinsky, P. Lavie and I. Pal ,Time-dependent effects of 5 mg melatonin on the sleep propensity function, *in:* 11th European Congress on Sleep Research, July 1992, Helsinki, Finland.J. Sleep Res. 1:Suppl. 1, 234 (1992).

22. A. Lewy, A. Saeeduddin, J. Latham Jackson and R. Sack, Melatonin shifts human circadian rhythms according to a phase-response curve, *Chronobiol. Internat.* 9: 380 (1992).

23. B. Claustrat, Presented at the Philipe Laudat Conference on Neureobiology of Circadian and Seasonal Rhythms: Animal and Clinical Studies, Strasbourg/Bischenbereg, France, (1991).

24. A.B. Dollins, I.V. Zhdanova, R.J. Wurtman, H.J. Lynch and M.H. Deng, Effect of inducing nocturnal serum melatonin concentrations in daytime on sleep, mood, body temperature, and performance, *Proc. Nat. Acad. Sci.* USA, 91:1824 (1994).

25. S. Deacon, J. English and J. Arendt, Acute phase-shifting effects of melatonin associated with suppression of core body temperature, *Neurosci. Lett.* (1994), in press.

26. A.L. Poulton, J. English, A.M. Symons and J. Arendt, Effects of various melatonin treatments on plasma prolactin concentrations in the ewe, *J. Endocrinol.* 108:287 (1986).

27. A.J. Lewy, T.A. Wehr, F.K. Goodwin, D.A. Newsome, and S.P. Markey, Light suppresses melatonin secretion in humans, *Science* 210:1267 (1980).

28. C. Bojkowski, M. Aldhous, J. English, C. Franey, A.L. Poulton, D.J. Skene and J. Arendt, Suppression of nocturnal plasma melatonin and 6-sulphatoxymelatonin by bright and dim light in man, *Horm. Metab. Res.* 19:437 (1987).

29. R.A. Wever, J. Polasek and C.M. Wildgruber, Bright light affects human circadian rhythms, *Pflugers Arch.* 396:85 (1983).

30. C.A. Czeisler, J.S. Allan, J.S. Strogatz, J.M. Ronda, R. Sandrez, C.D. Rios, W.O. Freitag, G.S. Richardson, R.E. Kronauer, Bright light resets the human circadian pacemaker independent of the timing of the sleep-wake cycle, *Science* 233:667 (1986).

31. D. Minors, J. Waterhouse and A. Wirz-Justice, A human phase-response curve to light, *Neurosci. Lett.* 133:36 (1991).

32. T. Sweeney, A.M. Strijkstra and G.A. Lincoln, Infusion of melatonin during the light phase does not alter the daily locomotor activity pattern in the Soay ram, Society for Research on Biological Rhythms, Florida, 4-8th May 1994, Abstr. 137.

33. S. Deacon and J. Arendt, Use of melatonin to adapt to phase-shifts. I. Melatonin counters sleep problems after a large advance shift in external time cues in spite of ambient light conditions, Society for Research on Biological Rhythms, Florida, 4-8th May 1994, Abstr. 147.

34. J. Arendt and S. Deacon, Use of melatonin to adapt to phase shifts II. Mood and performance after a large advance shift in external time cues, Society for Research on Biological Rhythms, Florida, 4-8th May 1994, Abstr. 148.

35. R.J. Strassman, C.R. Qualls, E.J. Lisansky and G.T. Peake, Elevated rectal temperature produced by all night bright light is reversed by melatonin infusion in man, *J. Appl. Physiol.*, in press.

36. A. Cagnacci, J. Elliott and S. Yen, Melatonin: a major regulator of the circadian rhythm of core temperature in humans. *J .Clin. Endocrinol .Metab.* 75:447 (1992).

37. J. Owen and J. Arendt, Melatonin suppression in human subjects by bright and dim light in Antarctica : time and season dependent effects, *Neurosci. Lett.* 137:181 (1992).

38. D.J. Kennaway and C.F. Van Dorp, Free running rhythms of melatonin, cortisol, electrolytes and sleep in humans in Antarctica, *Amer.J. Physiol.* R1137 (1991)

39. M. Dahlitz, B. Alvarez, J. Vignau, English J. Arendt and J.D. Parkes, Delayed sleep phase syndrome response to melatonin, *Lancet* i 337:1121 (1991).

40. J. Arendt, M. Aldhous and V. Marks, Alleviation of jet-lag by melatonin: preliminary results of controlled double-blind trial, *Br. Med. J.* 292:1170 (1986).

41. S. Folkard, J. Arendt and M. Clark, Can melatonin improve shift workers' tolerance of the night shift? Some preliminary findings, *Chronobiol. Internat.* 10:315 (1993).

42. J. Arendt, M. Aldhous, M. Marks, S. Folkard, J. English V. Marks and J. Arendt, Some effects of jet-lag and their treatment by melatonin, *Ergonomics*, 30:1379 (1987).

43. J. Arendt, M. Aldhous and J. Wright, Synchronisation of a disturbed sleep-wake cycle in a blind man by melatonin treatment, *Lancet* i:772 (1988).

44. M. Aldhous and J. Arendt, Melatonin rhythms and the sleep wake cycle in blind subjects.Proceedings of the 7th meeting of the European Society for Chronobiology,Marburg,Germany,1991, *J. Interdisciplinary Cycle Res.* 22:84 (1991).

THE INFLUENCE OF MELATONIN ON THE HUMAN CIRCADIAN CLOCK

Alfred J. Lewy,[1,2,3] Robert L. Sack,[1] Saeeduddin Ahmed,[1]
Vance K. Bauer,[1] and Mary L. Blood[1]

Sleep and Mood Disorders Laboratory
Departments of Psychiatry,[1] Ophthalmology[2] and Pharmacology[3]
Oregon Health Sciences University
Portland, Oregon 97201

INTRODUCTION

Now that it has been established that experimental administration of melatonin has circadian phase-shifting effects in humans, we can speculate about its endogenous function. More specifically, melatonin in physiologic amounts can produce effects that are described by a phase response curve (PRC) (Lewy et al., 1990a; Lewy et al., 1992; Lewy et al., 1991; Lewy et al., 1990b; Lewy et al., 1991; Zaidan et al., 1994) that is about 12 hours out of phase with the PRCs to light (Czeisler et al., 1989; Honma and Honma, 1988; Minors et al., 1991; Wever, 1989). Therefore, it appears that the most likely role of endogenous melatonin production in humans is to augment phase-shifting and entrainment of the endogenous circadian pacemaker (ECP) by the light-dark cycle. This is accomplished through the suppressant effect of light on melatonin production (Lewy et al., 1980).

EARLY STUDIES

In animals, melatonin has generally been associated with seasonal rhythms (in particular, reproductive cycles), although more recent studies have revealed circadian effects. The late Wilbur Quay found that, in response to a 12-hour shift in the light-dark cycle, young pinealectomized rats shifted faster (Quay, 1970), and older pinealectomized rats slower (Quay, 1972), than sham-controls. Subsequently, it was shown that melatonin implants lengthened free-running activity rhythms of house sparrows (Turek et al., 1976) and that single injections at the same time each day entrained them (Gwinner and Benzinger, 1978). When it was demonstrated that melatonin injections could entrain free-running Long-Evans hooded rats (Redman et al., 1983), interest in this aspect of melatonin significantly increased, even though human studies had begun a few years earlier.

Anecdotally (Arendt and Marks, 1982), as well as in some controlled human studies (Arendt et al., 1986), it was found that melatonin could affect circadian rhythms, particularly the sleep-wake cycle, or more specifically, sleepiness. Since pharmacologic doses of melatonin were taken at or near bed-time, two factors complicated speculation about the function of endogenous melatonin. First, a possible direct effect on sleep could not be ruled out. Although soporific properties have not been conclusively demonstrated (James et al., 1987), some individuals do report sleepiness which appears to be dose-dependent (Dollins et al., 1994; Hughes et al., 1994; Waldhauser et al., 1990). Therefore, any shifts in circadian phase noted above were not necessarily direct effects: an effect on sleep could indirectly result in a phase shift, in that when the eyes are closed darkness is superimposed upon the light-dark cycle; as the light-dark cycle shifts, so does the phase of the ECP.

A second confound is also related to the use of a pharmacologic dose. In addition to the putative dose-dependent soporific effect mentioned above, results from supraphysiologic levels do not necessarily suggest a physiologic function. Despite these problems, the early work of the Arendt (Arendt et al., 1987; Arendt et al., 1986; Arendt et al., 1985) and Claustrat groups (Mallo et al., 1988) did much to stimulate and maintain interest in a role for melatonin in the human circadian system.

THE MELATONIN PHASE RESPONSE CURVE (PRC)

In order to specifically avoid these confounds, we gave a physiologic dose (0.5 mg) of melatonin for four days, usually in the form of a split dose (described below) (Lewy et al., 1992). It was administered at all times of the day and night, not just in the evening. Therefore, for several reasons (some of which are described below), it does not seem likely that observed phase shifts were mediated through shifting the sleep-wake cycle.

In our studies, circadian phase position is assessed using the dim light melatonin onset (DLMO). Intra-individual variability of the DLMO is less than the sampling interval (30 minutes). Thus, we are able to reliably and consistently detect even small phase shifts.

The DLMO also provides a means for assessing circadian time (CT) of administration of exogenous melatonin. The pre-administration (baseline) DLMO is on average 14 hours after (bright) light onset. Since "lights on" is, by tradition, designated CT 0 (Johnson, 1990), the DLMO is designated CT 14. Between individuals, there is some variability in the clock time of the DLMO. Although the average time is 21.00 in individuals whose mean wake-up time is 07.00, the DLMO ranges from 18.00 to after midnight, which is most likely due to differences in sleep times (via changes in perceived light-dark cycles) and possibly to differences in the intrinsic period of the ECP. In any event, it is important to express scheduling of the phase-shifting stimulus (either bright light or melatonin) in CT rather than clock time. When expressed in CT, inter-individual differences in PRC phase diminish. PRCs from two subjects are plotted in Figure 1.

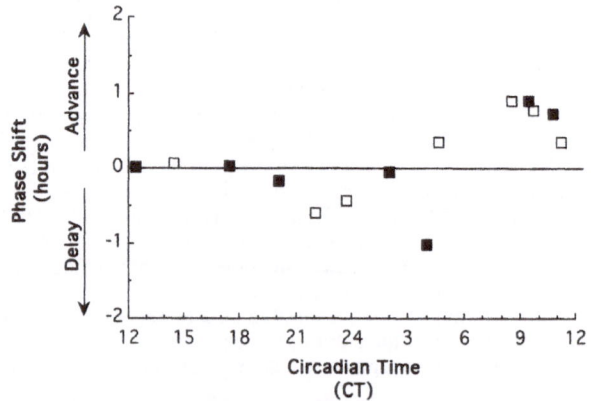

Figure 1. Phase shifts of the dim light melatonin onset (DLMO) as a function of circadian time (CT) of exogenous melatonin administration, for two subjects [JH (open-squares) and SE (closed squares)] who each participated in seven trials. Exogenous melatonin was administered at various times with respect to the time of endogenous melatonin production (CT 14 = baseline DLMO for each trial). When internal CT is referenced to the baseline DLMO, plots of the data for these two subjects nearly superimpose. On average, CT 0 = 07.00 clock time.

All thirty data points from nine individuals are plotted in Figure 2. The melatonin PRC appears to be about 12 hours out of phase with the PRCs to light, particularly the Czeisler group's PRC (Czeisler et al., 1989) when the scheduled bright light stimulus is expressed in CT (PRC #H/Hs-1 in Johnson, 1990).

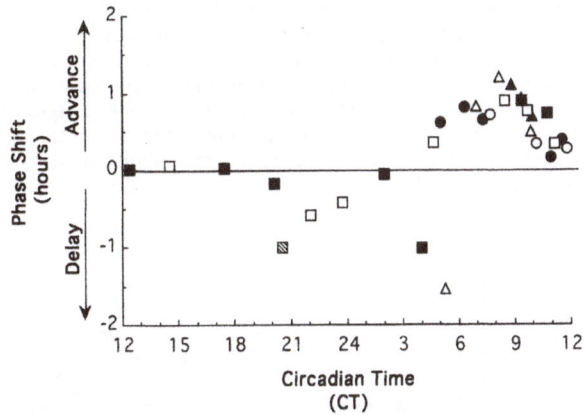

Figure 2. Phase shifts of the dim light melatonin onset (DLMO) as a function of circadian time (CT) for all nine subjects' 30 trials. For all but three trials (when melatonin was scheduled during sleep), a split dose was administered as two 0.25 mg capsules taken two hours apart. Administration time for the split doses is plotted as the mean time for example, capsules taken at CT 9 and CT 11 are shown as given at CT 10. From Lewy et al. (1992), with permission.

Several questions arise from this study. *Was there any evidence to suggest that phase shifts were the result of an effect of melatonin on the sleep-wake cycle?* Although subjects slept at home, they were asked to hold their sleep times constant and to record them in a daily log. Analysis of sleep data indicated that changes in circadian phase were not mediated by an effect of melatonin on the sleep-wake cycle. Trials in which sleep onset or offset was altered more than one hour were excluded from any further analysis (this occurred only once). To evaluate a systematic influence of sleep in the remaining data, phase shifts in sleep onset, offset and midsleep (after averaging sleep times for the first six days of each week) were plotted against DLMO phase shifts for the first eight trials (one trial per subject). No significant correlations were found. Next, one trial for each subject in the entire study was chosen randomly, and the analysis was repeated; again, no correlation was seen between changes in sleep times and DLMO phase shifts. Finally, after DLMO phase shifts from the first eight subjects' trials or the nine randomly-selected trials were compared to DLMO phase shifts minus sleep shifts, Wilcoxon's signed rank test was not significant. That observed phase shifts were generally more robust when melatonin was administered during the day further supports a direct circadian effect. Finally, it should be mentioned that holding the sleep-wake cycles constant (which was necessary so as to attribute solely to melatonin administration any resultant phase shifts) apparently caused an underestimation of the potential magnitude of melatonin's phase-shifting properties.

Why was melatonin administered for four days? This was done to maximize the possibility of achieving a consistent, measurable phase shift. Although we (Sack et al., 1991; Sack et al., 1987) and subsequently others (Arendt et al., 1988; Folkard et al., 1990; Palm et al., 1991; Sarrafzadeh et al., 1990) had shown that blind people had robust and consistent phase shifts in response to melatonin, we were doubtful that sighted people would respond as well, given that these individuals had to contend with a competing light-dark cycle. Since we were confined to a physiologic dose, we opted to give it for more than one day in order to maximize the possibility of obtaining measurable phase shifts. We restricted administration to four days, since a longer administration time might have produced too much of a shift in the phase of the melatonin PRC in the course of melatonin administration.

Why does the human PRC more closely resemble that of the lizard than that of a mammal? One explanation might be that both lizards and humans are diurnal. However, nocturnal and diurnal animals often have similar PRCs. A more likely reason is that melatonin was not administered for more than one day in the rodent studies (Armstrong and Chesworth, 1987). Recent data suggest that there is a phase delay zone where expected if mice are given melatonin for four days (M. Dubocovich, personal communication).

Why was a split dose used? Melatonin was divided between two doses given two hours apart in order to try to approximate the endogenous melatonin profile. A consolidated dose was used during certain hours, in order to avoid interference with sleep. The consolidated dose is now exclusively, since it seems no less capable of causing phase shifts. In order to be consistent with what is recommended in the field (Johnson, 1990) (Figure 3), we have re-plotted the original data using the time of administration of the first of the split doses as the reference for the zeitgeber stimulus rather than the average administration time. This appears as a one-hour advance in plotted administration time for the split doses.

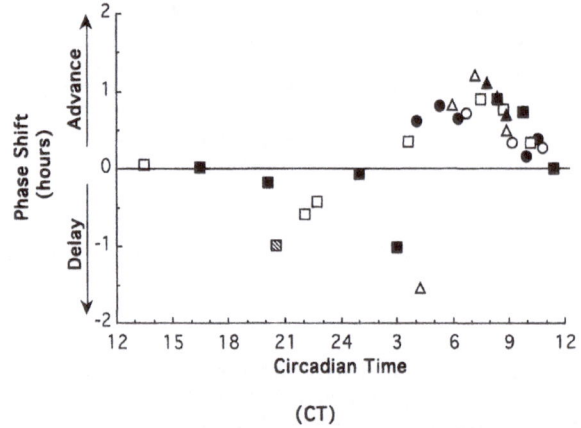

Figure 3. Data from Figure 2 re-plotted with the circadian time (CT) of melatonin administration recalculated using the time of the first 0.25 mg capsule. Therefore, for all trials using the split dose, time of administration is one hour earlier than in Figure 2.

Phase advances were observed between CT 3 and CT 11. Phase delays were found between CT 20 and CT 4. Zones of reduced responses are expected at the cross-over point between phase delays and phase advances (in a Type 1 PRC) and at the cross-over point between phase advances and phase delays (Johnson, 1990). A broad area of reduced responses was noted at the cross-over point between phase advances and phase delays (CT 11 - CT 18). Another cross-over point between phase delays and phase advances occurred at CT 3-4. Between CT 7 and CT 11, there was a linear relationship between magnitude of the phase advance and time of melatonin administration: the earlier the administration time, the greater the phase advance.

Our newest PRC data (not shown) are consistent with the original data. There continues to be remarkably little inter-individual variability in PRC phase, although response magnitude (which appears to increase in older people) varies somewhat. Indeed, the elderly respond as well or better to the phase-shifting effects of exogenous melatonin. Our new data indicate that advance responses continue linearly, slightly later than CT 14, thereby reducing the amount of "dead zone" (zone of reduced responses).

It is presumed that the time-dependency of phase-shift magnitude in the phase-advance zone will continue to be replicated. It is unclear why this is not as apparent in the phase-delay zone. Possible decreasing sensitivity of suprachiasmatic nuclei receptors (Reppert et al., 1988) as they are stimulated by endogenous melatonin during the night would provide a time-dependent change opposite to that expected from the shape of the delay zone of a typical PRC and could result in the observed square-wave appearance of this zone as indicated in Figure 3.

Other factors may also be important in modifying the classical PRC shape. Perhaps the offset of the exogenous melatonin pulse must be later than the endogenous offset to achieve a phase delay (and conversely, the onset of the melatonin pulse must precede the endogenous onset for phase advances to occur) and that for both advances and delays, the exogenous pulse must be continuous with the endogenous profile. Also, the time when orally-administered melatonin stimulates SCN melatonin receptors is somewhat later than ingestion. This last point, as well as our most recent data showing that the advance zone of the melatonin PRC extends slightly later than CT 14, are important when considering the function of endogenous melatonin (see below).

A FUNCTION FOR ENDOGENOUS MELATONIN PRODUCTION

A function for both the onset and offset of endogenous melatonin production in augmenting the entrainment and phase-shifting effects of the light-dark cycle on the ECP depends on an effect that works faster than the light-dark cycle's direct phase-shifting effect. While it is possible that the ECP shifts the same day, the DLMO and the melatonin profile clearly shift by transients. (If shown that the ECP shifts instantaneously, then this hypothesis may need to be revised somewhat.) A same-day suppressant effect is necessary in order to alter a crucial part of the melatonin profile sufficiently quickly so that there would still be some need for changing melatonin levels to help the light-dark cycle shift the ECP. Similarly, a same-day phase shift resulting from a change in melatonin levels is necessary: the Claustrat group has recently shown that exogenous melatonin can cause a phase shift within one day (Zaidan et al., 1994).

How same-day changes in melatonin levels would affect the ECP is explained by the melatonin PRC. If someone has not been getting up before dawn, an advance in wake-up time will cause the melatonin offset to occur earlier that same day thereby reducing melatonin stimulation of the melatonin PRC delay zone and thus causing a same-day advance. Also, there may be some suppressant effect of light at dawn, particularly under long photoperiods, such that a delay in sleep offset will delay the melatonin offset that same day, increasing melatonin stimulation of the delay zone. (Although we did not emphasize a suppressant effect of light at dawn in our clock-gate model (Lewy, 1983), we did not exclude this possibility.) Also, particularly under a photoperiod greater than 12 hours, an advance in dusk will increase melatonin stimulation of the advance portion of the melatonin PRC and cause a same-day phase advance of the ECP. Finally, our new data that reveal advance responses slightly after CT 14 make clear how a delay in dusk would result in removing melatonin from stimulating the advance portion of the melatonin PRC, thus causing a phase delay that same day.

Although a circadian phase-shifting role for endogenous melatonin is even more important after acute and substantial shifts in the light-dark cycle (as we experience now with shift work and after transmeridianal flight), whether or not our prehistoric ancestors were engaged in shift work, melatonin could have been important for small shifts in the light-dark cycle that result from the seasonal change in photoperiod or possibly through migratory behavior. Our newest PRC data now clarify how small changes in melatonin levels at dusk, as well as at dawn, can cause a same-day shift in the phase of the ECP through sunlight's suppressant effect. Some of these effects may be more robust during summer than during winter.

These effects depend on how sleep superimposes darkness on the light-dark cycle. Since bright light was not available in pre-modern times except from sunlight, it was not often that dusk was delayed much past the time it naturally occurred. That is, by going to sleep later, one can achieve a delay in (perceived) dusk, but only if one had been routinely going to sleep before sunset. This was probably not the case, since it is difficult to imagine our ancestors going to sleep while it was still light out. However, it is not difficult to imagine our ancestors staying asleep past sunrise. Under these circumstances, getting up earlier would cause an advance in the light-dark cycle and getting up later would cause a delay. Any change in sleep offset time would likely have a same-day effect on the time of offset of endogenous melatonin production, because of a suppressant effect of light at dawn, particularly under long photoperiods.

MELATONIN PRC AMPLITUDE AND SHIFT WORK

In addition to replicating previous studies indicating that night workers generally do not shift their circadian rhythms (Knauth et al., 1981; van Loon, 1963), our most recent night-work study shows that melatonin has robust phase-shifting effects when administered to a person who has experienced a 12-hour shift in their sleep-wake and perceived light-dark cycles (Sack et al., 1994). That is, subjects working a "7-70" shift often respond to melatonin with a shift of 1-2 hours per day (Figure 4), which is more than observed in our PRC studies. In the PRC protocols, the sleep-wake and light-dark cycles were held constant, so as to attribute any resultant phase shifts solely to exogenous melatonin administration. However, this experimental paradigm caused an underestimation of the potential magnitude of melatonin-induced phase shifts.

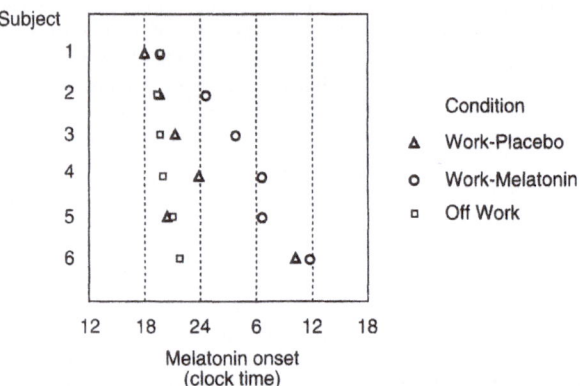

Figure 4. Dim light melatonin onsets (DLMO's) obtained at the end of the week in six night shift workers working on a "7-70" schedule (seven days of night work alternating with seven days off work). During the off week, two subjects received melatonin, two received placebo and two received no treatment; all six subjects had DLMOs (squares) that were in a normal phase indicating full adaptation to a conventional day-active schedule. During the run of night work, they were treated for a week with either 0.5 mg melatonin (circles) or placebo (triangles) given at bedtime in the morning after getting home from work (about 09.00). Treatment effects are indicated by the difference between placebo and melatonin treatment. No differences were observed in Subject #1 who did not respond to either treatment, and in Subject #6 who shifted on both treatments; however, in the other four subjects, melatonin appeared to enhance phase shifts and improve adaptation to the night work schedule. From Sack et al. (1994), with permission.

Indeed, the shifting of the sleep-wake (and ambient light-dark) cycles by 12-hours is reminiscent of the protocol employed by Czeisler and co-workers that was devised for their PRC to bright light study (Czeisler et al., 1989). It is noteworthy that their light PRC shows greater phase shifts than any found by other investigators (Honma and Honma, 1988; Lewy et al., 1987; Minors et al., 1991; Wever, 1989), in which the sleep-wake and ambient light-dark cycles were either held constant or allowed to free-run. Perhaps the Czeisler group is wrong about their claim of a Type 0 (high amplitude) light PRC, in that their greater phase shifts may have been due to 12-hour shifts in the sleep-wake and ambient light-dark cycles rather than to suppressing the amplitude of the ECP as they have speculated.

The phase-shift magnitudes obtained in both the Czeisler group's light PRC and our melatonin/shift work studies are similar. The question arises, as we originally pointed out (Lewy et al., 1994), could melatonin merely be "tipping the balance." That is, in helping shift-workers adapt to the night-work schedule, melatonin could simply be causing a slight phase shift barely sufficient to delay the cross-over point of the light PRC such that bright morning sunlight acts on the most robust portion of the phase delay zone rather than on the most robust portion of the phase advance zone (the latter might prevent a shift, whereas the former would promote one). This scenario calls into question the relative contributions of melatonin administration and bright light. Nevertheless, it should be pointed out that bright light was not specifically scheduled in the shift work study. Therefore, melatonin treatment appears to be as good as light treatment in helping shift workers adapt. Furthermore, we have evidence that at least one shift worker advanced when adapting to the 12-hour phase shift. It is difficult to imagine how melatonin administration could tip the balance in both directions, since the cross-over point of the light PRC is thought to occur somewhere between the middle and the end of the dark period.

DARKNESS AND MELATONIN PRODUCTION

Given that melatonin-induced phase shifts are greater after the sleep-wake and light-dark cycles have been shifted 12 hours, the possibility arises that melatonin administration and scheduled darkness add to (indeed, may potentiate) each other's phase-shifting effects. This suggests a second circadian role for endogenous melatonin for which there is a teleological argument. While exposure to sufficiently bright light during the night can suppress melatonin production, in almost all species darkness cannot induce melatonin production during the day. If when paired, melatonin enhances the phase-shifting effect of darkness (and vice versa), this can happen only at night and at the twilight transitions when endogenous melatonin is normally produced.

Therefore, a second circadian role for endogenous melatonin (that may or may not be identical to its phase-shifting effect as described by the melatonin PRC) is that melatonin may help the ECP distinguish between sporadic daytime darkness and nighttime darkness. While bright light was not available during the night in prehistoric times, darkness during the day could easily have been achieved by closing the eyes, going into a cave, covering the eyes, etc. As mentioned above, in almost all animals darkness cannot induce melatonin production during the day: perhaps this is to prevent sporadic daytime darkness from having a chaotic influence on ECP phase. However, it should be recognized that darkness and melatonin may be relatively less effective when scheduled during the night (at which time they normally occur), leaving the twilight transitions as the most important times under naturalistic conditions for zeitgeber effects of darkness and melatonin, or in the case of shift workers, daytime darkness enhanced by exogenous melatonin administration.

Perhaps this second circadian role for melatonin may explain why so many untreated night workers fail to shift their circadian rhythms (Figure 4): light during the nighttime work hours is probably not sufficiently bright to completely suppress endogenous melatonin production. Considering the pulses of sunlight exposure in the morning and evening as a skeleton photoperiod, the presence of endogenous melatonin production at the usual time may prevent recognition that the light-dark cycle is reversed, leaving bright morning sunlight as the same dawn signal it was prior to the shift in the sleep-wake cycle. Instead of perceiving morning light as "dusk," the night-worker's clock continues to interpret sunlight exposure at this time as dawn. That is, the ordinary-intensity light signal combined with endogenous melatonin production during the work night could prevent the ECP from interpreting sleeping during the day in total darkness (without a concomitant melatonin pulse) as the new "night cue." The darkness/sleep signal concomitant with exogenous melatonin treatment during the day might then override the nighttime ordinary-intensity light/endogenous melatonin signal as the more powerful night cue.

Of course, it cannot be ruled out that sleep per se contributes to the phase-shifting effects of sleeping in the dark. However, data from Australia (Dawson and Lack, 1988) and our group (Hoban et al., 1991) indicate that shifting the sleep-wake cycle, while holding the light-dark cycle constant, has minimal phase-shifting effects. Kronauer (Kronauer et al., 1993) has recently published similar results.

Further evidence for a potentiating effect of darkness and melatonin comes from a small pilot study (Figure 5). One of the subjects from our melatonin PRC study was asked to wear dark goggles for the duration of the exogenous melatonin pulse. The phase shift was enhanced considerably, even though darkness was scheduled at a time when it was not expected to cause a phase shift.

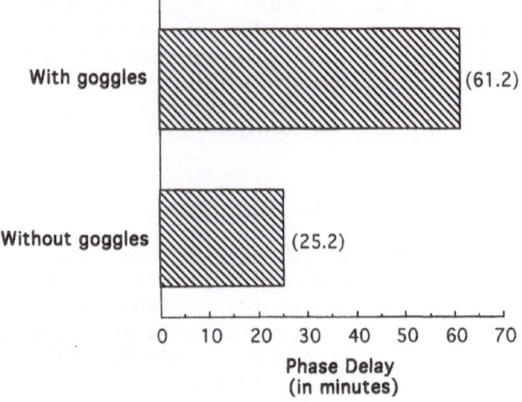

Figure 5. Phase delays (in min.) in a subject who took melatonin (0.5 mg) for four days on two separate occasions, once while wearing goggles and once without wearing goggles. Melatonin was administered at approximately the same circadian time (CT 1-2). The phase delay in the DLMO after wearing goggles was twice that achieved without the goggles. The subject had been exposed to 1-2 hours of sunlight before wearing goggles; therefore darkness at this time is not be expected to cause much of a phase shift.

CONCLUSIONS

In summary, it is clear that physiologic doses of melatonin can reliably produce changes in circadian phase, depending on when it is administered, as described by the melatonin PRC. This is not only significant for prescribing exogenous melatonin treatment, but it also has implications regarding the function of endogenous melatonin. Our newest PRC data are essentially similar to the original data. The new PRC does, however, reveal phase-advance responses that are slightly later. This finding, bolstered by the report that melatonin can cause phase shifts within one day (Zaidan et al., 1994), makes even more compelling our hypothesis that a function for endogenous melatonin is to augment phase-shifting and entrainment of the ECP by the light-dark cycle, that is, that acute changes in melatonin levels due to the suppressant effect of light can cause same-day phase shifts as described by the melatonin PRC.

ACKNOWLEDGMENTS

This monograph is in memory of Wilbur Quay, Ph.D. whose dedication and integrity, as well as prescient contributions, continue to inspire. We also wish to thank the nursing staff of the OHSU Clinical Research Center and to acknowledge the assistance of Neil L. Cutler, Rod J. Hughes, Jeanne M. Latham Jackson, Katherine H. Thomas, Clifford M. Singer, Angela J. McArthur, Rick S. Boney, Mary S. Cardoza, Neil R. Anderson, and Lynette K. Currie. Supported by Public Health Service research grants MH40161, MH00703, MH47089, MH01005, PO1 AG10794 and M01 RR00334 (OHSU GCRC), and the National Alliance for Research on Schizophrenia and Depression.

REFERENCES

Arendt, J., Aldhous, M., English, J., Marks, V., Arendt, J.H., Marks, M. and Folkard, S., 1987, Some effects of jet-lag and their alleviation by melatonin, *Ergonomics* 30:1379.
Arendt, J., Aldhous, M. and Marks, V., 1986, Alleviation of "jet lag" by melatonin: preliminary results of controlled double blind trial, *Br. Med. J.* 292:1170.

Arendt, J., Aldhous, M. and Wright, J., 1988, Synchronisation of a disturbed sleep-wake cycle in a blind man by melatonin treatment [letter], *Lancet* i:772.

Arendt, J., Bojkowski, C., Folkard, S., Franey, C., Marks, V., Minors, D., Waterhouse, J., Wever, R.A., Wildgruber, C. and Wright, J., 1985, Some effects of melatonin and the control of its secretion in humans, *in:* "Photoperiodism, Melatonin and the Pineal," D. Evered and S. Clark, eds., Pitman, London.

Arendt, J. and Marks, V., 1982, Physiological changes underlying jet lag, *Br. Med. J.* 284:144.

Armstrong, S.M. and Chesworth, M.J., 1987, Melatonin phase shifts a mammalian circadian clock, *in:* "Fundamentals and Clinics in Pineal Research," G.P. Trentini, C. de Gaetani and P. Pévet, eds., Raven Press, New York.

Czeisler, C.A., Kronauer, R.E., Allan, J.S., Duffy, J.F., Jewett, M.E., Brown, E.N. and Ronda, J.M., 1989, Bright light induction of strong (Type O) resetting of the human circadian pacemaker, *Science* 244:1328.

Dawson, D. and Lack, L., 1988, Can the position of the X oscillator be shifted while the sleep-wake cycle is held constant?, *Sleep Res.* 17:370.

Dollins, A.B., Zhdanova, I.V., Wurtman, R.J., Lynch, H.J. and Deng, M.H., 1994, Effect of inducing nocturnal serum melatonin concentrations in daytime on sleep, mood, body temperature, and performance, *Proc. Natl. Acad. Sci. U.S.A.* 91:1824.

Folkard, S., Arendt, J., Aldhous, M. and Kennett, H., 1990, Melatonin stabilises sleep onset time in a blind man without entrainment of cortisol or temperature rhythms, *Neurosci. Lett.* 113:193.

Gwinner, E. and Benzinger, I., 1978, Synchronization of a circadian rhythm in pinealectomized European starlings by injections of melatonin, *J. Comp. Physiol.* 127:209.

Hoban, T.M., Lewy, A.J., Sack, R.L. and Singer, C.M., 1991, The effects of shifting sleep two hours within a fixed photoperiod, *J. Neural Trans.* 85:61.

Honma, K. and Honma, S., 1988, A human phase response curve for bright light pulses, *Jap. J. Psychiat.* 42:167.

Hughes, R.J., Badia, P., French, J., Santiago, L. and Plenzler, S., 1994, Melatonin induced changes in body temperature and daytime sleep, *Sleep Res.* 23:496.

James, S.P., Mendelson, W.B., Sack, D.A., Rosenthal, N.E. and Wehr, T.A., 1987, The effect of melatonin on normal sleep, *Neuropsychopharmacol.* 1:41.

Johnson, C.H., 1990, "An Atlas of Phase Response Curves for Circadian and Circatidal Rhythms," Department of Biology, Vanderbilt University, Nashville, Tennessee.

Knauth, P., Emde, E., Rutenfranz, J., Kiesswetter, E. and Smith, P., 1981, Re-entrainment of body temperature in field studies of shiftwork, *Int. Arch. Occup. Environ. Health* 49:137.

Kronauer, R.E., Duffy, J.F. and Czeisler, C.A., 1993, Inversion of the sleep/wake cycle in a dim light environment has no significant effect on the circadian pacemaker, *Sleep Res.* 22:626.

Lewy, A., Sack, R. and Latham, J., 1990a, Circadian phase shifting of blind and sighted people with exogenous melatonin administration: evidence for a phase response curve., *Soc. Light Treatment Biol. Rhythms Abst.* 2:22.

Lewy, A.J., 1983, Biochemistry and regulation of mammalian melatonin production, *in:* "The Pineal Gland," R.M. Relkin, ed., Elsevier North-Holland, New York.

Lewy, A.J., Ahmed, S., Jackson, J.M.L. and Sack, R.L., 1992, Melatonin shifts circadian rhythms according to a phase-response curve, *Chronobiol. Int.* 9:380.

Lewy, A.J., Sack, R.L., Blood, M.L., Bauer, V.K., Cutler, N.L. and Thomas, K.H., 1994, *in:* "Circadian Clocks and Their Adjustment. Ciba Foundation Symposium 183," John Wiley, New York.

Lewy, A.J., Sack, R.L. and Latham, J., 1991, A phase response curve for melatonin administration in humans, *Sleep Res.* 20:461.

Lewy, A.J., Sack, R.L. and Latham, J.M., 1990b, Exogenous melatonin administration shifts circadian rhythms according to a phase response curve [Abstract 021]., *in:* "The Vth Colloquium of the European Pineal Study Group," Guilford, England.

Lewy, A.J., Sack, R.L. and Latham, J.M., 1991, Melatonin and the acute suppressant effect of light may help regulate circadian rhythms in humans, *in:* "Advances in Pineal Research," J. Arendt and P. Pévet, eds., John Libbey, London.

Lewy, A.J., Sack, R.L., Miller, S. and Hoban, T.M., 1987, Antidepressant and circadian phase-shifting effects of light, *Science* 235:352.

Lewy, A.J., Wehr, T.A., Goodwin, F.K., Newsome, D.A. and Markey, S.P., 1980, Light suppresses melatonin secretion in humans, *Science* 210:1267.

Mallo, C., Zaidan, R., Faure, A., Brun, J., Chazot, G. and Claustrat, B., 1988, Effects of a four-day nocturnal melatonin treatment on the 24 h plasma melatonin, cortisol and prolactin profiles in humans, *Acta Endocrinol. (Copenh.)* 119:474.

Minors, D.S., Waterhouse, J.M. and Wirz-Justice, A., 1991, A human phase-response curve to light, *Neurosci. Lett.* 133:36.

Palm, L., Blennow, G. and Wetterberg, L., 1991, Correction of non-24-hour sleep/wake cycle by melatonin in a blind retarded boy, *Ann. Neurol.* 29:336.

Quay, W.B., 1970, Precocious entrainment and associated characteristics of activity patterns following pinealectomy and reversal of photoperiod, *Physiol. Behav.* 5:1281.

Quay, W.B., 1972, Pineal homeostatic regulation of shifts in the circadian activity rhythm during maturation and aging, *Trans. N.Y. Acad. Sci.* 34:239.

Redman, J., Armstrong, S. and Ng, K.T., 1983, Free-running activity rhythms in the rat: entrainment by melatonin, *Science* 219:1089.

Reppert, S.M., Weaver, D.R., Rivkees, S.A. and Stopa, E.G., 1988, Putative melatonin receptors are located in a human biological clock, *Science* 242:78.

Sack, R.L., Blood, M.L. and Lewy, A.J., 1994, Melatonin administration promotes circadian adaptation to night-shift work, *Sleep Res.* 23:509.

Sack, R.L., Lewy, A.J., Blood, M.L., Stevenson, J. and Keith, L.D., 1991, Melatonin administration to blind people: phase advances and entrainment, *J. Biol. Rhythms* 6:249.

Sack, R.L., Lewy, A.J. and Hoban, T.M., 1987, Free-running melatonin rhythms in blind people: phase shifts with melatonin and triazolam administration, *in:* "Temporal Disorder in Human Oscillatory Systems," L. Rensing, U. an der Heiden and M.C. Mackey, eds., Springer-Verlag, Heidelberg.

Sarrafzadeh, A., Wirz-Justice, A., Arendt, J. and English, J., 1990, Melatonin stabilises sleep onset in a blind man, *in:* "Sleep '90," J.A. Horne, ed., Patenagel Press, Bochum.

Turek, F.W., McMillan, J.P. and Menaker, M., 1976, Melatonin: effects on the circadian locomotor rhythms of sparrows, *Science* 194:1441.

van Loon, J.H., 1963, Diurnal body temperature curves in shift workers, *Ergonomics* 6:267.

Waldhauser, F., Saletu, B. and Trinchard-Lugan, I., 1990, Sleep laboratory investigations on hypnotic properties of melatonin, *Psychopharmacology* 100:222.

Wever, R.A., 1989, Light effects on human circadian rhythms. A review of recent Andechs experiments, *J. Biol. Rhythms* 4:161.

Zaidan, R., Geoffrian, M., Brun, J., Taillard, J., Bureau, C., Chazot, G. and Claustrat, B., 1994, Melatonin is able to influence its secretion in humans: description of a phase-response curve, *Neuroendocrinol.* 60:105.

MELATONIN AND SLEEP

J David Parkes

University Department of Neurology
King's College Hospital and Institute of Psychiatry
London SE5

INTRODUCTION

The purpose of this review is two-fold. First, as an introduction to the role of melatonin in the circadian biological rhythm of sleeping and waking, I will discuss the normal sleep-wake cycle of man. Important issues here include the basic organisation of circadian mechanisms of sleeping and waking, the normal physiology of sleep and the biochemical systems involved. Sleep disorders in man and their present pharmacological management will be stressed, because of their importance in relation to circadian rhythm and melatonin research..

Second, the physiological and pharmacological effects of melatonin on this complex system will be considered from the clinical perspective. Important issues here include the physiological role of melatonin in the normal sleep-wake cycle as well as the use of melatonin in various sleep-wake disorders. These include different forms of insomnia, sleep problems in the elderly, and the delayed sleep phase syndrome.

The study of melatonin in human sleep disorders is in its infancy, and many detailed polysomnographic and clinical studies of melatonin and melatonin analogues remain to be done. However the initial results of investigations of the role of melatonin in normal and abnormal sleep present a major challenge for the members of this NATO workshop. A new class of pharmacological agent has emerged. Melatonin is a phase-setting hormone that can alter the time at which we fall asleep and wake as well as the timing of many other biological rhythms. The potential impact of melatonin and other phase-setting drugs in medicine as well as society is enormous.

This review will argue that melatonin does not act as a classic hypnotic but induces pre-sleep behaviour with some degree of tiredness and reduction in motor activity as preparation for sleep. This effect is seen with both low and high dosages of melatonin.

DEVELOPMENT OF THE SLEEP-WAKE CYCLE IN HUMANS

Considerable research has already been done on the acquisition of circadian rhythmicity and the role of the suprachiasmatic nucleus (SCN) and melatonin in many animal species although little is at present known in humans. In the human species there are major changes in sleep-wake pattern from infancy to old age. In the *30 week old fetus* there are regular cycles of sleeping and waking or at least rest and activity. *At birth* most of life is spent asleep. About half of this sleep time is occupied by the active phase of sleep, the neonatal equivalent of adult rapid eye movement (REM) sleep. *With development*, the percentage of each 24 hour cycle spent asleep as well as the percentage of time in active sleep declines.

In adults, the time spent asleep depends on ethnic, environmental, psychological, psychiatric and disease factors as well as on the day of the week, the nature of employment, the personality of the subject, the prevalent mood and the time of year. On average agricultural workers in Oxfordshire spend about one third of their life asleep and one tenth dreaming (Taub 1982). However there is a considerable variation in normal values. A few normal adult subjects sleep for only 2-3 hours each 24 hours, while others spend 10-12 hours asleep. There are no major waking behavioural correlates with sleep length (whether short or long) in normal subjects. It has been argued that many normal people sleep more, or at least spend more time in bed, that is necessary to maintain normal waking function.

In old age sleep usually deteriorates although a few 80 year old subjects sleep as well as young adults. With increasing age, the time taken to fall asleep (sleep latency) increases, total sleep time becomes shorter and night arousals more frequent. Sleep has a less refreshing quality and sleep disorders, in particular insomnia, sleep apnoea and sleep myoclonus (periodic movements in sleep) are more frequent with increasing age.

Deaths at night are more common than deaths during the day with a peak between 04.00 and 06.00. The likelihood of suffering a heart attack is greatest between 06.00 and 10.00 (Mitler et al, 1988). A common reason for the elderly dying painlessly and peacefully during their sleep is probably sleep apnoea. Diseases of many body systems may present during sleep rather than wakefulness as with ulcer pain, paroxysmal nocturnal haemoglobinuria, cluster headache and some forms of epilepsy.

Much of our knowledge of sleep mechanisms derives from animal studies. In all animal species that have been studied there is a greater prevalence of sleep in the young than in the old. Sleep-wake patterns show a great diversity in different species depending on age, environment, metabolic rate and body size. The largest living terrestrial animals, elephants, are, like humans, awake during the day and asleep at night. Elephants' total sleep time is a little less than that of humans, and they show a greater difference between winter and summer sleep.

Most mammalian species show non-rapid eye movement (NREM) - rapid eye movement (REM) sleep cycles. However REM sleep has not been identified in either the primitive species of echidna (the spiny anteater) or advanced dolphin species. A few animals, including hedgehogs, brown bats, hamsters and bears, hibernate, perhaps as an adaptation to energy conservation. Hibernation starts with a high proportion of NREM sleep followed by a dramatic decline in body temperature, heart rate and respiratory rate lasting a few hours to several weeks with eventual return to euthermia.

THE NORMAL SLEEP-WAKE CYCLE IN HUMAN ADULTS

In human adults, in contrast to many other mammalian species, sleep is usually consolidated in one single episode during each day-night cycle (with the exception of the Mediterranean siesta - this is usually taken at the lowest point of waking vigilance, commonly around 15.00). Night sleep consists of two main phases, NREM and REM sleep. (The rapid eye movements themselves are referred to as REMs). As stressed above, in infancy the equivalents of these phases are known as quiet and active sleep.

REM activity is suppressed in many conditions. However following REM suppression, a rebound of activity occurs the following night, and there must be a REM-memory system in the brain which measures and records this sleep phase.

In adults, sleep usually starts with NREM activity. However in infancy and a number of pathological states including the narcoleptic syndrome, sometimes following head injury or during recovery from metabolic coma, in the presence of depressive illness and following sleep deprivation, sleep may start with REM, not NREM activity.

NREM sleep is conventionally divided into four stages (1-4) characterised by a progressive increase in sleep depth and an increase in the strength of stimulus necessary to cause arousal. The occurrence of sleep spindles signals definite stage 2 sleep onset. Stages 3 and 4 NREM sleep (slow-wave sleep, delta sleep) comprise the "core sleep" of the early night together with the 2-3 initial REM periods. A variety of models, notably those of Hobson et al (1975) and Borbély et al (1983) have been developed to explain the brain processes that regulate these aspects of sleep homeostasis, and the interaction of sleep-wake and circadian systems.

In normal adults NREM sleep stages occupy the first 60-90 minutes of sleep followed by the first REM sleep cycle of the night. During each complete nocturnal sleep cycle there are about 4-6 NREM-REM cycles. Stage 3 and 4 NREM sleep is most intense during early sleep cycles. In contrast REM stages in the second half of the night last longer than those in the first half. Thus there is more REM activity in late than in early sleep and REM-linked sleep disorders may be more prominent in late sleep.

In NREM sleep, there are cycles of alternating electrical activation and inhibition over a 20-60 second time course. These electroencephalographic cyclic alternating patterns (CAPS) express the organised complexity of arousal-related phasic events. Frequent arousals, many or all of which may not be recalled on final waking, are particularly disruptive to sleep and may result in severe excessive daytime sleepiness.

The Rechtschaffen and Kales (1968) score system is the standard for sleep studies. Such scoring forms the basis for a number of different systems of automatic analysis which give an overview of sleep architecture. The system is being increasingly supplemented by additional physiological studies investigating, for example, motor activity or phasic events during sleep. Thus it may be important in the study of obstructive sleep apnoea to analyse changes in the QT interval, to determine the effects of sleep posture and to study anatomical changes in the upper airway during sleep.

ANATOMY OF SLEEPING AND WAKING

The control of sleeping and waking involves many brain structures including the basal forebrain area, the massa intermedia, intralaminar thalamic nuclei, mesencephalic and pontine reticular formation and raphe system nuclei as well as the locus coeruleus.

Forty years ago most researchers focused their attention on the brainstem. The reticular theory, which viewed sleep as a passive phenomenon, held a dominant position. Since then, neuro-anatomical studies of sleep have shown that the idea of a single "sleep centre" must be false.

Reticular activating system

From the time of Von Economo, who observed patients with encephalitis lethargica in 1917-27, it has been known that patients who died in a state of insomnia consistently had inflammatory lesions of the anterior hypothalamus whereas posterior hypothalamic lesions caused hypersomnia. Inflammation was usually patchy, and almost certainly areas of the hypothalamus were involved that were later recognised to contain the SCN. Many of these patients had a circadian disruption of both sleep-wake and locomotor activity although at the time, effects on other biological rhythms were not described.

From this work arose the idea that the brain had two separate systems to control sleep and waking. The experiments of Moruzzi and Magoun suggested there was a reticular formation in the brainstem which influenced the cortex through an activating system which then induced an arousal state. This system was called the ascending activating reticular system (AARS). The system became more complicated with the observation that electrical stimulation of the basal forebrain induced EEG synchronisation or sleep. The paramedian pre-optic areas which receive serotoninergic terminals originating from raphe neurons are responsible for the induction of synchronised and desynchronized sleep. One major problem at present is to understand how specific structures ascending from the brainstem, locus coeruleus and raphe modulate thalamic activity and hence EEG synchronisation and sleep.

Thalamus and hypothalamus

Various studies support the theory that sleep mechanisms, particularly REM sleep mechanisms, are localised in the brainstem but are subjected to a complex monitoring by hypothalamic and thalamic structures. The thalamus is the diencephalic structure most closely involved with the control of the sleep-wake cycle. Recent clinical studies of progressive fatal insomnia, where there is massive bilateral degeneration of the nucleus medialis dorsalis as well as anterior thalamic nuclei have stressed the importance of these regions for sleep control. Experimental investigations in animals suggest that the nucleus medialis dorsalis rather than the anterior thalamic nucleus is most closely involved. EEG synchronisation represents a transfer to the cerebral cortex of excitatory-inhibitory thalamic synaptic sequences. Although this thalamocortical synchronisation cannot be considered equivalent in all cases to behavioural sleep, this oscillatory phenomena must be important in sleep-wake control.

The participation of the hypothalamus is central to the organisation of the sleep-wake cycle as well as to circadian rhythmicity. This area may link the SCN central time clock mechanisms with thermoregulation control as well as with gene transcription mechanisms in the brain, and anterior forebrain mechanisms of sleep.

CIRCADIAN RHYTHMS AND SLEEP

The pattern of circadian rhythms in man is established in the first year of life and is then surprisingly stable. The first appearance of anything approaching normal adult sleep-wake, rest-activity cycles in human infants probably depends on the establishment of retinal connections to the hypothalamic suprachiasmatic nucleus. The study of human circadian function is intertwined with the study of the sleep-wake and locomotor rest-activity cycle. The most obvious of the biological functions in man that vary between maximum and minimum levels over a 24h period is the sleep-wake cycle.

The temperature rhythm is extremely stable. Over 24h, the temperature falls in the late afternoon, is lowest during the second third of the sleep period, and rises during morning awakening.

The normal development of an entrained 24h day after birth in human infants takes 3-4 months to establish. The initial 4 months are taken up by ultradian rest-activity cycles, waking about every 4h. This is then followed by a steady coalescence of waking periods. Proper entrainment to a 24h day usually occurs within 6-12 months of birth, although adult sleep-wake patterns are not achieved until puberty.

Do internal clocks run down in old age? Old people go to bed earlier than young people, wake up earlier and have a shorter total sleep time and more arousals at night. Perhaps the most striking circadian change in the all-night sleep pattern of aged humans is the extremely frequent interruption of sleep by long periods of wakefulness with more frequent shifts between sleep stages. Spindle activity in the sleep EEG becomes less frequent with age but this is less dramatic than the decrease in delta activity (slow wave: staged 3 and 4, corresponding to "deep" NREM sleep).

Shifts in human circadian sleep-wake, rest-activity, temperature and hormonal rhythms can be produced by many factors including prolonged sleep deprivation, large temperature changes, the administration of oral melatonin and a number of drugs. These drugs include monoamine oxidase A inhibitors and 5HT2 reuptake inhibitors, lithium and possibly sex hormones.

BIOCHEMISTRY OF SLEEPING AND WAKING

Much of the normal biochemistry of the sleep-wake cycle in man has been deduced from studies of disordered sleep, and in particular, of sleep in the narcoleptic syndrome.

There have been a few studies of human regional cerebral blood flow during sleep as compared with wakefulness. Townsend et al (1973), using [133]zenon inhalation techniques, and Sakai et al (1980), using the same technique found increases in cerebral blood flow between 3 and 41% during REM sleep but this has not always been confirmed. Looking at regional blood flow distribution during REM sleep, Madsen et al (1991) reported a decrease, not an increase in inferior frontal cortical areas.

Cerebral glucose utilisation during sleep and wakefulness in normal subjects is not homogenous. In stage 2 NREM sleep, cerebral glucose metabolism is slightly lower than during wakefulness with, in particular, a low thalamic glucose metabolism. During REM cerebral glucose utilisation is similar to that of wakefulness presumably because of reactivated neurotransmission and an increased need for ion gradient maintenance (Maquet et al, 1989a, 1989b). In the narcoleptic syndrome, where daytime sleepiness is characteristic, brain glucose metabolism may be reduced and amphetamines, which promote alertness, increase the incorporation of glucose into brain glycogen.

The role of many different neurotransmitter systems in sleep-wake control have been examined in detail in animals but surprisingly little is known with certainty in man. Most studies have looked at amine neurotransmitters. Neuronal discharges in *dopamine* neurones in the basal ganglia show, overall, little change in firing rate between sleeping and waking. In sleepy subjects with narcolepsy,. Montplaisir and his colleagues (1982) reported reduced levels of dopamine as well as the dopamine metabolite homovanillic acid, in the cerebrospinal fluid. Recent studies of D2 receptor systems in monozygous twin pairs discordant for the narcoleptic syndrome using raclopride to bind to basal ganglia D2 systems followed by positron emission tomography have shown normal raclopride uptake in both affected and non-affected co-twins. (Dahlitz et al, 1994)

Noradrenergic systems may be involved in the maintenance of vigilance. Many drugs which alter noradrenergic neurotransmission in the brain have profound effects on sleeping and waking. Both the MAO-A selective inhibitor, brofaramine, and the MAO-B inhibitor, selegiline, have vigilance stimulating effects. (Gillin et al 1976; Hohagen et al, 1993; Hublin el al, 1994; Schachter et al, 1979; Roselaar et al, 1987).

Phenylethylamine derivatives including amphetamine and methylphenidate enhance noradrenergic transmission in the central nervous system but have additional effects on other catecholamines.

Many studies have suggested a central role of 5HT mechanisms in REM sleep and, in particular, in sleep atonia which is most profound during this sleep phase. There are bulbospinal pathways which establish strong post-synaptic inhibition of spinal motoneurones during REM sleep as shown by absent tendon jerks, paucity of movement and, electrophysiologically, by reduced H-reflexes and F-responses. Changes in brain serotonin and serotonin metabolite concentration have been reported in both narcoleptic canines and

187

humans (Baker and Dement 1985, Miller et al 1990). In humans with narcolepsy low concentrations of 5-HIAA have been reported within the lumbar cerebrospinal fluid (Montplaisir et al 1982); however the majority of lumbar 5-HIAA is derived from spinal, not central, 5HT metabolism.

The *cholinergic* system plays an important part in REM sleep mechanisms but neither centrally active cholinergic agonists or antagonists have major effects on sleep length in man. In narcoleptic dogs, muscarinic cholinergic agonists elicit cataplexy but nicotinic agonists have little or no effect (Baker and Dement 1985, Mignot et al 1992; Kilduff et al 1986, Baruch et al 1987).

Fry et al (1986) and Sonka et al (1989) reported that *opiod* antagonists and partial agonists increased REM sleep latency and reduced sleepiness in normal control subjects. The importance of prostaglandins D2 and E2 in normal sleep mechanisms has recently been stressed but no changes in prostaglandin levels have been found in either narcoleptic dogs or humans (Nishino et al 1989, Wells et al 1992).

FAMILIAL AND GENETIC ASPECTS OF SLEEP-WAKE DISORDERS IN MAN

The importance of genetic control of sleep-wake and circadian homeostatic mechanisms has been explored by looking at genetic factors in human sleep-wake disorders. The distribution of total sleep time, the time of waking peak alertness, as well as respiratory sensitivity to hypoxia (which may determine certain clinical aspects of the common sleep disorder, sleep apnoea) may be inherited traits in man. "Larks" and "owls", short sleepers and long sleepers, may show a familial or possibly genetic grouping. Several sleep-wake disorders in man have a prominent genetic component.

One of the most interesting - although one of the most infrequent - of these, the *Smith-Magenis Syndrome* (SMS: with a deletion of genetic material at 17p 11.2) is characterised by developmental delay with learning difficulty. No definite circadian disorder has been described. It has been associated with absence of REM sleep; although present studies are very limited.

Major abnormalities of sleep-wake organisation, and possibly also disruption of circadian rhythms, are found in a childhood developmental disorder, the *Prader-Willi Syndrome*. Here there is a deletion of chromosome 15q with a mutation in the small nuclear ribonucleoprotein polypeptide N (SNRPN) gene with failure of normal hypothalamic development, reduced waking alertness and hypersomnia. Pathological changes have been described in the hypothalamus with an abnormal structure of the SCN.

In the *narcoleptic syndrome*, where sleep is markedly disrupted across a 24h cycle with fragmentation of REM sleep, a 98% association with an HLA gene (chromosome 6) DR2 DQw1 has been identified.

In 1986 Lugaresi et al reported 2 cases of a rapidly progressive familial disorder, named *fatal familial insomnia,* characterised clinically by untreatable insomnia, dysautonomia and motor signs and pathologically by selective atrophy of the anterior ventral and mediodorsal thalamic nuclei (Lugaresi et al, 1986). The sleep disorder is characterised by a marked reduction or loss of both slow wave and REM phases with, in the more advanced stages, complex hallucinations followed by terminal stupor and coma. The condition was subsequently shown to be a prion disease with a mutation in codon 178 of the prion protein gene. Several additional kindreds with the same mutation have been now been reported (Medori et al, 1992).

Familial and/or genetic factors are occasionally prominent in the *delayed sleep phase syndrome.* Up to one-third of subjects reported have a first degree family members with the same sleep-wake problem, or with the narcoleptic syndrome.

COMMON SLEEP-WAKE DISORDERS IN MAN

Daytime sleepiness

Daytime sleepiness (hypersomnia) is a common and serious complaint, although less common than insomnia. In a recent community survey in the United States (in Newhaven, Baltimore, St Louis, Durham and Los Angeles) 10.2% of the sample at the time of the interview described insomnia, and 3.2% described hypersomnia. Hypersomnia was most common among young and unemployed people.

The complaint of excessive daytime sleepiness includes inappropriate and undesirable sleep during waking hours; reduced motor and cognitive performance; unavoidable napping; sometimes - but not always - an increase in total 24 hour sleep time; and occasionally states of incomplete arousal with automatic behaviour and sleep drunkenness, slurred speech, impaired motor control, and difficulty in focusing. The disability caused by severe daytime sleepiness is comparable with that of severe epilepsy. Many hypersomniac patients are labelled dull, lazy, workshy, or stupid, and if they need treatment are considered to be drug addicts. They have considerable problems at school, work, and home. Daytime sleepiness is an important cause of industrial and road traffic accidents. Gaps of several years between the start of symptoms and the achievement of a definite diagnosis of the cause of the sleepiness are common.

Three common causes of excessive daytime sleepiness are obstructive sleep apnoea, the narcoleptic syndrome and idiopathic hypersomnia.

Obstructive sleep apnoea

Obstructive sleep apnoea is a very common cause of daytime sleepiness with a population prevalence in males of 1-3% of adults. A narrow upper airway collapses completely with sleep atonia, usually in the recumbent and prone posture, resulting in momentary airway occlusion, snoring and sleep hypoxia. This results in secondary daytime sleepiness as well as profound cardiopulmonary consequences with the development of pulmonary, and possibly systemic, hypertension. The condition may be lethal, particularly in combination with pre-existing other forms of cardio-respiratory disease.

The narcoleptic syndrome

The narcoleptic syndrome is characterised by excessive sleep, sleep episodes and cataplexy; and with or without sleep paralysis, hypnagogic hallucinations (pre-sleep dreams), and disturbed night sleep. An unequivocal history of cataplexy is necessary for a definitive diagnosis. Cataplexy (loss of muscle tone and paralysis of voluntary muscles) is caused by a sudden increase in arousal after being startled, surprised, excited or accompanying many sporting activities, emotion, and, in particular, laughter. This state resembles the physiological muscle atonia and paralysis of REM sleep although the subject is awake.

Idiopathic hypersomnolence

There are many common hypersomnolence syndromes, usually with prolonged night sleep, unremarkable sleep architecture, NREM sleep onset and periodic daytime sleepiness without cataplexy or sleep apnoea. These conditions include frequent snoring resulting in major disruption of night sleep with frequent micro arousals and secondary daytime sleepiness.

189

Sleep, movement and daytime drowsiness

Rhythmic leg jerking periodically during sleep with contraction of the anterior tibial muscle is a common and usually benign disorder which increases in incidence with age. The condition may be familial, and often is associated with restlessness of the legs (akathisia) during the pre-sleep period. Many sleep disorders are accompanied by this type of leg movement which is sometimes associated with daytime drowsiness though rarely with nocturnal sleep disruption. The diagnosis is established by electromyographic sleep studies.

INSOMNIA

Insomnia, the subjective complaint of poor quality sleep with, during the day. reduced wakefulness, tiredness, fatigue and poor function, is the most common sleep-wake disorder. Short-period insomnia, lasting only a few weeks, is usually due to life stress, psychological and environmental factors. In contrast, long-term persistent insomnia is often multi-factorial with a combination of medical, psychological, psychiatric and drug-related factors. In addition, genetic components may be important, particularly in familial forms of insomnia. An exact diagnostic formulation of the cause of insomnia can be difficult to achieve but it is essential to attempt this before considering treatment and aim for long rather than short term benefits in any subject with chronic insomnia. Drugs are never universal panaceas, and most treatment programmes rely on a combination of sleep hygiene, psychological support and drug treatment. There are few long-term surveys of benefit from any of these approaches (Ballinger 1990).

Drug treatment of chronic insomnia is often inappropriate. Benzodiazepines or alternatives such as chloral, dichloralphenazone, and chlormethiazole and cyclopyrolone derivatives such as zopiclone, although very effective as short-term sleep-promoting drugs are rarely if ever a universal panacea for chronic insomnia. If these drugs are used, short courses with frequent review are necessary. Once, or twice, weekly hypnotic drug treatment in a dose sufficient to give a good night's sleep is sometimes an effective strategy. Hypnotic drugs should not be given to those who are pregnant, alcoholic, have sleep apnoea or have to get up and function in the night.

DELAYED SLEEP PHASE SYNDROME

The delayed sleep phase syndrome is a form of insomnia with inability to fall asleep at conventional times but with normal sleep at a delayed clock time in respect of light-dark cycles, social, economic and family demands. The syndrome is the most common intrinsic disorder of human sleep-wake rhythmicity with a reported population prevalence rate of between 0.1 and 0.4%. In our clinic it accounts for about 2% of all subjects with a primary complaint of evening insomnia. The syndrome is familial in about one quarter of all subjects. It is more common in males than females and usually presents around puberty. The syndrome should be distinguished from a motivated sleep phase delay, associated with psychiatric problems, psychological factors or school avoidance behaviour in adolescence. The delayed sleep phase syndrome may be life-long. It appears to involve a number of biological rhythms coupled to the sleep-wake cycle including core temperature, alertness and melatonin rhythms.

The cause of the delayed sleep phase syndrome is uncertain but both genetic and environmental factors are involved. There is no evidence of intra-uterine or perinatal involvement and the reported incidence of breast feeding by mothers with affected offspring is normal as also is childhood development. Both paternal and maternal inheritance have been reported. One mechanism shown by the occasional progressive delay in sleep phase may involve the near 25-hour period of the endogenous pacemaker. Also, there may be a failure of normal entrainment mechanisms in the advance phase of the sleep-wake cycle.

ADVANCED AND NON 24-HOUR SLEEP PHASE SYNDROMES

In addition to the delayed sleep phase syndrome, a number of less common rhythmic and arrhythmic sleep-wake patterns have been described including the extremely uncommon advanced sleep phase syndrome. Non 24-hour sleep-wake patterns are sometimes described in blind or mentally retarded individuals. In these conditions the exact sleep-wake pattern may be determined largely by life-style, social and other time cues. Different patterns of sleep-wake phase may occur at different times in the same individual.

MELATONIN, SLEEPING AND WAKING IN HUMANS

Melatonin has a hypnotic effect. Early studies established that melatonin given by mouth, or intranasally, in the early afternoon or evening could induce sleep (Anton-Tay et al, 1971; Vollrath et al, 1981; Lerner and Nordlung, 1978). However there is still controversy about the physiological role of melatonin in the control of normal sleep and wakefulness as well as about the possible role of melatonin in human sleep-wake disorders.

Normal sleep is associated with a drop of body temperature. The lowest temperature occurs between 02.00 and 04.00. There is a variation of about 1 degree centigrade over the 24h cycle with a coupling of sleep-wake, dark-light, and temperature rhythms in the normal environment. However in conditions of temporal isolation, sleep-wake and body temperature rhythms may dissociate. When this occurs, sleep length may vary depending on the point on the temperature cycle at which sleep commences.

Melatonin 5mg by mouth causes a minor fall in core temperature of around 0.5°C. [All-night bright light, with wakefulness rather than sleep, causes an elevated rectal temperature in humans. In a group of 9 subjects in whom melatonin levels were suppressed by all-night sleep deprivation in bright light, infusion of melatonin lowered the elevated temperature. (Strassman et al, 1991)].

Over a 24-hour cycle, sleep itself and the changes in motor activity that accompany the sleep-wake cycle do not appear to directly influence the melatonin rhythm. However prolonged (several day) shifts in the sleep-wake cycle are associated with eventual changes in the melatonin rhythm.

Does the duration of the normal melatonin pulse influence normal sleep length? The duration of human melatonin secretion and the duration of sleep respond to changes in day length in humans as well as in many vertebrates. In man the duration of human melatonin secretion may be altered by artificial photoperiods with accompanying changes in the sleep-wake cycle. Wehr (1991) showed that the duration of nocturnal melatonin secretion was longer after exposure to a short photoperiod than to a long photoperiod, and the duration of the sleep phase was also considerably longer with exposure to short photoperiod.

Phase response curves: the sleep-wake cycle and melatonin

Melatonin shifts human circadian rhythms according to a phase-response curve. This is nearly opposite in phase with the PRCs for light exposure. Exogenous melatonin has been shown to delay the circadian rhythm of endogenous melatonin secretion when given in the morning and to advance it when given in the afternoon or early evening. (Lewy et al 1992).

Effects of melatonin on the sleep-wake cycle of normal subjects

Conflicting results have been reported of the effects of melatonin on sleep in normal subjects. Results vary with melatonin dosage and time of day of administration. Overall high dosages of melatonin (> 50mg by mouth) have a definite sedative and hypnotic effect. Low dosages have not invariably produced changes in sleep onset, maintenance or duration. Also, the results of subjective patient reporting of the effect of melatonin on sleep quality or subsequent wakefulness are not invariably in agreement with sleep laboratory studies.

Several behavioural and polysomnographic studies have looked at hypnotic effects of melatonin in young healthy subjects. In addition, the effects of night-time melatonin on next-day alertness have been determined. These studies show that melatonin does have sedative and hypnotic effects with reduction in sleep onset, reduced arousals and overall improvement in sleep efficiency (time spent asleep out of total time in bed). The initial stage of sleep (stage I) may be reduced. Waldhauser et al (1990) found additionally that wakefulness may "improve" after night sleep with melatonin.

In respect of the phase-response curve to melatonin, when melatonin (50mg) is given for one week, self-rated fatigue as measured by the Stanford Sleepiness Scale is increased during the 3 hours following melatonin intake in the morning but not when melatonin is given in the evening. However this study failed to show any cumulative effects of melatonin on sleep or other aspects of behaviour (Nickelsen et al 1989).

Phase response curves to the shift of human sleep-wake rhythm by melatonin have not yet been investigated in detail but as judged from studies in blind subjects, the use of melatonin opposite in phase with the PRC for light exposure may be critical. Initial studies in our laboratory combining light exposure at 07.00-08.00 with melatonin 5mg po at 19.00 have not shown sustained benefit in subjects with the delayed sleep phase syndrome.

Some evidence has been given that the minor hypnotic effect of l-tryptophan, a melatonin precursor, in man may be mediated by melatonin. The effects of tryptophan on vigilance, sleep and plasma melatonin were determined by double-blind administration of l-tryptophan 1, 3 and 5g both during the day and night. This resulted in the induction of sleep both during the day and night, and also a massive elevation of plasma melatonin levels. (Hajak et al, 1991).

Problems in the evaluation of the effects of melatonin on the sleep-wake cycle

One major problem in the evaluation of the effects of melatonin on the human sleep-wake cycle concerns the many environmental effects that may mask drug effects, These include body temperature, the rest-activity cycle and many different psychological, social and familial factors. Time of year and photoperiod may also compound the problem. In some investigations, phase response curves to melatonin have not been determined. Also, plasma levels of melatonin consequent upon oral treatment have not always been measured. Further problems in evaluation of the possible role of melatonin in both normal and abnormal sleep-wake patterns are the finding of both normal and abnormal sleep-wake cycles in subjects with undetectable levels of melatonin or melatonin metabolites in plasma, saliva and urine.

Melatonin profiles do not always match patterns of sleeping and waking in the blind.

Changes in sleep with advances in age in humans have not been clearly correlated with declining plasma melatonin levels, and the effect of the normal and progressive pineal calcification that occurs with age on pineal melatonin production and release is not known.

Melatonin may induce sleep. Does the sleep-wake state influence nocturnal melatonin excretion? This was investigated by Morris et al (1990). The 24-hour pattern of urinary 6-sulphatoxy melatonin excretion was monitored in a constant routine protocol, one with normal nocturnal sleep and one with continuous wakefulness. The sleep-wake state alone had no effect on nocturnal melatonin excretion but a correlation was present between the night-time percentage of the total 24-h melatonin output, and sleep length.

MELATONIN AND SLEEP-WAKE DISORDERS

Melatonin has been used in chronobiological disorders such as winter depression, in blind subjects with erratic sleep-wake cycles, in the delayed sleep phase syndrome and in groups of patients complaining of poor sleep quality, insomnia with accompanying waking fatigue and poor performance. Depressed mood has been correlated with decreased nocturnal melatonin in depressed patients.

Blind subjects

Several studies have looked at the action of melatonin in totally blind people with free-running melatonin rhythms. These are reviewed elsewhere in this volume. In some but not all such subjects daytime sleepiness and periodic insomnia are found. The importance of timing of melatonin administration to blind subjects may be critical. An 18 year old blind subject with a chronic sleep disturbance and excessive daytime sleepiness given melatonin 5mg po at 20.00, but not later, recovered normal alertness (Tzischinsky et al 1992). Overall, results suggest that the use of pulsed oral melatonin in such subjects may help to restore sleep-wake cyclicity at a conventional time.

Elderly

In elderly subjects, at least those in a hospitalised community, nocturnal melatonin profiles may be different from those living in the community with higher daytime plasma melatonin levels and an earlier nocturnal rise. It has been suggested that these changes may be related to the development of sleep disorders in elderly institutionalised people. (Baskett et al, 1991). The potential to improve daytime alertness and ameliorate night sleep in dementia, Alzheimer's disease, diffuse cerebrovascular disease and other conditions characterised by sleep disturbance in elderly people by a phase-setting drug is considerable but it must be admitted these benefits have not yet been achieved in clinical practice.

Insomnia

Chronic insomnia, as stressed above, is almost always multi-factorial. Any single pharmacological approach is most unlikely to be universally successful and the role of hypnotics in management needs careful definition. There have been few studies as yet of the actions of melatonin in such patients. Ten patients with persistent insomnia were given melatonin in 1mg and 5mg oral dosages in a double-blind design by James et al (1990). Subjects had no change in either the onset or duration of sleep although an increase in the latency to the onset of rapid eye movement sleep was noted at the 1mg dose. Patients reported less sleep on both melatonin 1 and 5mg but overall improvement in the subjective quality of sleep.

In a preliminary study, using much higher doses of melatonin (75mg by mouth), and measuring total sleep time and daytime alertness in thirteen 25-65 year old chronic insomniacs, melatonin at 22.00 daily for 14 consecutive days resulted in a significant increase in the subjective assessment of total sleep time and daytime alertness. Although the subjective assessment of total sleep time and daytime alertness improved with melatonin but not with placebo, only 6 of 13 subjects reported that melatonin caused an improvement in daytime feelings of well-being. (MacFarlane et al, 1991a, 1991b)

Melatonin in the delayed sleep phase syndrome

A number of treatment strategies have been used in the delayed sleep phase syndrome including chronotherapy with around-the-clock successive 1-2 hour delay in scheduled sleep onset time, photo-therapy (10,000 lux for 30 minutes at 08.00 for 14 days) and melatonin 5mg by mouth at 22.00. All these approaches are occasionally successful. The advance induced by melatonin 5mg by mouth given 4-6h before the time of expected sleep onset is at best partial (approximately 90 min) and may only be maintained during the treatment period (Vignau et al, 1993). However melatonin as well as melatonin analogues have been reported to cause phase advance in an animal model as well as in humans with the delayed sleep phase syndrome (Armstrong et al, 1993)

How does melatonin affect the sleep-wake cycle?

Present concepts of hypnotic and sedative drugs usually stress sleep-promoting and sleep-maintaining drug actions. However melatonin may have a subtly different effect here from that seen with "classic" hypnotics including benzodiazepines and cyclopyrolones. The present corpus of evidence indicates that melatonin has a direct effect on sleep tendency corresponding in many ways to "pre-sleep" behaviour observed in many animal studies. Extremely low dosages of melatonin may have near-comparable effects in this respect to very high (pharmacological rather than physiological) doses. Marczynski and his colleagues (1964) showed that melatonin injection into the hypothalamus of unrestrained cats resulted in pre-sleep behaviour comparable to that described in the 1920s as a consequence of electrical thalamic stimulation in the studies of Hess. In Hess's experiments the animal went to a corner of the room, curled itself up, groomed itself and took 15-30 minutes preparing for sleep. These classic experiments of Hess in cats were neglected however in the initial 30 year period of investigation of the action of melatonin on sleep and wakefulness. Major hypnotic effects and changes in sleep architecture were looked for but were seldom found, and the results with melatonin were highly dependent on the exact experimental condition. Thus we have virtually similar reports of sleep induction by melatonin with dosages between 0.1mg in non-sleep laboratory situations at midday (Dollins et al, 1994) as were found by Lieberman et al (!984) using 240mg in a sleep laboratory setting. Waldhauser et al (1990) described similar findings with dosages around 80mg po. It has often been necessary to use the "magnifying glass" approach of an artificial laboratory situation or a pathological disorder to determine any effect of melatonin on sleep. Thus, for example, normal subjects have been investigated just prior to the time of the naturally-occurring siesta or alternatively placed in a noisy, not quiet, sleep laboratory environment; or studies have focused on the delayed sleep phase syndrome.

It seems likely that the idea that melatonin is a "classic" hypnotic or sedative, like the benzodiazepines, is false. Dose-response relationships with melatonin and benzodiazepines are very different, there is a very different distribution of brain receptor sites, and different clinical phenomena are seen with melatonin and benzodiazepines. Melatonin, unlike benzodiazepines, has little effect on sleep architecture even in high dosage.

In addition to the pre-sleep behaviour enhancing actions of melatonin, melatonin may also possess an important phase-setting action on the sleep-wake cycle itself but further studies are necessary to separate these two different mechanisms.

REFERENCES

Anton-Tay F Diaz J Fernandez-Guardiola B On the effect of melatonin on human brain and its possible therapeutic implications Life Sci 1971 10 841-850

Armstrong SM McNulty OM Guardiola-Lemaitre B Redman JR Successful use of S20098 and melatonin in an animal model of the delayed sleep phase syndrome Pharmacol Biochem Behav 1993 46 45-49

Baker TL Dement WC Canine narcolepsy-cataplexy syndrome: evidence for an inherited monaminergic-cholinergic imbalance In: McGinty DJ Drucker-Colin R Morrison A Parmeggiani PL eds Brain mechanisms of sleep New York Raven Press 1985 199-234

Ballinger BR Hypnotics and anxiolytics Br Med J 1990 300 456-458

Baruch HL Kelwala S Kapen S The cardioacceleratory response to arecoline infusion during sleep in narcoleptic subjects and controls Sleep 1987 10 272-278

Baskett JJ Cockrem JF Todd MA Melatonin levels in hospitalised elderly patients: a comparison with community based volunteers Age Ageing 1991 20 430-434

Borbély AA Tobler I Croos G Sleep homeostasis and the circadian sleep-wake rhythm In: Chase M Weitzmann ED eds Sleep disorders: basic and clinical research Lancaster England: MTP Press 1983 227-44

Dahlitz MJ Vaughan R Leenders KL Khan N Parkes JD Twin studies in the narcoleptic syndrome Proc 12th Congress of the European Sleep Research Society Florence May 22-27 1994 In press

Dollins AM Zhdanova IV Wurtman RJ Lynch HS Deng MH Effects of induced nocturnal serum melatonin concentrations in daytime on sleep, mood, body temperature and performance Proc Natl Acad Sci USA 1994 91 1824-1828

Fry JM Pressman MR DiPhillipo MA Forst-Paulus M Treatment of narcolepsy with codeine Sleep 1986 9 269-275

Gillin JC Horwitz D Wyatt RF Pharmacologic studies of narcolepsy involving serotonin, acetylcholine and monoamine oxidase. In: Guilleminault C Dement WC Passouant P eds Narcolepsy New York Spectrum 1976 585-604

Hajak G Huether G Blanke J Blomer M Freyer C Poeggeler B Reimer A Rodenbeck A Schulz-Varszeggi M Ruther E The influence of intravenous l-tryptophan on plasma melatonin and sleep in men Pharmacol Psychiat 1991 24 17-20

Hess WR Stammganglion reizversuche, presented at a Frankfurt meeting of the German Physiological Society 27-30 September 1927 Ber Ges Physiol 1927 42 554-555

Hobson JA McCarley RW Wyzinski PW Sleep cycle oscillation: reciprocal discharge by two brainstem neuronal groups Science 1975 189 55-58

Hohagen F Mayer G Menche A Rieman D Volk S Meier-Ewert K Berger M Treatment of narcolepsy-cataplexy syndrome with the new selective and reversible MAO-A inhibitor brofaramine - a pilot study Journal of Sleep Research 1993 2 250-256

Hublin C Partinen M Heinonen EH Puukka P Salmi T Selegiline in the treatment of narcolepsy Neurology 1994 in press

James SP Sack DA Rosenthal NE Mendelson WB Melatonin administration in insomnia Neuropsychopharmacology 1990 3 19-23

Kilduff TS Bowersox SS Kaitin KI Baker TD Ciarenllo RD Dement WC Muscarinic cholinergic receptors and the canine model of narcolepsy Sleep 1986 9 102-106

Lerner AB Nordlung JJ Melatonin clinical pharmacology J Neural Trans 1978 Suppl 13 339-347

Lewy AJ Ahmed S Jackson JM Sack RL Melatonin shifts human circadian rhythms according to a phase-response curve Chronobiol Int 1992 9 380-392

Lieberman HR Waldhauser F Garfield G Lynch HJ Wurtman IJ Effects of melatonin on human mood and performance Brain Res 1984 323 201-207

Lugaresi E Medori R Montagna P et al Fatal familial insomnia and dysautonomia with selective degeneration of thalamic nuclei N Engl J Med 1986 315 997-1003

MacFarlane JG Cleghorne JM Brown GM Streiner DL The effects of exogenous melatonin on the total sleep time and daytime alertness of chronic insomniacs: a preliminary study Biol Psychiatry 1991 30 371-376

MacFarlane JG Cleghorne JM Brown GM Streiner DL The effects of exogenous melatonin on the total sleep time and daytime alertness of chronic insomniacs: a preliminary study Biol Psychiatry 1991 30 371-376

Madsen PL Holm S Vorstup S Freiberg L Lassen NA Wildschiodtz G Human regional cerebral blood flow during rapid eye movement sleep J Cereb Blood Flow Metab 1991 11 502-507

Marzynski TJ Yamaguchi N Ling GM Grodzinska L Sleep induced by the administration of melatonin to the hypothalamus of unrestrained cats Experentia 1964 20 436-473

Maquet P Dive D Salmno E Sadzot B Poirrier R Franco G Franck G Cerebral glucose utilisation during sleep in man J Cereb Blood Flow Metab 1989 9 S344

Maquet P Dive D Salmone E Von Frenckell R Franck G Stability of cerebral glucose utilisation in resting healthy human subjects measured by positron emission tomography and [18F]flurodeoxyglucose method European J Nucl Med 1989 15 418

Medori R Tritschler H-J Le Blanc A et al Fatal familial insomnia, a prion disease with a mutation at codon 178 of the prion protein gene N Engl J Med 1992 326 444-446

Mignot E Guilleminault C Dement WC Grumet FC Genetically defined animal models of narcolepsy, a disorder of REM sleep In Driscoll P ed Genetically defined animal models of neurobehavioural dysfunctions Boston Birkhauser 1992 89-110

Mitler MM Carskadon M Czeisler CA Dement WC Dinges DF Graeber RC Catastrophes, sleep and public policy: consensus report Sleep 1988 11 100-109

Montplaisir J, de Champlain J, Young SN, Missala K, Sourkes TL, Walsh J and Remillard G Narcolepsy and idiopathic hypersomnia: biogenic amines and related compounds in CSF Neurology 1982 32 1299-1302

Morris M Lack L Barrett J The effects of sleep-wake state on nocturnal melatonin excretion J Pineal Res 1990 9 133-138

Nickelsen T Demisch L Demisch K Radermacher B Schoffling K Influence of sub-chronic intake of melatonin at various times of the day on fatigue and hormonal levels: a placebo controlled, double blind trial J Pineal Res 1989 6 325-334

Nishino S Mignot E Fruhstorfer B Dement WA Hayaishi O Prostaglandin E2 and its methyl ester reduce cataplexy in canine narcolepsy Proc Natl Acad Sci USA 1989 86 2583-2487

Rechtschaffen A Kales A. A manual of standardized terminology, techniques and scoring system for sleep stages of human subjects 1968 NIH-NINDB Bethesda Maryland NIH Publication Number 204.

Roselaar SE Langdon N Lock CB Jenner P Parkes JD Selegiline in narcolepsy Sleep 1987 10 491-495

Sakai F Meyer JS Karacan I Derman S Yamamoto M Normal human sleep: regional cerebral haemodynamics Ann Neurol 1980 7 471-478

Schachter M Price PA Parkes JD Deprenyl in narcolepsy Lancet 1979 1 831

Sonka K Roth B Posmourova M The effect of naloxone on the symptoms of narcolepsy Agressologie 1989 30 93-96

Strassman RJ Qualls CR Lisansky EJ Peake GT Elevated rectal temperature produced by all-night bright light is reversed by melatonin infusion in man J Appl Physiol 1991 71 2178-2182

Taub JM Effects of scheduled afternoon naps and bedrest on day-time alertness Int J Neurosci 1982 16 107-127

Townsend RE Prinz PN Obrist WD Human cerebral blood flow during sleep and waking J Appl Physiol 1973 35 620-625

Tzischinsky O Pal I Epstein R Dagan Y Lavie P The importance of timing in melatonin administration in a blind man J Pineal Res 1992 12 105-108

Vignau J Dahlitz M Arendt J English J Parkes JD Biological rhythms and sleep disorders in man: the delayed sleep phase syndrome In: Light and Biological Rhythms in Man Wetterberg L (Ed) Pergamon Press Oxford 1993 261-271

Vollrath L Semm P Gammel G Sleep induction by intranasal application of melatonin Bioscience 1981 29 327-329

Waldhauser F Saletu B Trinchard-Lugan I Sleep laboratory investigations on hypnotic properties of melatonin Psychopharmacology Berl 1990 100 222-226

Wehr TA The durations of human melatonin secretion and sleep respond to changes in day length (photoperiod) J Clin Endocrinol Metab 1991 73 1276-1280

Wells P Dahlitz MJ Alexander G et al CSF cytokine concentrations in the narcoleptic syndrome J Sleep Res 1992 1 S 252

SEASONAL AFFECTIVE DISORDER , MELATONIN AND LIGHT

Lennart Wetterberg

Karolinska Institute
Institution of Clinical Neuroscience
Department of Psychiatry
St. Göran´s Hospital
S-112 81 Stockholm, Sweden

INTRODUCTION

Seasonal affective disorder (SAD) is characterized by changes of circadian and seasonal rhythms. The overall aim is to elucidate the mechanisms mediating the clinical effect of light treatment and study the covariation between biochemical rhythm markers, specially melatonin, and clinical outcome of light treatment. Light treatment has been given for 10 consecutive days and the patients treated with light 06-08 h in the morning or 18-20 h in the evening. Some subgroups of patients may be treatment responders and other patients non-responders to light therapy. Identifying clinical characteristics and biological markers for such subgroups is helpful in the choice of type of antidepressent treatment in clinical practice. In an Icelandic report the relationship between SAD and geographic latitude seemed more complex than previously thought and SAD may also be influenced by genetic adaptation (Magnusson and Axelsson, 1993).

Recent results demonstrate that bright artificial light confers certain physiological effects on human beings. It is now possible to study the normal regulation of biological rhythms and to investigate the rhythm changes that follow seasonal variations, and thereby maintain health.

The Pineal Gland and Its Hormones
Edited by F. Fraschini *et al.*, Plenum Press, New York, 1995

Biological rhythms of mood disorders

Treatment with light in some forms of depressive disorders rests on the hypothesis that light therapy at specific time periods of the day will normalize disturbed diurnal rhythms. Derangement of circadian pacemakers has been suggested in some forms of mood disorders such as recurrent brief depression.

Inner biological rhythms may be in harmony with the circadian and seasonal rhythms in the environment or deranged as in certain diseases (Wetterberg, 1990). More than 25 years ago Franz Halberg proposed that periodic depressions might result from so called free running rhythms, explained as "the continuance of an endogenous bioperiodicity consistently different from any known enviornmental schedule, i.e. from its usual synchronizer or usual pacemaker rhythm". Such "free running" rhythms have been hypothesized to be caused by seasonal alterations in light intensity and/or the ratio between the light period and the period of darkness over the 24 hours.

Light is a form of energy that can pass from one body to another without the need of any material substance. The light waves are electromagnetic in character and occupy a part of a range of wavelengths of the electromagnetic spectrum. The visible portion of the spectrum consists of wavelengths from 380 to 760 nm. We discriminate between different wavelengths within the range by the sensation of colours. Violet and blue corresponds of course to the short wavelengths, yellow and green to the middle, and red to the long visible range. Rhythm regulation by light in humans also involves the non-visual effects of light which recently has become an area of great research interest. Even low levels of light may affect the circadian secretion of hormones. Individual response to light varies in humans from the completely blind without a retina to healthy seeing persons.

Ocular mechanisms of photic effects

Ocular mechanisms mediating photic effects are dependent on many factors influencing light transmission, such as gaze behaviour relative to the light source, the age of the ocular lens which modifies the balance of wave lengths reaching the retina, the modulation of different degrees of pupillary dilation, and the structure and function of the retinal field in its different parts. Recently, Brainard et al. (1993) have shown that near-ultraviolet radiation traditionally considered outside the visual spectrum may in fact stimulate visual responses in children and young adults at least up to age 25 years.

In patients who have free-running rhythms, light may stabilize the 24-hour rhythms and by this mechanism have a therapeutic effect. One method of stabilizing an unstable circadian rhythm is to treat the patient with light of a specific intensity, for a specific length of time, at a specific period of the day. In several trials patients with seasonal affective disorder, in particular of the winter type in the northern hemisphere, have been successfully treated with light.

Light synchronizes disturbed biological rhythms

Sunlight strongly synchronizes circadian rhythms in humans with an active-awake phase during the day and one of rest-sleep during the night. The light which enters the eye has both visual as well as non-visual effects. Poor lighting conditions at work can result in complaints of ocular fatigue and reduced visual acuity. Now, it is equally apparent that the non-visual effects of light are also important for health and well-being.

The biological clock mechanism

Already at the beginning of this century it was demonstrated that some rhythms, for instance variations in body temperature with lower temperature at night compared to daytime are influenced by the light-dark cycle as reported by Simpsom and Galbraith in 1906. Certain anatomical structures in the mammalian brain such as the sumprachasmatic nucleus of the hypothalamus serve as regulators of endogenous rhythms. This so-called biological clock has been transplanted from one strain of hamsters with a short diurnal rhythm to another strain, with the result that the shorter rhythm from the host strain was transferred to the recipient (Ralph et al. 1990). This shows that the cells, which come from the superchiasmatic nucleus in the midbrain, function as a biological clock in the regulation of circadian rhythms. An inner biological clock governing temperature control was demonstrated already in the 1950's by Pittendrigh (1954, 1979).

A disturbed diurnal rhythm is a common symptom in depression. Sleep deprivation is a non-pharmacological treatment mode that has been investigated during recent decades. The effects of complete 40 h sleep deprivation have been evaluated (Roy-Byrne et al. 1984). Attempts have also been made with partial sleep deprivation (Schilgen and Tolle, 1980) as well as REM-sleep deprivation (Vogel et al. 1980). Sleep deprivation has, however, not been compared to placebo, and there have been difficulties in the assessment of whether the effect is symptomatic or truly antidepressant. Hypothetically, the effect has been considered to operate by a chronobiological mechanism irrespective of whether the effects are antidepressant or symptomatic. Support for this assumption includes the observation of changes in body temperature rhythm during this treatment (Pflug et al. 1981).

Acute exposure to light at night reduces the nocturnal decline of core body temperature and inhibits the secretion of the pineal hormone melatonin. Results show that the elevation of core body temperature induced by nocturnal exposure to bright light can be reversed completely by circumventing the decline of serum melatonin levels with concurrent oral administration of melatonin. It is thus established that melatonin as the mediator of the effect of light on core body temperature provides a rationale for the use of oral melatonin as an aid in the re-entrainment of the body temperature rhythm in desynchronized conditions (Cagnacci et al. 1993). For review about light and biological rhythms see also Wetterberg (1994).

Light therapy may be more practical than sleep deprivation since its requires less staff, as patients subjected to sleep deprivation have to be kept awake throughout an entire night as even short naps may nullify the therapeutic effect. Another practical aspect is that the antidepressant effect of sleep deprivations may cease following the first night after the treatment (Wehr et al. 1988). A third factor, which has lowered the interest in sleep deprivation, is that treatment with light for some hours in the morning is simpler to administer and has proven as effective as sleep deprivation, and rests on the same hypothetical mechanism of normalizing a disturbed diurnal rhythm.

Melatonin as a circadian rhythm marker

According to an earlier report both light and darkness affect the production of the hormone melatonin in the pineal body (Wetterberg, 1978). A coupling between the rhythm of melatonin and the adrenal cortical hormone cortisol as well as a possible rhythm disturbance of these two hormones in patients with depression was reported by Wetterberg et al. (1979). We have since had an active interest in biological rhythm disturbances in mood disorders. Further early investigations, which pointed to the importance of light for synchronizing circadian rhythms in general, were published by Pittendrigh (1954) and Wever et al. (1983) whose diurnal rhythm studies were performed in volunteers who spent prolonged periods in isolation. The system that generates the diurnal rhythms of melatonin includes signal transmission of light impulses via the retina to the hypothalamus and via the upper cervical ganglion to the pineal body. Formation of melatonin from the pineal body is rhythmically regulated by alterations in light/darkness that affect noradrenaline concentrations through nervous pathways from the eye to the hypothalamus and superchiasmatic nuclei, via the thoracic spinal cord and upper cervical ganglion to the pineal body. The nerve endings release noradrenaline at beta-adrenergic receptors which stimulate the transformation of ATP to cyclic AMP and, thereby, increase the protein synthesis in the pineal body. The activity of two enzymes, N-acetylserotonin transferase (NAT) and hydroxyindole-O-methyl transferase (HIOMT), increases in darkness, and melatonin is secreted into the blood. Blood levels of melatonin reach their highest values during the night at about 02.00 h. Melatonin formation decreases during daytime and exposure to bright light. Light treatment in depression hypothetically acts as a regulator of rhythms and normalizes a disturbed diurnal rhythm. According to an alternative hypothesis light therapy results in a general increase in biogenic amine related receptor sensitivity in the central nervous system which results in an antidepressant effect.

The first placebo controlled investigation of light treatment in depression was performed by Kripke et al. (1981) in San Diego, California, USA. They treated male patients with nonseasonal depression with red (placebo) or bright white light in the mornings (Kripke, 1981). The following year Lewy et al. (1982) described treatment with intensive light administered mornings and evenings for so called winter depression. Seasonal affective disorder has not been accepted as a separate diagnosis but is included in

DSM-III as a specfic type of depression namned "Seasonal Pattern". Apart from its appearance during winter, winter depression is characterized by an increase in appetite in certain patients with weight gain, augmented need of sleep, and a feeling of loss of energy (Kräuchi et al. 1990).

Light treatment has been used in several psychiatric departments world wide during the last decade with evidently beneficial results, especially in patients with depressions that develop during winter months (Avery et al. 1990, Czeisler et al. 1987, Kasper et al. 1990, Terman et al. 1989, Wirz-Justice et al. 1993)). These studies were directed at optimising treatment and to ascertain the most suitable duration of therapy, the best time of day for treatment, the spectral composition of the light as well as its intensity, and also to elucidate which symptoms and/or hormonal parameters may be used for prediction of therapeutic efficacy.

A pronounced effect of light treatment on melatonin rhythm has been reported by Terman at al. (1988) and Lewy et al. (1987) and is in accordance with the work by Czeisler et al. (1987) who reported that bright light also phase shifted serum cortisol and body temperature. Several other studies about the effect of light on melatonin in seasonal affective disorders has recently been reviewed by Thalén at al. (1993).

A Swedish study of light therapy

Parts of an ongoing Swedish investigation were recently reported (Wetterberg et al. 1991, Kjellman et al. 1993). The study was open and did not include placebo controls. Because of the publicity given to light treatment during recent years, many patients who want to receive light treatment are aware of previous results that have been obtained with different forms of light and belive that bright white light is the most effective. In a subset of fifty-six Swedish patients (41 women, 15 men) who fulfilled the criteria of an episode of major depression according to DSM-III-R participated. Before entry a clinical examination was performed and the patient history was scrutinised.

At the commencement of the study the patients were asked to rate their own rating of depression with the aid of Beck's depression scale. The degree of depression was determined by two psychiatrists who used the Hamilton Depression Scale (HDS) and that part of the Comprehensive Psychopathological Rating Scale (CPRS), which measures signs and symptoms of depression. Blood samples were drawn from an indwelling catheter in an antecubital vein before and after treatment. A "light test" was performed in a room with 350 candela (cd)/m^2. Analyses of the pineal hormone melatonin were performed before and after exposure to bright light at 22.00 - 23.00 h in the light test (Beck-Friis et al. 1985). A lower level of serum melatonin at 23.00 than at 22.00 h was considered as normal result of light exposure for one hour in this investigation.

Light treatment was administered either as morning treatment at 06.00 - 08.00 or as evening treatment 18.00 - 20.00 h in a room specifically designed for this purpose. In a room of this design all surfaces will receive a luminance of 350 cd/m^2. Luminance is expressed as the impression of amount of light which the eyes receive from a source of light, e.g. a bulb, or a light reflecting surface, e.g. the surfaces of the room. The strength of the light in the room is about 1.500 lux per m^2 at the height of 0.8 m above the floor. The adjustable electric fittings were equipped with 36 W low-energy fluorescent tubes of a "full-spectrum" type, 4.000 K, Ra 88 or 6.500 K, Ra 94. (K=Kelvin; Ra=Randering average = an index of colour reproduction). Furniture and patients were clothed in white to reduce absorption of white light.

Results of light therapy

The treatment was considered positive as the ratings on one or both of the instruments used for clinical evaluation of degree of depression, HDS and CPRS, decreased by at least 50% after a ten day period with light treatment. The evaluations were performed by two psychiatrists, and there was a high correlation between the results obtained by the two scales. Of the 56 patients who received light treatment, 29 improved within ten days according to the above criteria. Morning treatment was given to 38 patients, and 24 of these improved as compared to 5 of 18 who received evening treatment. Of 42 patients with seasonal depression 27 improved as compared with 2 of 14 of patients with non-seasonal disease. The most favourable results were obtained in patients with winter depression where 22 out of 30 showed amelioration. Four out of five patients with winter depression showed improvement when subjected to light treatment 6.00 - 8.00 a.m.. Most patients with winter depression were given maintenance treatment 1 to 3 times weekly after the initial therapy. Out of 20 patients with winter depression who received morning light treatment and who showed improvement, 19 patients had lower serum melatonin levels at 23.00 than at 22.00 h. The same pattern was, by contrast, observed in only one of five patients who did not improve.

Light treatment in depression seems to confer favourable effects, and the best results are obtained in patients with depressive episodes that occur during the winter months. The two observations of marked differences in therapeutic response in patients with seasonal depression as compared with those who suffer from nonseasonal depression as well as a more favourable effect when light treatment is provided in the mornings rather than in evenings, speak against placebo effects and non-specific therapeutic effects due to overall hospital care as the full explanation of the treatment effects.

Further evaluation of light treatment needed

The light therapy requires further evaluation for optimization before general recommendation. Previous reports have shown that there are subgroups of patients with mood disorders with different profiles of biological markers (Wetterberg et al. 1990). The whole research field would gain, if valid and reliable biological markers for subgroups of depression were found and treatment response could be predicted by objective markers.

Prevention of recurrent seasonal depression with light therapy

Phototherapy has an evident place in future studies of depressive conditions, and particularly so in patients with winter depressions who seem to respond most favourably. An interesting field of study is whether light therapy also can be used as a preventive measure against this apparently specific form of depression. Not all patients who suffer from seasonal affective disorder, winter type, become depressive every winter (Rosenthal et al. 1984). In an English study, 2/3 of winter depressives diagnosed in the summer became depressed in the following winter while 1/3 did not (Thompson, 1989). Meesters and coworkers (1991) administered light at the development of the first signs of a winter depression. No patient in a group of ten treated in this way developed any signs of depression during the rest of the winter season, while five of seven patients from a control group became depressed and needed treatment during the winter season. Light treatment seems to cause specific brain metabolic patterns which are currently being studied by positron emission tomography (PET) (Cohen et al. 1992). The functional abnormalities found in patients with winter depression could possibly be explained by a dampened behaviour. Any causal relationship between altered behaviour and brain changes found, must however wait for results of future studies. Such studies should, in my view, include the therapeutic as well as preventive effects of light therapy as well as different serotonergic drugs in controlled longitudinal studies in patients with subtypes of depression.

CONCLUSIONS

A topic of interest is to compare the possible preventive effects of light treatment with antidepressant drugs in longitudinal controlled investigations. In a single case study Wirz-Justic at al. (1992) reported therapeutic and possibly preventive effects of light therapy as well as treatment with the antidepressant drug citalopram. They concluded that central serotonergic mechanisms may be involved in the therapeutic effect of light treatment in SAD. As in the psychopharmacological literature, the apparent efficacy of placebo has varied widely. The importance of placebo controlled studies of phototherapy in winter depression has been discussed by Eastman et al. (1993).

The intensity and duration of light, and time of day of light exposure in prevention of depression need more studies. The data, so far, does not support the hypothesis that high expectations are linked to the depression profile in SAD or low expectations to the melancholic depression (Terman, 1993). Given the increasing use of light to treat and prevent mood disorders, it seems useful from the clinical perspective to establish the interaction between the mechanisms in the retina of the human eye and the pineal gland hormone melatonin which may carry markers for biological vulnerability of SAD as well as mediate the therapeutic effects of light.

REFERENCES

Avery, D.H., Khan, A., Dager, S.R. , Cox, G.B., and Dunner D.L., 1990, Bright light treatment of winter depression: morning versus evening light. *Acta Psychiatr Scand* 82:335.

Beck-Friis, J., Borg, G., Wetterberg, L., 1985, Rebound increase of nocturnal melatonin levels following evening suppression by bright light exposure in healthy men: relationship to cortisol levels and morning exposure. In: Wurtman ,R.J., ed. *The Medical and Biological effects of Light.* Ann NY Acad Sci, 453:371.

Brainard, G.C., Gaddy, J.R., Barker, F.M., Hanifin, J.P., and Rollag, M.D., 1993, Mechanisms in the eye that mediate the biological and therapeutic effects of light in humans. In: Wetterberg, L., ed. *Light and Biological Rhythms in Man.* Pergamon Press, London, 29.

Cagnacci, A., Soldani, R., and Yen S.S.C., 1993, The effect of light on core body temperature is mediated by melatonin in women. *J Clin Endocrinol Metab* 76:1036.

Cohen, R.M., Gross, M.,,Nordahl, T.E., Semple, W.E., Oren, D.A., and Rosenthal N., 1992, Preliminary data on the metabolic brain pattern of patients with winter seasonal affective disorder. *Arch Gen Psych* 49:545.

Czeisler, C.A., Kronauer, R.E., Mooney , J.J., Andersson, J.L. and Allan, J.S., 1987, Biological rhythm disorders, depression and phototherapy. *Psychiatr Clin North Am* 10:687.

Eastman, C.I., Young, M.A., and Fogg L,F., 1993, A comparison of two different placebo-controlled SAD light treatment studies. In: Wetterberg, L., ed. *Light and Biological Rhythms in Man.* Pergamon Press, London, 371.

Kasper, S. Rogers, S.L.B. Madden, P.A. Joseph-Vanderpool J.R., and Rosenthal N.E., 1990, The effects of phototherapy in the general population. *J Affect Disord* 18:211.

Kjellman, B., Thalén, B-E., and Wettereberg, L., 1993, Light treatment of depressive states; Swedish experience at latitude 59° North. In: Wetterberg, L., ed. *Light and Biological Rhythms in Man.* Pergamon Press, London, 351.

Kripke, D,F., 1981, Photoperiodic mechanisms for depression and its treatment. In: Perris, C., Struwe, G., Jansson, B, *Biological Psychiatry.* Elsevier-North Holland Biomedical Press, 1249.

Kräuchi, K. ,Wirz-Justice, A., and Graw, P., 1990, The relationship of affective state to dietary preference: winter depression and light therapy as a model. *J Affect Disord* 20:43.

Lewy , A.J., Sack, R.L., Miller, L.S., Duncan, C.C., Jacobsen, F.M., and Wehr, T.A., 1987, Antidepressant and phase-shifting effects of light. *Science*, 235,352.

Lewy, A.J., Kern, H.A., Rosenthal, N.E., Wehr, T.A., 1982, Bright artificial light treatment of a manic-depressive patient with a seasonal cycle. *Am J Psychiatry* 139:1496.

Magnusson, A., and Axelsson, J., 1993 The prevalence of seasonal affective disorder is low among descendants of Icelandic emigrants in Canada. *Arch Gen Psychiatry, 50:*947.

Meesters ,Y., Lambers, P.A., Jansen, J.H.C., Bouhuys, A.L. Beersma, D.G.M., van den Hoofdakker, R.H., 1991, Can winter depression be prevented by light treatment ? *J Affect Disorders* 23:75.

Pflug, B., Johnsson, A., and Ekse, A.T., 1981, Manic-depressive states and daily temperature. Some circadian studies. *Acta Psychiatr Scand* 63:277.

Pittendrigh, C.S., 1954, On temperature independence in the clock-system controlling emergence in Drosophila. *Proc Natl Acad Sci USA* 40:1018.

Pittendrigh, C.S., 1979, Some functional aspects of circadian pacemakers. In: Suda ,M., Hayaishi, Nakagawa H. eds. *Biological rhythms and their central mechanism.* Amsterdam; Elsevier/North-Holland Biomedical Press, 3.

Ralph, M.R., Foster, R.G., Davis, F.C., and Menaker, M. ,1990, Transplanted superchiasmatic nucleus determines circadian period. *Science* 247:975.

Rosenthal, N.E., Sack, D.A., Gillin, J.C., Lewy, A.J., Goodwin F.K., Davenport, Y., Mueller P.S., Newsome D.A., and Wehr, T.A., 1984, Seasonal affective disorder. A description of the syndrome and preliminary findings with light therapy. *Arch Gen Psychiatry* 41:72.

Roy-Byrne, P.P., Uhde, T.W., Post, R.M., 1984, Antidepressant effects of one night's sleep deprivation: Clinical and theoretical implications. In: Post RM. Ballenger, J.C. eds. *Neurobiology of Mood Disorders.* Baltimore: Williams & Wilkins, 817.

Schilgen, B. ,Tolle, R., 1980, Partial sleep deprivation as therapy for depression. *Arch Gen Psychiatry* 37:267.

Simpsom, S., Galbraith, J.J., 1906, Observations on the normal temperature of the monkey and its diurnal ariation, and on the effect of changes in the daily routine on the variation. *Trans R Soc Edinb* 45:5.

Terman ,M., 1993, Problems and prospects for use of bright light as a therapeutic intervetion. In: Wetterberg, L, ed. *Light and Biological Rhythms in Man.* Pergamon Press, London, 421.

Terman ,M., Terman, J.S., Quitkin, F.M., McGrath, P.J., Stewart, J.W., and Rafferty, B., 1989, Light therapy for seasonal affective disorder: A review of efficacy. *Neuropsychopharmacol* 2:1.

Terman, M., Terman, J.S., Quitkin, F.M., Cooper, T.B., Lo, E.S., Gorman, J.M., Stewart, J.W., and McGrath, P.J., 1988, Response of the melatonin cycle to phototherapy for seasonal affective disorder, *J Neural Transm,* 72, 147.

Thalén , B-E., Kjellman, B.F., and Wetterbertg, L., 1993, Phototherapy and melatonin in relation to seasonal affective disorder and depression. In *Melatonin; Biosynthesis, physiological effects, and clinical applications.* Ed. Yu, H-S., and Reiter, R.J., CRC Press, London. 495.

Thopmson ,C., 1989, The syndrome of seasonal affective disorder. In: Thompson, C. ,Silverstone ,T. ,eds. *Seasonal affective disorder.* CNS, London, 37.

Vogel, G.W., Vogel, F., McAbee, R.S., Thurmond, A.J., 1980, Improvement of depression by REM sleep deprivation. New findings and a theory. *Arch Gen Psychiatry* 37:247

Wehr, T.A., Rosenthal, N.E., and Sack, D.A., 1988, Environmental and behavioral influences on affective illness. *Acta Psychiatr Scand* 77 (Suppl): 44.

Wetterberg, L., 1978, Melatonin in humans: physiological and clinical studies, *J Neural Transmission* (Suppl) 13:289.

Wetterberg, L., 1990, Chronobiology in psychiatry. In: Stefanis, C.N., Soldatos, C.R., and Rabavilas, A.D., eds. *Psychiatry: A World Perspective.* Excerpta Medica. 2:776.

Wetterberg, L., 1994, Light and biological rhythms. *J Int Med,* 235:5.

Wetterberg, L., Beck-Friis, J., Aperia, B., and Petterson, U., 1979, Melatonin/cortisol ratio in depression. *Lancet* 2:1361.

Wetterberg, L., Beck-Friis, J., and Kjellman, B.F., 1990, Melatonin as a marker for a subgroup of depression in adults. In: Shafii, M. ,Shafii, S.L., eds. *Biological rhythms, mood disorders, light therapy and the pineal gland.* Am Psych Press, Washington, 69.

Wetterberg, L., Kjellman, B., Thalén, B-E., Beck-Friis, J., Freund-Levi, Y., Friberg ,Y., Nordlund, A-L., Sparring Björkstén, K., Wiberg, B., Zettergren, M., and Wibom, R., 1991, Light treatment in depression. *Läkartidningen* , 88:310.

Wever, R.A., Polasek J., and Wildgruber, C.M., 1983, Bright light affects human circadian rhythms. *Pflugers Arch* 396:85.

Wirz-Justice, A., Graw, P., Kräuchi, K., Haug, H-J., Leonardt, G., and Brunner, D.P., 1993, Effect of light on unmasked circadian rhythms in winter depression. In: Wetterberg L, ed. *Light and Biological Rhythms in Man.* Pergamon Press, London, 385.

Wirz-Justice, A., Van der Velde P., Bucher, A., and Nil, R., 1992, Comparison of light treatment with citalopram in winter depression: a longitudinal single case study. *Internat Clin Psychopharmacol* 7:109.

MELATONIN ACTION ON HUMAN BREAST CANCER CELLS: INVOLVEMENT OF GLUTATHIONE METABOLISM AND THE REDOX ENVIRONMENT

David E. Blask and Sean T. Wilson

The Mary Imogene Bassett Research Institute
Cooperstown, NY 13326-1394 USA

INTRODUCTION

In spite of the evidence that melatonin inhibits breast cancer growth *in vivo* and *in vitro* (*1,2*), very little is known regarding the mechanism(s) of action by which this indoleamine exerts its oncostatic action (*3*). Recently, some progress has been made in elucidating melatonin's mechanism of action at the cellular level using the estrogen receptor (ER) + human breast cancer cell line MCF-7. In this cell line, physiological concentrations of melatonin retard cell growth by delaying the progression of cells from G_0/G_1 to S phase of the cell cycle (*4*). Furthermore, melatonin not only blocks estrogen-stimulated MCF-7 cell growth, but it also down-regulates the expression of ER protein as well as ER mRNA (*5,6*) while transiently increasing the expression of *c-fos* (*7*). Additionally, melatonin inhibits the mitogenic action of other peptide growth factors such as prolactin and epidermal growth factor (*8,9*). Also, the antiproliferative action of physiological levels of melatonin appears to be restricted to ER+ human breast cancer since ER- cells have little or no response to this indole (*10*).

Owing to its high degree of lipophilicity and its ability to influence a variety of mitogenic response pathways in MCF-7 cells, melatonin may act at multiple sites including the intracellular compartment (*11*). In fact, melatonin itself may be an intracellular free-radical scavenger (*12*) and have the ability to increase the levels of glutathione (GSH) (*13,14*), another important free-radical scavenging molecule found in abundance within both normal and cancer cells (*15*). Glutathione is important in the proliferation of normal cells (*16*) and has been shown to be involved in mediating the cytotoxic effect of the anticancer drug taxol on MCF-7 cells grown *in vitro* (*17*). It is against this background that we have been investigating the role of GSH in the mechanism(s) of melatonin's oncostatic action on MCF-7 cells as well melatonin's influence on the production of GSH in this cell line. Moreover, we have been examining the relationship between melatonin and glutathione S-transferase (GST), an important enzyme involved in GSH conjugation reactions and multidrug resistance (*18*), in ER- human breast cancer cells as well as in MCF-7 cells that have developed resistance to melatonin.

THE ROLE OF GSH AND GST IN CANCER THERAPY

Glutathione is the major non-protein thiol found within mammalian cells and serves as one of the intracellular environment's most important redox molecules. It functions in a variety of reductive processes essential for DNA synthesis, protein metabolism (i.e., synthesis, degradation and folding), enzyme regulation as well as drug and hormone metabolism via conjugation reactions involving GST. As a potent antioxidant, GSH, through its redox cycling, provides cells with an impressive degree of protection against oxidative stress, free radical damage and other types of toxicity (*15*). Through a complex metabolic interrelationship, both GSH and GST have an important impact on determining the therapeutic responsiveness or resistance of cancer cells to anticancer drugs (*19*).

THE GSH REDOX CYCLE AND THE EXPRESSION OF GSH AND GST IN BREAST CANCER CELLS

The vast majority (99.5%) of GSH molecules within cells exist in the reduced state while the remainder are in the oxidized form (GSSG). The synthesis of GSH occurs intracellularly via two consecutive reactions first involving γ-glutamylcysteine synthetase (GCS) (rate-limiting enzyme) which catalyzes the formation of an amide linkage between cysteine and the γ-carboxyl of glutamate followed by the GSH synthetase (GS)-mediated reaction of glycine with the cysteine carboxyl of γ-glutamylcysteine to form γ-glutamylcysteinylglycine (GSH) (Fig. 1). Although GSH exists in a number of metabolic forms that are dynamically interchangeable, under steady-state conditions most of the GSH is in the reduced form which can be readily oxidized by glutathione peroxidase (GSH-Px) to GSSG, under conditions of oxidative stress (Fig. 1). The low levels of intracellular GSSG are then maintained primarily by its redox cycling back to reduced GSH in a reaction catalyzed by glutathione reductase (GSH-R) in the presence of reducing equivalents from nicotinamide adenine dinucleotide phosphate (NADPH) generated from the hexose monophosphate shunt (Fig. 1) (*20*). Additionally, a family of enzymes collectively known as glutathione-S-transferases (GST) are involved in the conjugation of GSH with a variety of exogenous substances as well as endogenous compounds including hormones (*21*).

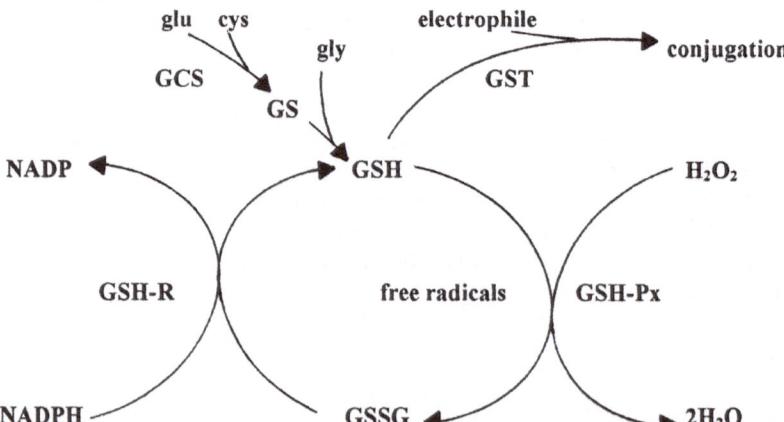

Figure 1. Glutathione metabolism and redox cycle. GSH, reduced glutathione; GSSG, oxidized glutathione; GST, glutathione S-transferase; GSH-Px, glutathione peroxidase; GCS, γ-glutamylcysteine synthetase; GS, glutathione synthetase; GSH-R, glutatione reductase; NADPH, nicotinamide adenine dinucleotide phosphate; glu, glutamate; cys, cysteine; gly, glycine.

The overexpression of GST, particularly the isoenzyme GSTπ, has been implicated in the development of multidrug resistance in human breast cancer cells. That is, those breast cancer cells phenotypically expressing high levels of GST exhibit a reduced response or fail to respond to a variety of anticancer drugs. For example, MCF-7 cells that have developed resistance to the chemotherapeutic agent adriamycin (Adr^R MCF-7) and express very high levels GST, are cross-resistant to other chemotherapeutic agents as well (22). As compared with wild-type (WT) MCF-7 cells, Adr^R MCF-7 cells express 40 to 50-fold more GST activity and significantly lower GSH levels (23,24). Therefore, GST, together with multidrug resistance-associated membrane proteins, may be part of a more complex multidrug resistance system operating in breast cancer cells (25).

EXPERIMENTAL MANIPULATION OF GSH AND GST AND THE RESPONSE TO ANTICANCER AGENTS

The advent of pharmacological agents that alter the metabolism of GSH have provided a useful approach for determining the role of GSH and GST in governing the response of cancer cells to anticancer drugs. For example, buthionine sulfoximine (BSO), a very specific and effective inhibitor of GCS and thus GSH synthesis, has been shown to deplete cancer cells of GSH and make them more sensitive to certain classes of anticancer compounds, particularly alkylating agents (26). On the other hand, it was recently shown that the cytotoxic effects of taxol on MCF-7 cells *in vitro* were completely blocked by inhibition of GSH synthesis with BSO (17). Thus, a reduction in the levels of GSH in some cancer cells can actually decrease their sensitivity to antineoplastic agents (19).

Another pharmacological agent that has been used to alter the sensitivity of cancer cells to chemotherapeutic drugs is the diuretic ethacrynic acid (EA). Ethacrynic acid is a potent inhibitor of GST activity and as such, has been used to enhance the antineoplastic effects of cytotoxic drugs on tumor cells which express high levels of GST and have acquired drug resistance (27,28). Moreover, EA forms a conjugate with GSH which is a more potent inhibitor of GST activity than EA alone (29).

We have taken advantage of the ability of BSO and EA to alter various aspects of GSH metabolism in order to study the role of this important biochemical pathway in mediating melatonin's mechanism of action on MCF-7 cell growth in culture. Specifically, we have determined the effects of BSO on melatonin's antiproliferative effect on MCF-7 cell growth as well as melatonin's effects on GSH in these cells. Furthermore, we have determined whether EA increases the sensitivity of human breast cancer cells, relatively resistant to melatonin, to the antiproliferative effects of this molecule.

EFFECTS OF GSH SYNTHESIS INHIBITION ON MELATONIN'S ONCOSTATIC ACTION ON MCF-7 HUMAN BREAST CANCER CELL GROWTH

As alluded to above, BSO has been shown to eliminate the cytotoxic effects of taxol on MCF-7 cells in culture by blocking the synthesis of GSH and causing a depletion in the intracellular levels of this endogenous antioxidant (17). We have used a similar strategy to determine whether the response of MCF-7 cells to the oncostatic effects of physiological levels of melatonin are altered when GSH synthesis is inhibited with BSO. Additionally, we have determined melatonin's effects on intracellular levels of total GSH (reduced GSH + GSSG) in the absence or presence of BSO.

The first hint that GSH was involved in melatonin's inhibitory mechanism of action on MCF-7 cell growth came from experiments in which the simultaneous exposure of cells to both melatonin (1 nM) and BSO (1 μM) completely blocked melatonin's inhibitory effect (*14*). Additionally, we determined that exogenously added GSH (1 μM) reverses the melatonin-blocking effect of BSO on cell growth, further suggesting that GSH is a critical component of the biochemical pathways mediating melatonin's oncostatic action on this cell line (Table 1). We found this to be an interesting result since native GSH apparently does not readily re-enter most cells due to its extracellular breakdown by γ-glutamyl transpepitdase (GGT). Perhaps enough GSH escaped extracellular degradation and entered the cells to reconstitute intracellular GSH concentrations sufficiently (see below) to restore the response to melatonin.

Table 1. Effects of melatonin (MLT) (1 nM), buthionine sulfoximine (BSO) (1 μM), glutathione (GSH) (1 μM) and vehicle (CTL) either alone or in combination on MCF-7 cell growth (% Δ from plating density) and intracellular total GSH levels (nM/10^6 cells) following 5 days of culture in DMEM with 10% FBS.

Treatment	Cell Growth	GSH
CTL	892.3 ± 23.5	13.1 ± 1.9
MLT	253.5 ± 10.9*	45.9 ± 7.4*
MLT + BSO	855.8 ± 30.4	1.0 ± 0.6
MLT + BSO + GSH	367.3 ± 38.6*	19.8 ± 4.1*
BSO	850.3 ± 20.5	1.7 ± 0.7
GSH	871.8 ± 32.3	10.5 ± 5.8

Data are expressed as the mean ± SE. * $p < 0.05$ vs MLT + BSO

Having established that GSH is involved in melatonin's action, we also tested whether melatonin itself altered the intracellular content of GSH and whether such a change would be negated by treatment with BSO. A recent report by Kothari and Subramanian (*13*) indicated that pharmacological doses of melatonin increase concentrations of GSH in both liver and mammary gland tissue from female rats treated with the mammary tumor-inducing carcinogen dimethylbenzanthracene (DMBA). Thus, the *in vitro* MCF-7 cell system provided us with the opportunity to determine whether physiological levels of melatonin would act directly on the GSH-generating pathway.

Following five days of continuous incubation with melatonin (1 nM), cell proliferation is not only inhibited but the intracellular content of GSH in MCF-7 cells increases anywhere from two- to four-fold over the levels in control cells (Table 1). When doses of melatonin from 10^{-13}M to 10^{-5}M are tested, a bell-shaped pattern of GSH levels results with 10^{-9}M melatonin eliciting the maximal GSH-stimulatory response in addition to the maximal growth-inhibitory effect (unpublished results). Moreover, this melatonin-induced increase in GSH is completely blocked by BSO (Table 1) suggesting that melatonin increases the synthesis and/or activity of GCS. Although augmented GSH levels might alternatively be due to a melatonin-induced decrease in the degradation of GSH, this seems unlikely since GSH degradation occurs extracellularly and is catalyzed by GTT and dipeptidases bound to the external surface of the plasma membrane (*26*).

In order to determine whether the increase in GSH is a primary response to melatonin or a secondary result of melatonin-inhibited cell growth, we tested the effects of tamoxifen on GSH since this non-steroidal antiestrogen is an effective inhibitor of MCF-7 cell growth. Although tamoxifen inhibited cell growth by more than 50%, no change in GSH levels occurs in comparison with control cells indicating that the melatonin-induced increase in GSH levels is not a phenomenon merely secondary to cell growth inhibition. Furthermore, the fact that BSO does not block the antiproliferative effects of tamoxifen on this cell line (unpublished results) also indicates that these compounds may ultimately inhibit MCF-7 cell growth by very different mechanisms of action .

EFFECTS OF INHIBITION OF GST ACTIVITY ON MELATONIN'S ACTION ON MELATONIN RESISTANT HUMAN BREAST CANCER CELLS

Our previous work demonstrated melatonin's oncostatic action on breast cancer growth *in vitro* appears to be restricted to ER+ breast cancer cell lines since ER- cells (i.e., Hs578t, BT-20, MDA-MB-231, MDA-MB-364) fail to respond to this indole (*10*). This suggested that the ER content and/or other components of the estrogen-response pathway are critical determinants of melatonin's action on breast cancer cell growth. However, another feature common to a number of ER- cell lines, particularly the Hs578t line, is that they express high levels of GST activity (*24*). Additionally, high levels of certain GST isozymes have been found in ER-poor primary breast cancers as compared with ER-rich tumors (*30*). Thus, we became interested in the possibilty that the refractoriness of ER- cell lines to the oncostatic action of melatonin might be related to a more general phenomenon of multidrug-resistance due to the over-expression of GST rather than or in addition to the lack of an estrogen-response system in these cells.

In order to test this hypothesis we determined whether EA, an inhibitor of GST, would affect the sensitivity of the ER- human breast cancer cell line Hs578t to melatonin. In our laboratory, this cell line not only exhibits little or no response to melatonin but expresses high levels of GST activity (7.0 - 9.0 nM/min/mg protein) as compared with our melatonin-responsive MCF-7 cells in which GST activity is undetectable (unpublished results). The exposure of Hs578t cells to melatonin (1 nM) for five days resulted in only a 24% inhibition of cell proliferation. However, in the presense of EA (1 µM), melatonin inhibited cell proliferation by 85% (Table 2). Interestingly, when BSO was added to the cultures, the inhibitory action of melatonin in combination with EA was completely blocked suggesting that GSH is involved in mediating the oncostatic action of melatonin in EA-treated cells. This postulate is further supported by the fact that melatonin causes a three-fold stimulation of GSH in EA-treated cells whereas BSO completely prevents this effect (Table 2). We have documented a similar effect of EA on the sensitivity of MLT^R MCF-7 cells to melatonin which have spontaneously developed resistance to melatonin's inhibitory effects over several passages (unpublished results).

SUMMARY AND CONCLUSIONS

There is little question that the intracellular metabolism of GSH, particularly through its synthetic and conjugation pathways, represents an important biochemical determinant of the responsivenss or resistance of cancer cells to therapeutic agents (*19*). The ability of

Table 2. Effects of melatonin (MLT) (1 nM), ethacrynic acid (EA) (1 μM), buthionine sulfoximine (BSO) (1 μM) and vehicle (CTL) either alone or in combination on MCF-7 cell growth (% Δ from plating density) and intracellular total GSH levels (nM/10^6 cells) following 5 days of culture in DMEM with 10% FBS.

Treatment	Cell Growth	GSH
CTL	289.8 ± 50.3	8.8 ± 0.9
MLT	226.8 ± 12.4	11.1 ± 0.7
MLT + EA	43.8 ± 13.8*	29.3 ± 3.1*
MLT + EA + BSO	233.8 ± 16.4	7.1 ± 0.3
BSO	281.0 ± 19.7	4.6 ± 0.2
EA	291.8 ± 32.0	8.8 ± 1.1
EA + BSO	309.0 ± 25.6	4.5 ± 0.3

Data are expressed as the mean ± SE. * $p < 0.05$ vs MLT and MLT + EA + BSO

melatonin to easily enter cells makes the various components of GSH metabolism easily accessible to this unique molecule. Furthermore, we have taken advantage of two important pharmacological tools, namely BSO and EA, to inhibit the synthesis and conjuation of GSH, respectively, in addition to measuring GSH itself in order to study the potential involvement of this endogenous antioxdiant in the intracellular mechanism(s) mediating melatonin's oncostatic action on human breast cancer cells. From our studies, it appears that like the anticancer agent taxol (*17*), a physiological concentration of melatonin requires sufficient GSH in order to inhibit MCF-7 cell growth under the conditions of our experiments. The fact that melatonin induces a robust increase in GSH levels in mammary cancer cells is also consistent with a report on the ability of pharmacological doses of melatonin to increase mammary gland and liver tissue levels of GSH in rats either treated or not treated with the mammary carcinogen DMBA (*13*). Additionally, we apparently can convert ER- Hs578t and MLT^R MCF-7 cells, which exhibit little or no response to melatonin, to a considerably more responsive phenotype by inhibiting GST with EA. Even more remarkable is the fact that GSH is also required for inhibition of the growth by and increases in response to melatonin in these EA-treated cells. As interesting as these data are, they do not reveal the exact nature of the role played by GSH and GST in mediating melatonin's oncostatic effect on human breast cancer cell proliferation.

It is tempting to speculate that the high levels of GSH stimulated by melatonin are responsible for its antiproliferative effects. However, it seems unlikely that GSH itself mediates melatonin's oncostatic effect since GSH alone added to MCF-7 cell cultures does not inhibit cell proliferation. Rather, it appears that GSH is in some way permissive for melatonin's oncostatic action and thus needs to be present within the cell at some threshold level. This could explain why GSH depletion by BSO is effective in blocking the inhibitory action of melatonin as well as the ability of moderate intracellular GSH repletion (to control levels) to restore the inhibitory effect of melatonin in BSO-treated cultures. If such is the case, why then does melatonin generate seemingly such unnecessarily high levels of GSH? One might speculate that the "overproduction" of GSH in response to melatonin serves as a mechanism by which melatonin sustains an above threshold level of GSH to permit the maintainence of its oncostatic effect for the duration of the culture period. Does melatonin directly or indirectly increase the synthesis of GSH or does it inhibit its efflux from the cell? Regardless of the exact scenario, it is clear from our experiments thus far that melatonin and GSH must be present together in order for a growth inhibitory response to be manifested in WT MCF-7 human breast cancer cells.

Like GSH in MCF-7 cells, GST also appears to play an important role in the resistance of ER- Hs578t cells to the inhibitory effects of melatonin. However, the converse situation occurs in that high levels of GST appear to impart these cells with resistance to melatonin whereas the pharmacological inhibition of GST activity converts them to the MCF-7 phenotype with respect to inhibition of cell proliferation and the induction of increased GSH levels. That high GST expression in ER- breast cancer cells is responsible for their refractoriness to melatonin is further supported by the fact that our responsive MCF-7 cells are devoid of GST activity. It appears that the presence of significant GST levels and/or activity, through as yet an unknown mechanism, prevents melatonin from interacting with GSH and its synthetic pathway. Could GST be acting as an intracellular binding protein for melatonin as has been demonstrated for other hormones including estradiol? If such were the case, then GST might act as an intracellular "melatonin sponge" and diminish the availability of melatonin for interaction with GSH. Our experimental results strongly indicate that GSH is permissive for melatonin's oncostatic action at the cellular level, while GST seems to be inhibitory.

Thus, the responsiveness or resistance of breast cancer cells to the inhibitory action of melatonin may ultimately depend on a critical ratio of GSH:GST which may be specific for the cell type examined as well as the culture conditions employed. Regardless of the exact mechanism(s) involved, the redox state of the breast cancer cell, as governed by GSH and GST metabolism, appears to be a critical determinant of melatonin's oncostatic action. Whether this principle extends to other cancer cells affected by melatonin awaits further investigative efforts.

REFERENCES

1. D.E. Blask, D.B. Pelletier, S.M. Hill, A. Lemus-Wilson, D.S. Grosso, S.T. Wilson and M.E. Wise, Pineal melatonin inhibition of tumor promotion in the N-nitroso-N-methylurea model of mammary carcinogenesis: potential involvement of antiestrogenic mechanisms *in vivo*, *J. Cancer Res. Clin. Oncol.* 117:526 (1991).

2. S.M. Hill, and D.E. Blask, Effects of the pineal hormone melatonin on the proliferation and morphological characteristics of human breast cancer cells (MCF-7) in culture, *Cancer Res.*, 48:6121 (1988).

3. D.E. Blask, Melatonin in oncology, *in*: "Melatonin Biosynthesis, Physiological Effects, and Clinical Applications," H.S. Yu, and R.J. Reiter, eds., CRC Press, Boca Raton, 447 (1993).

4. S. Cos, D.E. Blask, A. Lemus-Wilson, and A.B. Hill, Effects of melatonin on the cell cycle kinetics and "estrogen-rescue" of MCF-7 human breast cancer cells in culture, *J. Pineal Res.*, 10:36 (1991).

5. T.M. Molis, M.R. Walters, and S.M. Hill, Melatonin modulation of estrogen receptor expresssion in MCF-7 human breast cancer cells, *Int. J. Oncol.* 3:687 (1993).

6. T.M. Molis, Y. Cockerham, and S.M. Hill, Regulation of ER mRNA expression by melatonin in MCF-7 human breast cancer cells, *74th Ann. Mtg. Endocr. Soc.*, Abs. 1030, 309 (1992).

7. K.M. Hull, T.J. Maher, and K.L. Jorgenson, Melatoin increases c-*fos* mRNA in MCF-7 human breast cancer cells, *Soc. Neurosci.*, 19:Abs. 773.5, 1896 (1993).

8. A. Lemus-Wilson, D.E. Blask , and P.A. Kelly, Direct inhibitory effect of melatonin on prolactin receptor-mediated growth of MCF-7 human breast cancer cells, *74th Ann. Mtg. Endocr. Soc.*, Abs. 1297, 376 (1992).

9. S. Cos, and D.E. Blask, Melatonin modulates growth factor activity in MCF-7 human breast cancer cells, *J. Pineal Res.*, (1994) in press.

10. S.M. Hill, L.L. Spriggs, M.A. Simon, H. Muraoka, and D.E. Blask, The growth inhibitory action of melatonin on human breast cancer cells is linked to the estrogen response system, *Cancer Lett.*, 64:249 (1992).

11. R.J. Reiter, Melatonin: multifaceted messenger to the masses, *Lab. Med.*, 25:438 (1994).

12. R.J. Reiter, D.X Tan, B. Poeggler, A. Menendez-Pelaez, L.D. Chen, and S. Saarela, Melatonin as a free radical scavenger: implications for aging and age-related diseases, *Ann. N.Y. Acad. Sci.*, 719:1 (1994).

13. L. Kothari, and A. Subramanian, A possible modulatory influence of melatonin on representative phase I and phase II drug metabolizing enzymes in 9,10-dimethyl-1,2-benzanthracene induced rat mammary tumorigenesis, *Anti-Cancer Drugs*, 3: 623 (1992).

14. D.E. Blask, S.T. Wilson, and A.M. Lemus-Wilson, The oncostatic and oncomodulatory role of the pineal gland and melatonin, *in*: "Advances in Pineal Research", vol. 7, G.J.M. Maestroni, A. Conti, and R.J. Reiter, eds., John Libbey, London, 235, (1994).

15. A. Meister, Glutathione, ascorbate, and cellular protection, *Cancer Res.*, 54:1969s (1994).

16. J.P. Shaw, and I.N. Chou, Elevation of intracellular glutathione content associated with mitogenic stimulation of quiescent fibroblasts, *J. Cell. Physiol.*, 129:193 (1986).

17. J.E. Leibmann, S.M. Hahn, J.A. Cook, C. Lipschultz, J.B. Mitchell, and D.C. Kaufman, Glutathione depletion by L-buthionine sulfoximine antagonizes taxol cytotoxicity, *Cancer Res.*, 53:2066 (1993).

18. S. Simon, and M. Schindler, Cell biological mechanisms of multidrug resistance in tumors, *Proc. Natl. Acad. Sci. USA*, 91:3497 (1994).

19. B.A. Arrick, and C.F. Nathan, Glutathione metabolism as a determinant of therapeutic efficacy: a review, *Cancer Res.*, 44:4224 (1984).

20. A. Meister, and M.E. Anderson, Glutathione, *Ann. Rev. Biochem.*, 52:711 (1983).

21. D.J. Waxman, Glutathione S-transferases: role in alkylating agent resistance and possible target for modulation chemotherapy - a review, *Cancer Res.*, 50:6449 (1990).

22. R.D.H. Whelan, L.K. Hosking, A.J. Townsend, K.H. Cowan, and B.T. Hill, Differential increases in glutathione S-transferase activities in a range of multidrug-resistant human tumor cell lines, *Cancer Commun.*, 1:359 (1990).

23. G. Batist, R. Schecter, A. Woo, D. Greene, and S. Lehnert, Glutathione depletion in human and in rat multi-drug resistant breast cancer cell lines, *Biochem. Pharmacol.*, 41:631 (1991).

24. D.K. Armstrong, G.B. Gordon, J. Hilton, R.T. Streeper, O.M. Colvin, N.E. Davidson, Hepsulfam sensitivity in human breast cancer cell lines: the role of glutathione and glutathione S-transferase in resistance, *Cancer Res.*, 52:1416 (1992).

25. R.A. Kramer, J. Zakher, and G. Kim, Role of the glutathione redox cycle in acquired and *de novo* multidrug resistance, *Science*, 241:694 (1988).

26. A. Meister, Glutathione deficiency produced by inhibition of its synthesis, and its reversal; applications in research and therapy, *Pharmac. Ther.*, 51:155 (1991).

27. K.D. Tew, A.M. Bomber, and S.J. Hoffman, Ethacrynic acid and piripost as enhancers of cytotoxicity in drug resistant and sensitive cell lines, *Cancer Res.*, 48:3622 (1988).

28. J.H.T.M. Ploemen, A. Van Schanke, B. Van Ommen, and P.J. Van Bladeren, Reversible conjugation of ethacrynic acid with glutathione and human glutathione S-transferase P1-1, *Cancer Res.*, 54:915 (1994).

29. Y. Takamatsu, and T. Inaba, Inhibition of human hepatic glutathione S-transferase isozymes by ethacrynic acid and its metabolites, *Toxicol. Lett.*, 62:241 (1992).

30. A.F. Howie, W.R. Miller, R.A. Hawkins, A.R. Hutchinson, and G.J. Beckett, Expression of glutathione S-transferase B_1, B_2, Mu and Pi in breast cancers an their relationship to oestrogen receptor status, *Br. J. Cancer*, 60:834 (1989).

TREATMENT OF HUMAN METASTATIC MALIGNANT MELANOMA WITH HIGH DOSE ORAL MELATONIN

William A. Robinson, Lyndah Dreiling, Rene Gonzalez and Carol Balmer

Melanoma Research Clinic
Division of Medical Oncology
Department of Medicine
University of Colorado Health Sciences Center
Denver, Colorado USA 80262

INTRODUCTION

Cutaneous malignant melanoma (MM) is the most rapidly increasing cancer in Caucasians throughout the world(1,2). The cause of the current epidemic is increasing exposure to ultraviolet light in the form of sunshine that has occurred over the past 50 years from changes in clothing habits and lifestyle (3,4). Most MM develop in pre-existing nevi, primarily in sun exposed sites, in fair skinned, blue eyed, brown haired, upper middle class Caucasians who work indoors but spend large amounts of time outdoors in leisure time activities (5,6). Unlike cancers beginning in the epithelial skin, which usually remain localized, MM spreads rapidly from the primary site to other parts of the body via the blood and lymphatic systems. Early surgical excision may be curative but once the disease has spread beyond regional lymph nodes no form of therapy has been shown to be of consistent and long lasting benefit. Based on current figures 20 to 30% of all persons currently diagnosed with MM will develop metastatic disease, and most if not all, will die as a direct consequence. Numerous forms of therapy have, or are, being tried including chemo and immunotherapy, vaccines and recently gene therapy (7,8,9). All have some limited effect but most patients eventually develop brain metastases and no therapy has been shown to significantly prolong life when this occurs (10). With this background we have begun to seek new and novel agents for the treatment of metastatic MM. Melatonin has been previously examined as a possible anti-neoplastic agent by a number of investigators, including ourselves (11,12,13). In-vitro melatonin has been shown to suppress the growth of cancer cells, including MM (12). Likewise in animal models of MM, administration of melatonin has been shown to delay disease progression and death (13). Melatonin, both alone and with other agents, has been used to treat a variety of human cancers with variable results, but generally low response rates (11,14). We previously reported on a phase 1, dose escalation trial of oral melatonin in patients with metastatic cutaneous and ocular MM(11). In this initial trial there was little toxicity, even at very high daily doses of oral melatonin, and occasional patients appeared to have minor responses with partial regression of metastatic disease. No dose response curve was noted in this study. Patients receiving very high doses (700mg per M2) were no more likely to respond than those receiving much lower doses. We therefore elected to carry out a further study using a fixed dose of oral melatonin (200mg orally per day) in a larger group of patients with metastatic MM to determine more precisely the possible therapeutic benefit and toxicity. The results of this study are reported here.

METHODS AND MATERIALS

All of the patients entering the present study were seen and followed in the Melanoma Research Clinic of the University of Colorado Health Sciences Center between 1990 and 1994. All had histologically documented Stage 3 or Stage 4 metastatic MM. Stage 3 designates the presence of regional lymph node involvement only while Stage 4 indicates the presence of metastatic disease in areas beyond the regional lymph nodes. Patients with both cutaneous and ocular primary sites, considered to have a life expectancy of 3 months or longer, were eligible for the study provided they met all other entry criteria. Prior to entry into the study all patients had a full evaluation to determine the extent of metastatic spread including a complete physical examination, blood cell counts, chemistry panel, chest radiograph, CT scan of the abdomen and pelvis and MRI brain scans. Other tests were performed as clinically indicated. Prior to treatment all patients gave written informed consent as approved by the Human Subjects Committee of this institution. The first 7 patients were admitted to the Clinical Research Center for measurement of serum FSH, LH and melatonin levels for 24 hours before and after treatment. All other patients were treated and followed in the Clinical Research Unit outpatient clinic on a weekly basis for 4 weeks and then monthly, or as the clinical situation dictated. At each visit a physical examination was performed and every 4 weeks the blood counts and chemistry studies were repeated. Radiographs and scans were repeated every 3 months unless the clinical situation required more frequent study. In addition to the routine studies noted 10 patients had monthly measurement of skin reflectance to determine possible changes in skin color. Controls were spouses or family members who were exposed to the same lighting conditions, but were not necessarily of the same phenotype. Reflectance readings were taken monthly from the forehead, upper inner arm and upper back to give data from both sun and non sun exposed sites. The results of these studies have been reported in detail elsewhere (15). Melatonin was purchased from Regis Chemical Co. and prepared in our pharmacy as 50mg gelatin capsules. Patients were asked to take one 50mg capsule at 6 hourly intervals for a total daily dose of 200mg in all patients regardless of body size. No attempt was made to control the time of administration although most patients preferred a 6-12-6-12 dosing schedule. No other anti-cancer therapy was given during the duration of melatonin treatment. In order to be evaluable for therapeutic response patients had to have taken a minimum of 2 months of continuous melatonin. Response criteria were those used in standard cancer chemotherapy trials. A complete remission was defined as total disappearance of all measurable disease for two months or longer, a partial remission as a reduction by 50% or greater of all measurable disease for at least two months and stable disease was defined as a less than 50% reduction or no change for the same period.

RESULTS

Table 1 shows the clinical characteristics of the 22 patients initially entered on the study. There were 12 males and 10 females with a wide age range, from 15 to 83 years, and a median age of 53. All were Caucasian reflecting the phenotypic basis of the disease rather than investigator selection. Nineteen patients had cutaneous primaries (86%) while 3 (14%) had MM originating in the eye. Most patients had multiple areas of metastases, as shown in Table 2. The most common site of involvement was the brain where metastases were present in 14 patients (64%), followed by lung (45%) and liver (32%). Two patients initially had only lymph node and/or subcutaneous metastases and had refused other treatment.

Table 1. Clinical characteristics of the 22 patients treated

Sex	Number	Percent	Age (years)
Male	12	55	Range 15-83
Female	10	45	Median 53

Side Effects

No patient withdrew from the study due to side effects from the melatonin. In general it was well tolerated, as we have noted previously, even when given orally in this very large dose. Two patients reported mild diarrhea or other gastrointestinal complaints, 3 had mild drowsiness that they thought was from the melatonin and a single patient had itching without a clinical skin rash. Most of the patients in the study were moderately to severely ill due to the underlying disease and vital organ involvement. While many had mild changes in blood counts none of these could be directly attributed to melatonin. Further no patient had a perceived change in libido, sexual function, potency or menstruation that was felt to be due to melatonin.

Table 2. Primary and metastatic sites of melanoma in the study group

Primary Site	Number	Percent	Sites of metastases	Number	Percent
Cutaneous	19	86	Brain	14	64
Ocular	3	14	Lung	10	45
			Liver	7	32
			Subcutaneous	7	32
			Lymph node	6	27

Special Studies

The results of the special studies done as part of this therapeutic trial have been reported in detail elsewhere (15,16). Reflectance measurements for changes in skin color were done over long periods of time on patients and appropriate normal controls, usually a spouse or family member (15). Readings were taken from both sun exposed and non sun exposed skin sites to control for the effect of differing exposure to UV light among subjects. Measurements were done on the forehead, upper inner arm and back at monthly intervals for as long as one year. No changes in skin color which could be attributed to the melatonin were recorded. Likewise neither the patients or family members reported any subjective change in skin color. We were also concerned in this, and our previous study, about the possible effects of melatonin on retinal pigmentation. Although this was not measured directly due to the complexity of the techniques required, no patient reported significant changes in visual acuity, light sensitivity or external pigmentation. In addition no changes were noted in retinal pigmentation on opthalmascopic examination. Finally in this regard no subjective or objective changes were noted in hair color during the time of melatonin administration.

Prior to beginning melatonin therapy serum levels of melatonin were measured at hourly intervals for 24 hours in 8 patients (16). Typical diurnal curves were recorded with peak nocturnal levels nearing 100pg as has been recorded for normal controls. Thus the melanoma patients in the present study did not appear to have abnormal baseline values either in quantity or rhythm. Following the oral administration of melatonin in the doses noted serum values rose to one hundred times or greater that of baseline values with total ablation of the normal diurnal/nocturnal rhythm. Long term measurements of serum levels were not carried out to determine if this was sustained over the duration of treatment. There appeared to be modest differences in absorption and metabolism among individual patients, but these were not examined in detail. The details of these studies will be found in a previous publication (16).

As noted above there were no subjective changes in libido or sexual function. Previous investigators have evaluated the possible effects of melatonin on the hypothalamic gonadal axis in man and animals because of it's presumed role in sexual maturity and ovulatory regulation. In the present and previous study we measured serum LH and FSH levels at 15 minute intervals for 24 hours each before, and after, melatonin administration. The results of these studies will be the subject of a subsequent publication, but in summary showed no major changes in either the amplitude or number of FSH or LH pulsations after the administration of melatonin, as compared to pre therapy. There did appear to mild decreases in the levels both hormones but there was a wide range of variability which could not be explained on the basis of sex, age, menstrual function or the extent of MM. It should be noted that these studies were done on two sequential days at the start of melatonin treatment. It may be that greater changes would have been noted had the post therapy studies been carried out at a later date, and after longer treatment. The clinical facts, however do not suggest that this would be the case.

General Effects

It was difficult to evaluate the effects of melatonin administration on appetite and general well being in this study group. Most were moderately to severely ill on entry, and as noted frequently had vital organ involvement by MM. We did not, however, detect any significant changes in these parameters using either subjective or objective criteria. There was no perceptible weight gain or loss, change in muscle mass or appetite. Mental function was subjectively unaltered without increased drowsiness, change in sleeping habits or alertness. One patient, with known brain metastases, had increased headache but this was not relieved when melatonin was discontinued and not reported by similar patients.

Therapeutic Effect and Patient Survival

The therapeutic benefit of melatonin, in this as well as the previous study, was small. Details are shown in Table 3. Two patients were ineligible for evaluation because of rapidly progressive disease and were taken off treatment, or died, before the two month period required for therapeutic evaluation. Two patients had partial remissions of local lymph node metastases which lasted from 4 to 6 months. Both patients however developed widespread metastases involving other organs while taking melatonin and died. A single patient each with lymph node, subcutaneous and brain metastases had stabilization of disease, but all eventually had local progression and further systemic metastases. The 15 remaining evaluable patients had no objective response. There was no apparent correlation between age, sex, menstrual status or previous anti-cancer therapy and response to melatonin. The only important factor, in this small group, appeared to be the extent of disease. There were no responses in patients with widespread metastases, liver, lung involvement or multiple brain metastases. No patient with an ocular primary lesion had a response in the present study, although the numbers were small.

The median survival for the entire group of 22 patients was 5.5 months which is also similar to that reported in our previous study. Only 1 patient remains alive with continued, slowly progressive disease at 50 months follow-up. The remaining 21 patients have all died of progressive MM with brain metastases being the area most responsible for death. This survival is similar to that expected for a similar group of untreated patients.

Table 3. Therapeutic response to melatonin in the 20 evaluable patients

Response	Number	Percent	Sites of disease
Complete remission	0	0	
Partial remission	2	10	Lymph node
Stable disease	3	15	Lymph node, brain, SQ
No response	15	75	Liver, lung, brain

The median survival for the entire group of 22 patients was 5.5 months which is also similar to that reported in our previous study. Only 1 patient remains alive with continued, slowly progressive disease at 50 months follow-up. The remaining 21 patients have all died of progressive MM with brain metastases being the area most responsible for death. This survival is similar to that expected for a similar group of untreated patients.

DISCUSSION

This study confirms our previous report that melatonin can be safely given in large doses to humans over long periods of time with few side effects, but little or no, therapeutic activity against metastatic malignant melanoma. The lack of side effects was, to us, both surprising and unexpected. Based upon previous animal and human data we had expected, at a minimum, some changes in mental status, sleep patterns, skin pigmentation, and sexual function. None of these were noted to any significant degree in either this or the previous study. This was not the result of lack of absorption or adequate serum levels of melatonin. As shown in a previous publication (16) levels one hundred times baseline were achieved in most patients and were sustained over the 24 hour observation period. Further even with the mild side effects noted there did not appear to be any correlation between dose and possible side effects. In patients in the preceding dose escalation study these were as likely at low doses as at very high doses. We have speculated that this lack of observed physiologic effect may have resulted from the timing of administration of melatonin in the both studies and that a schedule which more closely resemble the normal diurnal pattern might have had more effect. The objective in the present study was however to maintain high levels over long periods of time to obtain direct anti-cancer activity. This, in retrospect, may not have been the optimal way to test this unique agent, but is the means by which most new experimental cancer treatments are carried out. It should be noted in this regard however that some patients did not take their capsules on a 6 hourly basis, choosing instead to take the entire dosage in the evening hours before retiring, and this did not seem to result in any fewer or greater side effects.

This study is the first to look at the effect of large doses of melatonin on changes in human skin color over a long period of time using an objective measurement of pigmentation in both sun and non-sun exposed sites. The full details of these studies have been reported elsewhere (15). Changes in pelage and coat color have been reported in animals kept under various lighting conditions and after the administration of melatonin. One study in man also suggested in a single hyperpigmented patient that melatonin administration led to lightening of skin color (17). Our studies were all conducted in fair skinned Caucasians who are the phenotypic group most likely to develop MM. Whether our findings would be confirmed in other phenotypic types and with other dosing schedules must await further study.

This is also one of the first and most comprehensive studies in humans of changes in the hormones of the hypothalamic gonadal axis after the administration of large doses of melatonin. Although these were not normal subjects, and the timing of administration was not physiologic, the lack of significant effects on the levels of FSH, LH and sexual function was surprising. Others have suggested roles for melatonin in human sexual maturity and as a possible contraceptive when used in conjunction with other agents (18,19). Certainly as used here melatonin does not appear to suppress gonadal function in either males or females and if it is to have a role in the above further studies will be necessary to determine more precisely it's normal physiologic function and mechanism of action in this area.

Finally we were prepared, as were the patients in this study, for significant effects on sleep patterns, mood and affect. We anticipated that we would upset the normal biorhythms presumed to be regulated by, or in conjunction with, melatonin secretory patterns. As noted however this was not the case. Only mild and transient drowsiness was reported by a few patients and there were no instances of major sleep disorders regardless of the timing of melatonin administration. We can only speculate that this was the result of the continuous dosing schedule used although we have considered other possibilities as well. It may be that the neuro-physiologic effects of melatonin require close cellular interaction within the brain which cannot be mimicked by external administration, or that it's effects need to be more carefully timed with other phyisologic biorhythms for the anticipated effects to be seen. Regardless, the present studies clearly show that large doses given continuously do not have major neurologic effects in man as judged clinically.

The lack of significant therapeutic benefit of melatonin in metastatic MM was not unexpected, but disappointing, particularly in view of the minimal number of side effects observed. It is clear from both this and the previous study that a very small number of patients do appear to have minor objective responses to melatonin, but these are in general short lived and do not appear to prolong overall survival. There are those that would argue that the number of responses seen in this study were no greater than those reported to occur spontaneously in untreated MM patients. In our experience however spontaneous regressions are very rare and do not occur with even the very low frequency seen here. Since this was not a randomized trial with a placebo treated control group this question cannot be answered. Such trials are obviously quite difficult to do in very ill and desperate patients such as those studied here.

If one accepts that some of the responses noted in our studies were the result of the administration of melatonin, which we believe to be the case, then the question arises as to the possible mechanism of action in this regard. We have discussed this issue in detail in our previous paper and have considered the possibilities of direct action of melatonin on MM cells and various indirect mechanisms including effects on the immune or cytokine systems. All of the above have been speculated on and discussed by other authors as well (12,13).Our initial reason for evaluating melatonin in MM was based on it's possible direct effect on the melanocyte cell lineage. We, and others, have shown that MM cells have high affinity melatonin receptors (20) but have not studied the relationship of receptor number to response rates, tumor stage or other prognostic factors. Others have also noted that melatonin, in vitro, increases the number of estrogen receptors on a breast cancer cell line (21) raising the possibility that it may act through another intermediate endocrine pathway. It is noteworthy, in this regard to note that the anti-estrogen, tamoxifen, has significant therapeutic benefit in metastatic MM when used in conjunction with conventional chemo and immunotherapy. We have done a preliminary study in 4 patients examining estrogen receptor levels in MM tumor tissue before, and after, the administration of melatonin and have not observed any increase in this small sample. Nevertheless we have speculated that a combined regimen of melatonin and tamoxifen might be warranted in future studies.

Even the most potent agents, when used alone, have only a 10 to 15 per cent response rate in advanced cutaneous MM and even lower rates are observed with metastatic ocular MM. More recent therapeutic trials have used intense combinations of a number of chemotherapy drugs, interferon, IL-2 and tamoxifen with reported response rates of 30-50%(7,8). Again however most of these are short lived with most patients eventually succumbing to brain metastases, for which there is no therapy of long term benefit. Any further trials of melatonin in advanced metastatic MM should be done in conjunction with one or more of the above agents. We do not feel that further studies with melatonin alone are warranted in this group of patients. Further we believe that if melatonin therapy trials are to conducted in MM that they should be done in patients with early stage disease and should take into account the normal biorhythms of melatonin secretion. In the our studies the patients most likely to respond to melatonin were those with early stage disease, usually involving the lymph nodes only, and these are the group who should be targeted for any future studies.

As reported elsewhere in this book a large number of melatonin agonists have been recently synthesized and studied. Several of these have much more potent effects on a variety of physiologic functions than the parent compound. The role of these as anti-neoplastic agents has not however been investigated in detail. Based upon our knowledge that most MM cells contain high affinity receptors, to which many of these agents bind, they may offer a new therapeutic avenue for study in MM and other human cancers containing melatonin receptors.

The current epidemic of MM will continue for many years to come. Unlike many other cancers therapeutic advances have not kept pace with the growing problem. The ultimate answer to MM is prevention by through education about the dangers of sun exposure and the importance of early diagnosis. Unfortunately many people around the world have already sown the seeds, and will reap the harvest of hours spent in the sun. We are in desperate need of effective treatment and although it was not found in the studies done, and reported here, we hope that the basic knowledge gained will further us on that difficult path.

SUMMARY

We carried out a therapeutic and toxicity trial of high dose oral melatonin in humans with advanced metastatic cutaneous and ocular malignant melanoma (MM). Twenty two patients were initially entered on the study and 20 were evaluable for response. All patients had biopsy proven disease and were evaluated extensively prior to treatment to define and quantify metastases. Melatonin was given to all patients in a dosage of 50mg 4 times daily. Special studies, in addition to evaluation for tumor response, included monthly measurement of skin pigmentation using reflectance, and 24 hour measurements of serum FSH and LH levels before and after melatonin. This dosage of melatonin was well tolerated. No patient withdrew from the study due to side effects. There were no significant changes in skin or hair pigmentation. Serum FSH and LH levels were slightly lower in some patients after melatonin administration but there were no significant changes in libido, sexual function or menstruation. Some patients reported mild drowsiness, but this was not common, sustained, or associated with changes in mental status or sleep patterns. The overall therapeutic response was small, and similar to that reported in a previous study. Two patients had partial regression of lymph node metastases and 3 others had stabilization of disease for short periods. None of the remaining 15 evaluable patients had objective responses. The median survival for the entire group was 5.5 months with only 1 patient still alive, with disease, at 50 months. We conclude that melatonin given in this dose and schedule has little therapeutic benefit in metastatic MM, but can be given safely and with minimum side effects to humans at the doses used here.

REFERENCES

1. S.J. Hoffmann, J.J. Yohn, D.A. Norris, C. Smith, and W.A. Robinson, Cutaneous malignant melanoma, *Current Problems in Dermatology* V(1):1 (1993).
2. J.M. Elwood, Recent developments in melanoma epidemiology, *Melanoma Research* 3:149 (1993).
3. A.M. Goldstein and M.A. Tucker, Etiology, epidemiology, risk factors, and public health issues of melanoma, *Current Opinion in Oncology* 5:358 (1993).
4. A.J. Sober, F.A. Lew, H.K. Koh, and R.L. Barnhill, Epidemiology of cutaneous melanoma. An update, *Dermatologic Clinics* 9:617 (1991).
5. R.M. Rifkin, M.R. Thomas, T.I. Mughal, J.S. Kaur, L.U. Krebs, and W.A. Robinson, Malignant melanoma - profile of an epidemic, *Western Journal of Medicine* 149:43 (1988).
6. W.A. Robinson, J. Ferguson, and J.F. Robinson, Malignant melanoma: The dark side of the sun, at 40° north, *Transactions of the Menzies Foundation (Melbourne, Australia)* 15:121 (1989).
7. J.M. Richards, N. Mehta, K. Ramming, and P. Skosey, Sequential chemoimmunotherapy in the treatment of metastatic melanoma, *J. Clin. Onc.* 10:1338 (1992).
8. L.E. Flaherty, W.A. Robinson, B.G. Redman, R. Gonzalez, S. Martino, M. Kraut, M. Valdivieso, and A. Rudolf: Phase II study of dacarbazine and cisplatin in combination with outpatient administered interleukin-2 in metastatic malignant melanoma, *Cancer* 71:3250 (1993).
9. G.J. Nabel, E.G. Nabel, Z.Y. Yang, B.A. Fox, G.E. Plautz, X. Gao, L. Huang, S. Shu, D. Gordon, and A.E. Chang, Direct gene transfer with DNA-liposome complexes in melanoma: Expression, biologic activity, and lack of toxicity in humans, *Proc. Natl. Acad. Sci. USA* 90:11307 (1993).
10. K. Brega, W.A. Robinson, K. Winston, and W. Wittenberg, Surgical treatment of brain metastases in malignant melanoma, *Cancer* 66:2105 (1990).
11. R. Gonzalez, A. Sanchez, J.A. Ferguson, C. Balmer, C. Daniel, A. Cohn, and W.A. Robinson, Melatonin therapy of advanced human malignant melanoma, *Melanoma Research* 1:237 (1991).
12. A. Slominski and D. Pruski, Melatonin inhibits proliferation and melanogenesis in rodent melanoma cells, *Exp. Cell Res.* 206:189 (1993).
13. L.R. Stanberry, T.K. Das Gupta, and C.W. Beattie, Photoperiodic control of melanoma growth in hamster: Influence of pinealectomy and melatonin, *Endocrinology* 113:469 (1983).
14. P. Lissoni, S. Barni, G. Tancini, A. Ardizzoia, G. Ricci, R. Aldeghi, F. Brivio, E. Tisi, F. Rovelli, R. Rescaldani, et al, A randomised study with subcutaneous low-dose interleukin 2 alone vs interleukin 2 plus the pineal neurohormone melatonin in advanced solid neoplasms other than renal cancer and melanoma, *Br. J. Cancer* 69:196 (1994).
15. D.B. McElhinney, S.J. Hoffman, W.A. Robinson, and J. Ferguson, Effect of melatonin on human skin color, *J Invest Derm* 102:258 (1994).
16. M.A. Kane, A. Johnson, A.E. Nash, D. Boose, G. Mathai, C. Balmer, J.J. Yohn, and W.A. Robinson, Serum melatonin levels in melanoma patients after repeated oral administration, *Melanoma Research* 4:59 (1994).
17. J.J. Nordlund and A.B. Lerner, The effects of oral melatonin on skin color and on the release of pituitary hormones, *J. Clin. Endo. & Metab.* 45:768 (1977).
18. F. Waldhauser, G. Weiszenbacher, H. Frisch, U. Zeitlhuber, M. Waldhauser, and R.J. Wurtman, Fall in nocturnal serum melatonin during prepuberty and pubescence, *Lancet* 1:362 (1984).
19. B.C. Boordouw, R Euser, R.E. Verdonk, B.T. ALberda, F.H. de Jong, A.C. Drogendijk, B.C. Fauser and M. Cohen, Melatonin and melatonin-progestin combinations alter pituitary-ovarian function in women and can inhibit ovulation, *J. Clin. Endo. & Metab.* 74:108 (1992).
20. R.A. Helton, W.A. Harrison, K. Kelley, and M.A. Kane, Melatonin interactions with cultured murine B16 melanoma cells, *Melanoma Research* 3:403 (1993).
21. S.T. Wilson, D.E. Blask, and A.M. Lemus-Wilson, Melatonin augments the sensitivity of MCF-7 human breast cancer cells to tamoxifen in vitro, *J. Clin. Endo. & Metab.* 75:669 (1992).

CONTRIBUTORS

AHMED S., Sleep and Mood Disorders Laboratory, Depts of Psychiatry, Ophtalmology and Pharmacology, Oregon Health Sciences University, Portland, Oregon 97201, USA

ARENDT J., School of Biological Sciences, University of Surrey, Guildford, Surrey, GU2 5XH, U.K.

BALMER C., Melanoma Research Clinic, Division of Medical Oncology, Department of Medicine, University of Colorado Health Sciences Center, Denver, Colorado, USA 80262

BARRETT P., Molecular Neuroendocrinology Group, Rowett Research Institute, Greenburn Road, Bucksburn, Aberdeen AB2 9SB, Scotland, UK

BAUER V.K., Sleep and Mood Disorders Laboratory, Depts. of Psychiatry Ophtalmology and Pharmacology, Oregon Health Sciences University, Portland, Oregon 97201, U.S.A.

BENLOUCIF S., Department of Molecular Pharmacology and Biological Chemistry, Northwestern University Medical School, Chicago, IL 60611, U.S.A.

BLASK D.E., The Mary Imogene Bassett Research Institute, Cooperstown, NY 13326-1394, USA

BLOOD M.L., Sleep and Mood Disorders Laboratory, Depts. of Psychiatry, Ophtalmology and Pharmacology, Oregon Health Sciences University, Portland, Oregon 97201, U.S.A.

CAIGNARD D.H., I.R.I.S., 6, Place des Pléiades, 92415 Courbevoie Cedex, France

CANGUILHEM B., Neurobiologie des Fonctions Rythmiques et Saisonnières, CNRS-URA 1332, Université Louis Pasteur, Strasbourg, France,

CAPSONI S., Department of Pharmacology, University of Milan, Via Vanvitelli, 32, 20129 Milan, Italy

CARDINALI D.P., Department of Physiology, Faculty of Medicine, University of Buenos Aires CC 243, 1425 Buenos Aires, Argentina

DELAGRANGE P., I.R.I.S., 6, Place des Pléiades, 92415 Courbevoie Cedex, France

DEMARTINI G., Department of Pharmacology, University of Milan, Via Vanvitelli, 32, 20129 Milan, Italy

DREILING L., Melanoma Research Clinic, Departement of Medicine, University of Colorado, Health Science Center, Denver, Colorado, USA 80262

227

DUBOCOVICH M.L., Department of Molecular Pharmacology and Biological Chemistry, Northwestern University Medical School, Chicago, IL 60611, USA

EBLING F. J.P., Department of Anatomy, University of Cambridge, Downing Street, Cambridge CB2 3DY, U.K.

ESPOSTI D., Institute of Human Physiology II, Via Mangiagalli, 32, 20133 Milano, Italy

FISZMAN M., Department of Physiology, Faculty of Medicine, University of Buenos Aires, CC 243, 1425 Buenos Aires, Argentina

FRASCHINI F., Department of Pharmacology, Chair of Chemotherapy, University of Milan, Via Vanvitelli, 32, 20129 Milan, Italy

GOLOMBEK D.A., Department of Physiology, Faculty of Medicine, University of Buenos Aires, CC 243, 1425 Buenos Aires, Argentina

GONZALEZ R., Melanoma Research Clinic, Division of Medical Oncology, Department of Medicine, University of Colorado Health Science Center, Denver, Colorado, USA 80262

GUARDIOLA-LEMAITRE B., I.R.I.S., 6, Place des Pléiades, 92415 Courbevoie Cedex, France

HASTINGS M.H., Department of Anatomy, University of Cambridge, Downing Street, Cambridge CB2 3DY, U.K.

IACOB S., Department of Molecular Pharmacology and Biological Chemistry, Northwestern University Medical School, Chicago, IL 60611, U.S.A.

LEWY A., Sleep and Mood Disorder Laboratory, Depts. of Psychiatry, Oftalmology and Pharmacology, Oregon Health Sciences University, Portland, Oregon 97201, U.S.A.

KANTEREWICZ B.A., Department of Physiology, Faculty of Medicine, University of Buenos Aires, CC 243, 1425 Buenos Aires, Argentina

KIRSCH R., Neurobiologie des Fonctions Rythmiques et Saisonnières, CNRS-URA 1332, Université Louis Pasteur, Strasbourg, France

KRAUSE D.N., Department of Pharmacology, College of Medicine, University of California, Irvine, Irvine, CA 92717, U.S.A.

LUCINI V., Department of Pharmacology, University of Milan, Via Vanvitelli, 32, 20129 Milan, Italy

MARIANI M., Geriatric Inst. Redaelli, Vimodrone, Italy

MASANA M.I., Department of Molecular Pharmacology and Biological Chemistry, Northwestern University Medical School, Chicago, IL 60611, U.S.A.

MASSON-PEVET M., Neurobiologie des Fonctions Rythmiques et Saisonnières, CNRS-URA 1332, Université Louis Pasteur, ,Strasbourg, France

MAYWOOD E.S., Department of Anatomy, University of Cambridge, Downing Street, Cambridge CB2 3DY, U.K.

MICK G., Unité 94, Institut National de la Santé et de la Recherche Médicale, Bron - France

MØLLER M., Institute of Medical Anatomy, University of Copenhagen, Denmark

MORGAN P.J., Molecular Neuroendocrinology Group, Rowett Research Institute, Greenburn Road, Bucksburn, Aberdeen AB2 9SB, Scotland, UK

PARKES J.D., University Department of Neurology, King's College Hospital and Institute of Psychiatry, London SE5, U.K.

PEVET P., Neurobiologie des Fonctions Rythmiques et Saisonnières, CNRS-URA 1332, Université Louis Pasteur, Strasbourg, France

PHANSUWAN-PUJITO P., Department of Anatomy, Faculty of Medicine, Srinakarinwirot, University of Prasarnmit, Bangkok, Thailand

PITROSKY B., Neurobiologie des Fonctions Rythmiques et Saisonnières, CNRS-URA 1332, Université Louis Pasteur, Strasbourg, France

REDMAN J.R., Psychology Department, Monash University, Clayton, Victoria, 3168 Australia

REITER R. J., Department of Cellular and Structural Biology, University of Texas Health Science Center, 7703 Floyd Curl Drive, San Antonio, Texas 78284, U.S.A.

RENARD P., I.R.I.S., 6, Place des Pléiades, 92415 Courbevoie Cedex - France

ROBINSON W.A., Melanoma Research Clinic, Division of Medical Oncology, Department of Medicine, University of Colorado Health Science Center, Denver, Colorado, U.S.A. 80262

ROSENSTEIN R.E., Department of Physiology, Faculty of Medicine, University of Buenos Aires, CC 243, 1425 Buenos Aires, Argentina,

SACK R.L., Sleep and Mood Disorder Laboratory, Depts. of Psychiatry, Ophtalmology and Pharmacology, Oregon Health Science University, Portland, Oregon 97201, U.S.A.

SAAVEDRA J.M., Laboratory of Clinical Science, National Institute of Mental Helath, Bethesda, MD 20892-1514, U.S.A.

SCAGLIONE F., Department of Pharmacology, Chair of Chemotherapy, University of Milan, Via Vanvitelli, 32, 20129 Milan, Italy

STANKOV B., Department of Pharmacology, Chair of Chemotherapy, University of Milan, Via Vanvitelli, 32, 20129 Milan, Italy

STEHLE J., Klinikum der Johaun Wolfgang Goethe-Universitat, Zentrum der Morphologic, AG Neurobiologic, 60590 Frankfurt am Main, Germany

VISWANATHAN M., Laboratory of Clinical Science, National Institute of Mental Helath, Bethesda, MD 20892-1514, USA

VIVIEN-ROELS B., Neurobiologie des Fonctions Rythmiques et Saisonnières, CNRS-URA 1332, Université Louis Pasteur, Strasbourg, France

WETTERBERG L., Karolinska Institute, Institution of Clinical Neuroscience, Department of Psychiatry, St. Goran's Hospital, S-112 81 Stockholm, Sweden

WILSON S.T., The Mary Imogene Bassett Research Institute, Cooperstown, NY 13326-1394, USA

INDEX

Immune system, 224
In situ hybridization, 2, 3, 5, 14, 15
Insomnia, 183, 186, 189, 190, 193
Ionizing radiation, 28

Jet lag, 139, 169

Lagomorphs, 51
LH see Luteininsing hormone
Light, 35, 64, 69-72, 99, 125, 126, 148, 166, 169, 173-194, 199-205
Light/dark cycle, 64, 69-72, 148, 149, 166
Locomotor activity, 148, 150, 151
Luteininsing hormone, 85, 221, 224

Malignant melanoma, 219-225
Mammals, 1, 7, 21, 33, 39, 40, 62, 86, 91, 98, 110, 120, 122, 184, 201
Mannitol, 25
Marsupials, 51
MCF-7 cells, 209, 211, 212, 214
Melatonin
    actions, 21-29, 83, 85, 86, 98, 101, 102, 123-125, 209-212
    activity, 141
    administration, 156, 157, 161, 167-169,
    analogs, 83, 89, 91, 131-134, 139-151, 155, 159, 162, 183, 194
        agonist/antagonist activity, 143, 144
        chemistry, 139-145
    and cerebral blood flow, 80
    anti-proliferative action, 169, 209, 219
    as free radical scavanger, 21, 25, 27
    biological activity, 33, 83-85, 158
    metabolism, 159
    receptors, 49-55, 61-72, 75-84, 83-91, 125, 127, 131-134, 136, 151, 177
        binding affinity, 142, 155, 158, 162
        distribution, 50-55, 83
        localization, 61
        pharmacology, 83, 86, 89, 91
        regulation, 67
        subtypes, 62, 65, 142, 151
    sedative effects, 149
    synthesis, 7, 17, 18
    therapy
        and sexual function, 221, 225
        and skin pigmentation, 221, 223, 224
Membrane polarisation, 85
Menstrual cycle, 166
5-Methoxytryptamine, 25
Muscarinic receptor, 4

Neuronal discharge activity, 146, 147
Neurotrasmitter receptor, 1-4, 6, 17, 62, 63, 120-133
Non-photic zeitgeber, 169

Norepinephrine, 80

Ovariectomy, 80
Oxidative stress, 21-23, 26
Oxigen free radicals, 22-29

Pertussis toxin, 77, 80, 89
Pharmacokinetics, 107, 156-160
Pharmacological effects, 107, 108, 115, 183
Phase response curve, 69, 109, 111, 114, 147, 155, 169, 173-194
Phase shifting, 71, 107, 108, 111, 115, 123, 147, 148, 167-169, 173-194
Phospholipases A2, 85
Phospholipases C, 85, 89
Phospholipases D, 85
Photoperiodic, 33-35, 42, 43, 85, 96-99, 101-103, 120-132, 166, 167, 177, 191, 192
Photoperiodism, 13, 35, 139, 166
Pineal gland, 165
Pinealectomy, 83, 165
Pituitary gland, 83, 85, 86
    Pars distalis of, 52-55, 85, 91
    Pars tuberalis of, 21-33, 52-55, 85-103, 131-134, 140, 143-145, 150
PRC see Phase response curve
Primates, 80, 96, 98, 101, 102
Prolactin, 85, 167

Rabbit, retina, 145
Rat, 51, 75, 80, 85, 89, 109-115, 121-126, 133, 146-150, 161
    spontaneously hypertensive, 80
Regulation, 4-6, 13-19, 26, 40, 41
Reproduction, 166
Reticular system, 185, 186
Retina, 14, 62, 63, 83, 91, 107, 108, 122, 200, 202, 205
Rodents, 1, 51, 96, 101, 107, 125

S- 20098, 145, 146, 148, 149, 150
S- 20242, 150, 151
SAD see Seasonal affective disorder
Safrole, 21
SCN see Suprachiasmatic nuclei
Suprachiasmatic nuclei, 13-16, 34-39, 40-44, 51-55, 61-71, 77, 101, 103, 107, 108, 115, 120-122, 126, 139, 145-150, 176, 177, 183, 185-188, 201, 202
Seasonal affective disorder, 138, 155, 166-168, 199, 202-205
Seasonality, 166, 169
Sheep, 85, 139-153, 166, 168
Shift work, 139, 150, 166-169, 177-179
Signal, 13, 35, 39, 127, 131, 185
Signal transduction, 85